Nanotechnology in Cement-Based Construction

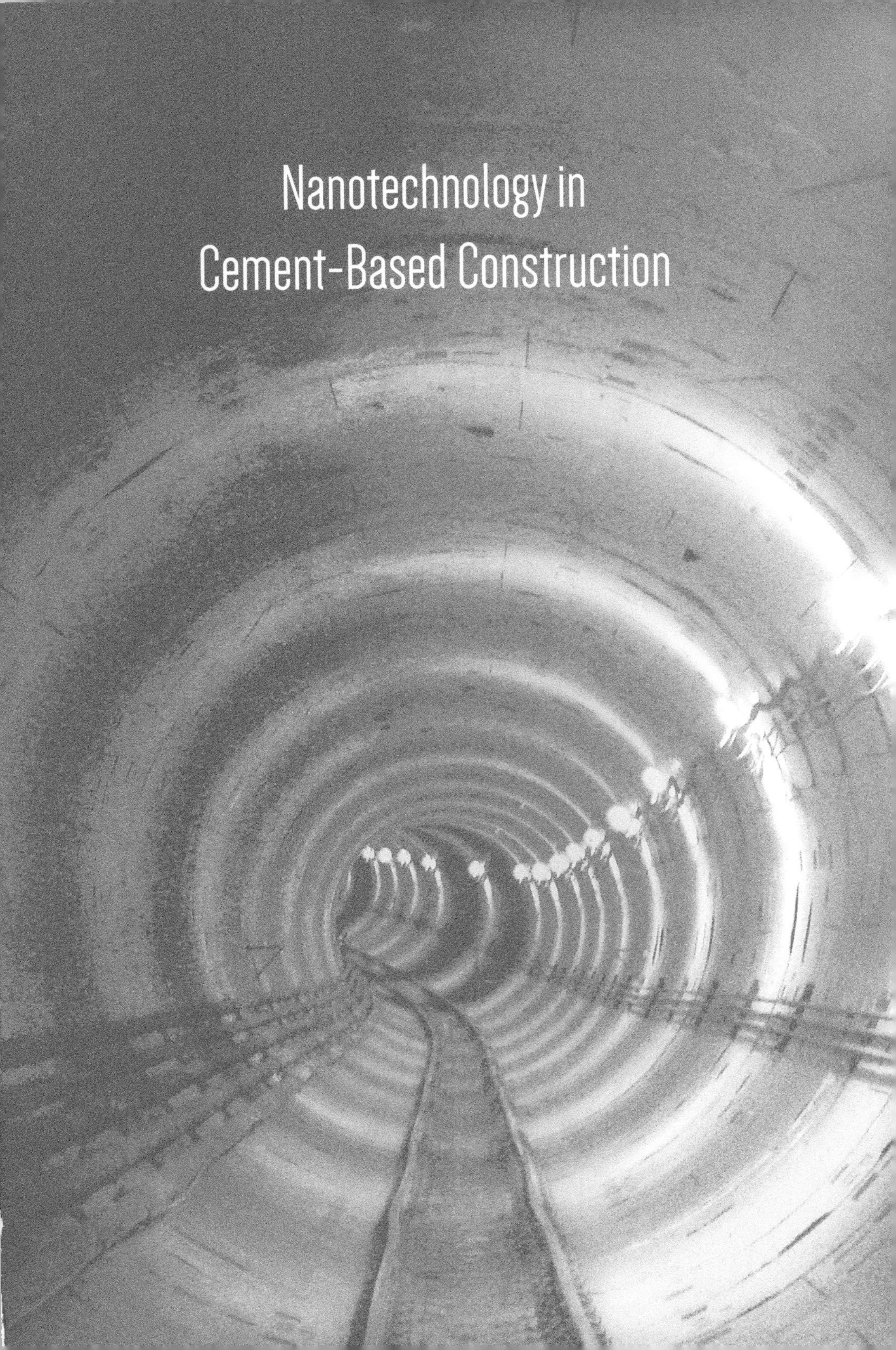

Nanotechnology in Cement-Based Construction

edited by

Antonella D'Alessandro
Annibale Luigi Materazzi
Filippo Ubertini

JENNY STANFORD
PUBLISHING

Published by

Jenny Stanford Publishing Pte. Ltd.
Level 34, Centennial Tower
3 Temasek Avenue
Singapore 039190

Email: editorial@jennystanford.com
Web: www.jennystanford.com

British Library Cataloguing-in-Publication Data
A catalogue record for this book is available from the British Library.

Nanotechnology in Cement-Based Construction

ISBN 978-981-4800-76-1 (Hardcover)
ISBN 978-0-429-32849-7 (eBook)

Contents

Preface

In recent years, developments in the field of nanotechnology have impressively increased because of extensive scientific and technological research. Now they involve, among others, branches like science, medicine, and engineering.

While many books are now on the market, the demand for specialized references that go in depth into single topics still remains.

In this general frame, this book focuses on the application of nanotechnology to cement-based materials for civil engineering applications.

It is aimed at giving both scholars and practitioner engineers up-to-date information about recent developments and perspectives in this field, with special reference to the topics that seem susceptible to major developments in the near future.

In fact, while in the past, nanoinclusions into cement-based composites had the main purpose of increasing the mechanical strength, today it is possible to give materials many interesting and useful properties.

This book is divided into two parts.

The first part, "Advanced Cement-Based Composites," comprises nine chapters and presents novelties in the production of cementitious nanocomposites. Special attention is paid to the types of nanoinclusions, to novel techniques to mix the components, and to the analysis of the properties that can be achieved by paste, mortar, or concrete if they are added with nanoinclusions.

The second part, "Innovative Applications of Advanced Cement-Based Nanocomposites," is devoted to the analysis of the new properties that can be given to nanocomposites.

Among them, very promising is the capability of sensing mechanical strain, which can lead to the production of embedded sensors for dynamic monitoring of structures. They can measure internal forces as well as detect damage.

Moreover the use of phase-changing materials gives the nanocomposite the capability of storing and releasing heat, with evident possibility of application in the field of heating and cooling apartments, houses, and industrial plants.

Finally, they can collect energy from the environment (energy harvesting), as well efficiently shield electromagnetic waves.

Annibale Luigi Materazzi
2020

Part I

Advanced Cement-Based Composites

Chapter 1

Nanoinclusions for Cementitious Materials

Antonella D'Alessandro
Department of Civil and Environmental Engineering, University of Perugia,
Via G. Duranti 93, 06125 Perugia, Italy
antonella.dalessandro@unipg.it

The development of nanotechnology allows the progress of several disciplines. New devices permit us to investigate materials and objects in depth and detect characteristics and properties at the nanoscale. Moreover, the production of always more novel nanomaterials with enhanced capabilities determines the nano-engineering of composite devices with potentialities not possible before. In the field of engineering, the availability of new micro- and nanoparticles, above all the carbon-based ones, produces multi-functional composite materials by adding inclusions in the original matrix. Concrete and cementitious materials result particularly suitable for a nano- or micromodification due to the composite nature that characterizes them and the presence of micro- and nano-elements in their structure. The application of cement-based materials doped with nanofillers in concrete structures, such as buildings, bridges, and pavements, enables us to improve their performance, to detect degradations and damages during the service

Nanotechnology in Cement-Based Construction
Edited by Antonella D'Alessandro, Annibale Luigi Materazzi, and Filippo Ubertini

ISBN 978-981-4800-76-1 (Hardcover), 978-0-429-32849-7 (eBook)
www.jennystanford.com

life, and consequently to increase the safety of the users of such structures and infrastructures.

1.1 Introduction

Different types of inclusions for cementitious materials, with various characteristics and properties, are available in the market and their number is constantly growing. The most adopted particles in the field of engineering are the carbon-based ones and oxides, such as nanosilica and nanotitania. Their applicability is strictly related to the good dispersion of the filler into the matrix in order to obtain a homogeneous material. Fillers can be particles or fibers: both types exhibit benefits and disadvantages. Indeed, fibrous particles possess a higher aspect ratio, which is the ratio between length and diameter, related to enhanced electrical and mechanical performances but also related to the difficulty in obtaining a good dispersion and to a greater tendency to be damaged during the mixing [1].

This chapter is aimed at describing the main particles suitable for cement-based materials, the main dispersing techniques of nanoinclusions in a cementitious matrix, and the issues related to the effects on health and on the environment related to their use.

1.2 Dispersion of Nanoinclusions in a Cementitious Matrix

An optimal dispersion of inclusions in a material determines (i) the reliability and stability of the sensing and mechanical properties, (ii) the enhancement of the filler efficiency, and (iii) the decrease of the mechanical energy for mixing [2]. The dispersability of an inclusion depends on its hydrophilicity—because the cement-based materials are water based—and on its predisposition of producing agglomerations, due to the mutual forces between the elements forming the fillers. There are three main methods used to disperse nanoinclusions in a cement matrix: using dispersants, by mechanical mixing, and using sonication (Fig. 1.1). Dispersants can be sur-

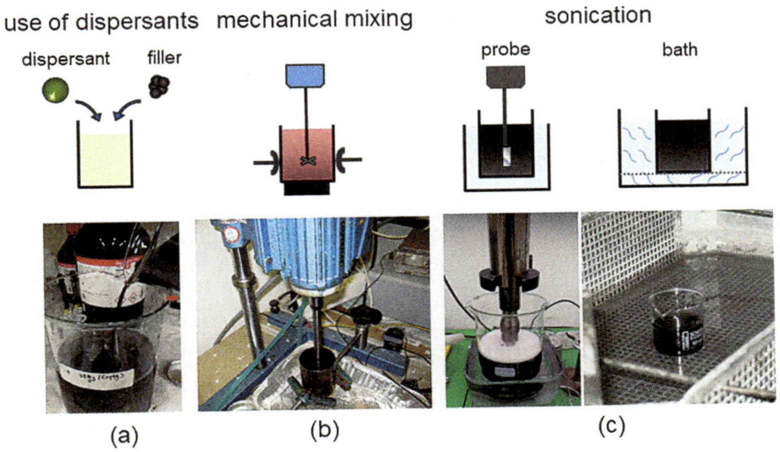

Figure 1.1 Dispersion methods (a) using dispersants, (b) through mechanical mixing, and (c) by sonication.

factants and mineral admixtures (Fig. 1.1a). The former possess a particular chemical composition able to disperse by wetting, electrostatic repulsion, or steric hindrance effect [1]. The mineral admixtures act with gradation, adsorption, and separation effects. However, some of the effective surfactants are not compatible with cementitious materials and with their applications. Mechanical methods, based on milling at high speed, are not effective with all types of particles and may produce damages in fibrous fillers due to the shear energy released during the mix (Fig. 1.1b). Often mechanical mixing is adopted in combination with ultrasonication, with an embedded probe, or through the immersion in a sonicated bath (Fig. 1.1c). Effective results also are obtained through the combination of organic ligands, including polycarboxilate-based water reducers, with probe sonication [3].

Chemical methods use covalent surface modification through surface functionalization of nanoparticles in order to improve their chemical compatibility with the matrix, enhancing their wettability and reducing their tendency to agglomerate. However, they result in aggressive methods because the procedures might introduce structural defects and changing in peculiar properties that could affect their capabilities [4].

An alternative innovative approach was proposed by Cwirzen et al.: the nanoparticles, as carbon nanotubes (CNTs) or carbon nanofibers (CNFs), are grown on the surface of cement using a modified chemical vapor deposition method [5].

1.3 Nanoinclusions for Cement-Based Materials

Concrete and cement-based materials are composite materials with components and pores at the nanoscale, too (Fig. 1.2). So, they are particularly suitable for nanomodification. The addition of nanofillers may enhance the concrete properties, acting on the hydration products or producing a filler effect. For example, the calcium-silicate-hydrate (C–S–H) gel, which determines the mechanical and physical properties—shrinkage, elasticity, creep, porosity, etc.—of cement pastes, can be modified in order to enhance durability. Besides, micro- or nanosize particles may improve the resistance of the composites through the filler effect or may provide new multifunctional capabilities, as thermal or electrical conductivity. There are various types of fillers for cementitious materials, from dimensions of millimeters up to nanometers (Fig. 1.3). When the dimensions of a material decrease, important changes occur, for example, in electronic conductivity, optical absorption,

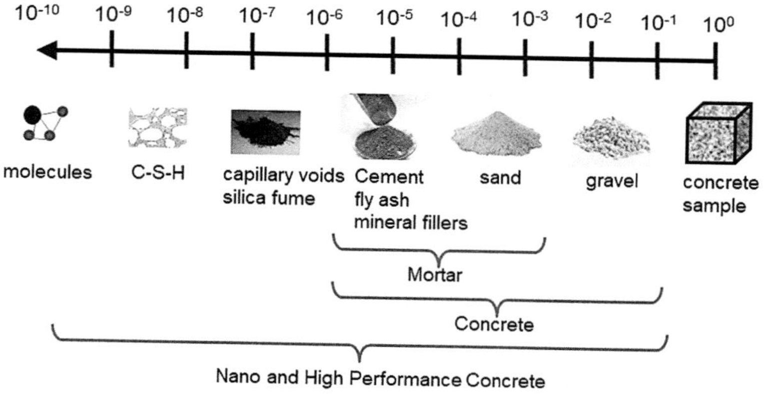

Figure 1.2 Dimensions of constituents of concrete and definition of main cement-based structural materials.

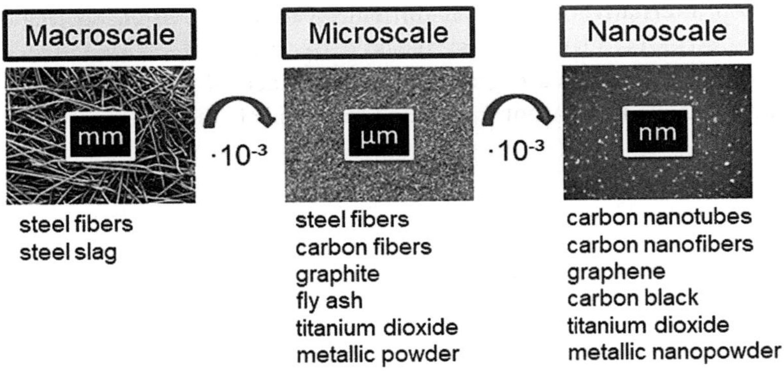

Figure 1.3 Classification of the main fillers for concrete at the macro-, micro-, and nanoscale.

mechanical resistance, and chemical reactivity. The same materials can demonstrate enhanced properties at the nanosize level. The two main reasons of such behavior are the higher specific surface and the quantum effects that increase the reactivity [6]. Nanomaterials for cement-based materials can be organized into three main categories: carbon-based nanofillers, metallic nanoinclusions, and other noncarbon nanoinclusions.

Nanoparticles have costs 100 to 1000 times higher than Portland cement or others constituents of cement-based materials: this occurrence must be considered in the material design in relation to enhancement of the structural efficiency deriving from the use of such nanoparticles. So a key issue with the use of nanofillers in cementitious materials is the choice of the optimal amount of addition [7].

The main concrete properties influenced by the presence of nanoparticles are workability, cohesiveness, content of entrained air in concrete, setting times (for fresh concrete), hydration reactions, compressive strength, tensile/flexural strength, modulus of elasticity, and durability (for hardened concrete) [8].

1.3.1 Carbon-Based Inclusions

The nanoparticle fillers most used for engineering applications are carbon-based ones. Carbon is the only element with stable allotropes from the zero to the third dimension. Carbon-based fillers include CNTs, CNFs, carbon black (CB), graphene nanoplatelets (GNPs), and graphene oxide (GO) [6]. These ones are characterized by different morphology and aspect ratio and can be classified as 0D (nanoparticles), 1D (nanofilaments), and 2D (nanosheets). Generally, their reinforcement effect into composite materials proportionally increases with an increase in the aspect ratio.

1.3.1.1 Carbon nanotubes

CNTs are rolled sheets of graphene arranged in cylinders. They can be single-walled CNTs (with one rolled sheet) or multiwalled CNTs (with more concentric cylinders). Single-walled CNTs have internal diameter of 0.4–2.5 nm and a length of tenths of microns; multiwalled CNTs have an average inner diameter of 5–10 nm and a length of 0.1–10 microns (Fig. 1.4a). They can be considered as 1D fillers. Both types of CNTs are used to strengthen composite materials and to provide materials with conductive and thermal capabilities. Carbon atoms were arranged in a hexagonal pattern with strong covalent bonds, type σ–σ, while the tubes were connected through van der Waals forces, which determine the formation of agglomerates during their dispersion into cementitious matrices. Such undesirable structures can be limited through the use of a proper dispersion technique. CNTs appear as a black powder, with a very high specific surface area (of the order of 100–250 m^2/g) and a very low apparent density. The important properties of CNTs make them suitable for applications in different fields. Their strength-to-weight ratio higher than other materials allows us to obtain light-strength composites, as those used for spacecraft. In cement-based materials they are adopted, above all, for increasing mechanical performance [9, 10] and for smart monitoring applications [11–13]. CNTs can be metals, semiconductors, or small-gap semiconductors, depending on the orientation of the graphene sheets. They possess enhanced mechanical capabilities: they are highly resilient, their Young's modulus is roughly 1.2 TPa, and their tensile strength is about a hundred times higher than that of steel [6].

Their piezoresistive capability is used to realize composites with self-monitoring properties. Moreover, nanocement (made of CNTs within the cementitious material) has the potential to create tough, durable, high-temperature, fire-resistant coatings [14]. The literature shows positive mechanical performances of doped cement-based materials with volume contents lower than 0.01% of CNTs [15, 16], related to a total porosity reduction [17]. Also single-walled CNTs determine a nucleation effect in cement-based composites [18].

Moreover, the rheological behavior of cementitious materials with CNTs is different from that of normal cement-based materials, showing a higher suspension viscosity [9].

1.3.1.2 Carbon nanofibers

CNFs are 1D nanofillers as CNTs but with greater dimensions and a different internal structure. They are stacked according to three different patterns: cups, cones, or plates. They can be produced by low heat treatment with temperatures up to 1500°C, which partially chemically graphitize the vapor-deposited carbon on the surface and allow us to obtain ordered structures. Their average diameter and length vary from 70 to 200 nm and 50 to 200 μm, respectively (Fig. 1.4b). They have a tensile strength slightly less than that of CNTs, but their cost is about 50 times lower, making them particularly appropriate for scalable applications [19]. Their mechanical and conductive properties make them suitable for realizing cementitious materials with high mechanical resistance [20] and electrical capabilities for applications in de-icing [21], self-sensing [22], and electromagnetic shielding [23].

1.3.1.3 Graphene nanoplatelets

GNPs generally are chemically exfoliated from natural graphite. They consist of small layers of graphene with an overall thickness of about 3–10 nm (Fig. 1.4c). In the scientific literature, these elements are often named "graphite nanoplatelets" [24]. GNPs are excellent electrical and thermal conductors as a result of their pure graphitic composition. Unlike other nanofillers presented before, they are 2D, resulting in the achievement of isotropic nanomodified composites [25]. Graphene is one of the strongest materials: its tensile strength is approximately 100–300 times higher than that of steel. So, its possible applications are an interesting topic in the field of engineering. Added to cement matrices, graphene nanomaterials provide enhanced mechanical [26] and smart capabilities, resulting in multifunctional composites [27]. Graphene is an expensive material with respect to other carbon-based nanoparticles, and it is highly hydrophobic, so a dispersion procedure is needed to obtain homogeneous composites. Cementitious materials doped with hybrid GNPs show enhanced strength, fracture resistance, and higher durability than high-performance concrete (HPC) [28]. Also, such nanomaterials significantly contribute to arrest crack generation.

1.3.1.4 Carbon black

CB is composed of pure elemental carbon in the form of colloidal spherical particles in an amorphous molecular structure (Fig. 1.4d). The row material is produced by incomplete combustion of organic materials such as petroleum or coal. CB particles are able to provide electrical conductivity both in water and in solvent-borne systems. Electron tunneling is one of the factors influencing the electrical conductivity of such nanoparticles, a property that comes from the capacity of electrons to jump the gaps between the closed CB pigments. This amorphous carbon allotrope represents a cheaper alternative to provide electrical properties to cement-based materials. The reduced size of CB and its electrical conductivity make it an economical method for protecting steel rebars from corrosion [29]. Despite being less effective than carbon fibers, the literature [30, 31] reports a good self-sensing behavior of cement matrices containing CB: an increase in the compressive stress leads to a linear decrease in the fractional change of the electrical resistance [32]. Other researchers [33] have measured both the mechanical and the piezoresistive performances of mortar specimens doped with CB. They found that the addition of CB with a content of approximately 4% w/b (weight-to-binder ratio) is

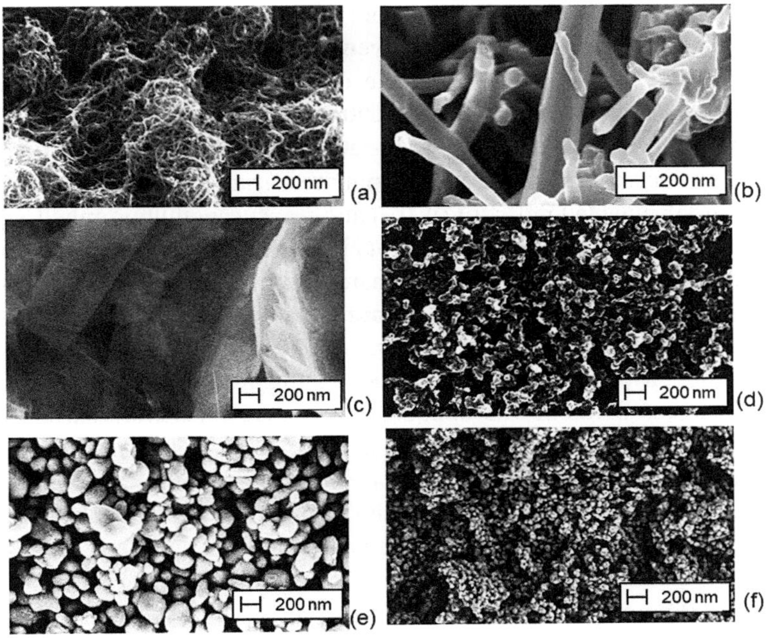

Figure 1.4 Micrographs obtained by scanning electron microscopy of different fillers: (a) multiwalled CNTs, (b) CNFs, (c) GNPs, (d) CB, (e) titania, and (f) nanonickel particles.

favorable for improving the compressive and tensile strengths of the composites.

1.3.1.5 Graphene oxide

GO consists of graphite oxidized with oxygen molecules that separate the carbon layers [34]. Carboxyl, hydroxyl, and epoxy functional groups confer high water solubility to GO but, on the other hand, degrade its electrical properties [35]. GO can compete economically with CNTs because of its ease of production and its higher solubility in aqueous solutions. It acts as a nucleation area for C–S–H gel, promoting cement hydration [36]. A 0.05 wt% of GO increases the compressive strength of cement paste by 15%–33% and the flexural strength by 41%–59% [37]. Also, the presence of GO enhances the ductility of cementitious materials.

1.3.2 Metallic Nanoinclusions

1.3.2.1 Nano-TiO_2

Titanium dioxide nanoparticles (titania, TiO_2) are white pigments with reflective properties. Nano-TiO_2 is the most widely used photocatalyst filler in construction materials and the second most used nano-oxide particle [7]. It has been successfully applied in the production of self-cleaning concrete, as in the construction of the Dives in Misericordia Church in Rome. This nanoparticle is able to break down organic pollutants, volatile organic compounds, and bacterial membranes through a photocatalytic reaction. Its hydrophilic capability provides self-cleaning properties to materials where it is dispersed or applied. The resulting concrete surfaces remain with a white color [14]. Nano-TiO_2 possesses a high density (about 3.9 g/cm^3) [7], which determines a higher hardness and has a direct influence on the properties and durability of cement-based materials where it is dispersed. Literature studies show that compressive strength gains about 10% and 20% for filler contents of 0, 5, and 10 wt% [38]. The packing effect of nano-TiO_2 in cementitious materials reduces permeability and the chloride diffusion coefficient [39]. Also, nano-TiO_2 accelerates cement hydration and increases the reaction total heat, indicating a nucleation effect of nano-TiO_2 [38, 40].

1.3.2.2 Nano-Fe_2O_3

An enhancement of compressive strength of composites with these nano-oxides is observed in cement-based materials containing 0.5%, 1.25%, and 2.5% of these nanoparticles [16]. The capillary permeability decreases for concrete with nano-Fe_2O_3. Literature studies also show that it is able to confer self-sensing properties to cementitious materials where it is dispersed [42].

1.3.2.3 Silver nanoparticles

Nanosilver particles are able to inhibit the growth of bacteria, with a disinfection effect toward infection, odor, itchiness, and sores. The strength of such particles is their capability to be distributed uniformly due to their very small size. Their application results in the realization of smart surfaces and coatings [14].

1.3.2.4 Nano-Al_2O_3

Aluminum oxide nanoparticles react with calcium hydroxide in the hydration of calcium silicates. This pozzolanic reaction is related to the amount of available surface area. So, the addition of alumina improves the mechanical properties of concretes, such as tensile and flexural strength. The alumina could also substitute cement up to 2.0% as a binder of cementitious materials, with an optimal replacement amount of 1.0% [7, 9], even if good results have been obtained for higher amounts [41]. Nano-Al_2O_3, or nanoalumina, is also identified as a retarder of cement hardening [43]. The capillary permeability decreases for concrete with nano-Al_2O_3. It is also able to determine a significant increase in the elastic modulus (up to 143% with an addition of 5%) but with a limited effect on compressive strength [44].

1.3.2.5 Nano-ZnO

Nano-ZnO possesses semiconducting and piezoelectric dual properties, which can confer to materials, such as cement-based ones, enhanced electrical properties. Used for concrete manufacturing, it also improves the processing time and the resistance of concrete against water [14].

1.3.2.6 Nano-ZrO_2

Zirconium oxide nanoparticles, or zirconia nanopowder, have dimensions between 5 and 100 nm. They are white and possess high-surface-area particles (specific surface area of 25–50 m^2/g). Nano-ZrO_2, or nanozirconia, shows good properties with regard to aesthetics (translucency), mechanical strength (hardness, flexibility, durability), and chemical resistance (basically inert) and is a notable insulator [14]. Small amounts of nano-ZrO_2, up to 2.5%, enhance the compressive strength of nanocementitious composites [25].

1.3.2.7 Nano-MgO

Nano-MgO is a newly explored chemical nanofiller for cementitious materials. It possesses properties as a shrinkage compensator, and it is more effective than other expansion agents [46].

1.3.3 Noncarbon Nanoinclusions

1.3.3.1 Nano-SiO_2

Nano-SiO_2, or nanosilica, is the main studied oxide nanofiller for cement-based materials that could be delivered in powdered or liquid suspensions. Its real density is about 2.25–2.54 g/cm^3 [6, 7], and it leads to the cheapest cost–volume relationship for oxide nanoparticles. A pozzolanic reaction is directly related to its high surface area.

The use of nano-SiO_2 in concretes significantly increases the compressive strength by filling the pores between large fly ash and cement particles. Compressive strength could increase by about 18% due to the reaction of nanomaterials with calcium hydroxide crystals of the interfacial zone between cement paste and aggregates, producing C–S–H gel. Flexural strength could increase up to 8%. Also the filling action of nano-SiO_2 enhances the strength of the composites through the resulting denser microstructure [47]. The optimal percentage of addiction is around 3% because greater amounts may reflect negatively to the performance of the composite due to the agglomeration of the fillers. Other studies [41] have indicated that the optimum nano-SiO_2 volume content is about 2%, and the compressive strength decreases after this filler quantity. Permeability shows a reduction of about 45% for concrete with 3 wt% of nano-SiO_2 [48], with a decrease of carbonation velocity and chloride diffusion coefficient [49, 39]. These properties and the durability enhancement could be related with two important microstructural aspects of cementitious nanomaterials: the packing effect and the pozzolanic reaction of the nano-SiO_2 filler [14]. Moreover, nano-SiO_2 is able to decrease the setting time of cementitious materials when compared to silica fume (microsize), can reduce bleeding water and avoid segregation, and is helpful in facilitating the use of recycled materials in the cement. The addition of nano-SiO_2 to cement-based materials can also decrease the degradation of the calcium-silicatehydrate reaction caused by calcium leaching into water, preventing water penetration and so enhancing the durability. Nano-SiO_2 is a cement-hardening accelerator; it improves workability and resistance to water penetration [50]. Furthermore, nano-SiO_2 is one of the most economical, the most studied, and, also, the most consumed nanofiller in cement worldwide [51].

1.3.3.2 Nano-$CaCO_2$

Nano-$CaCO_3$ accelerates the hardening process of concrete and reduces the shrinkage, although high humidity of the environment is needed during curing of the samples to improve their durability [52]. Small amounts of such a nanofiller, up to 2.5%, enhance the compressive strength of nanocementitious composites [53]. Higher percentages improve microhardness and the elastic modulus [54], which produce also a nucleation effect.

1.3.3.3 Nanoclay

Another pozzolanic material used in the cement matrix is nanomontmorillonite, commonly called nanoclay. It consists of a three-layered 2D structure of aluminum inserted between two layers of silicon. It seems promising in enhancing mechanical properties, resistance to chloride penetration, and self-compacting ability and in reducing the permeability and shrinkage [55] of cementitious materials. Nanoclay also acts as a plasticizer and reduces permeability [56]. Compressive strength of cementitious materials could increase by about 11% due to the reactions in the interfacial zone between cement paste and aggregates, producing C–S–H gel. Flexural strength could increase by 4% [47].

1.3.3.4 Cement nanoparticles

Cement nanoparticles are binder nanoparticles of C_2S, C_3S, C_3A, and C_4AF obtained by sol-gel processing and heating treated by a flame spray reactor and oven heating. They possess a high surface area, which determines a notable water demand, and consequently, the nanocomposites they create generally have higher water/binder ratios and so greater porosity. Also, ultrasonic treatment induces a positive impact on compressive strength [7].

1.4 Safety of Nanomaterials

The risks related to the use of nanomaterials essentially occur by exposures during the preparation of nanoparticles and their possible accidental release. The toxicity of nanocomposites depends on a number of factors such as particle size, surface area, crystallinity, surface chemistry, and particle agglomeration tendency [57, 58].

The main risks of nanoparticles can be summarized as follows:

- Aggregation: Leads to poor corrosion resistance, high solubility, and deterioration with consequent issues in the structure maintenance [59, 60].
- Reactivity or charge: Chemical structure and charge-related critical functional groups could be critical for specific functionality and for bioavailability of nanomaterials [61].
- Impurity: Nanoparticles can interact with impurities due to their high reactivity and generate undesirable features.
- Contaminant dissociation: This refers to contamination of residual impurities in the nanoparticles.
- Size: It affects the reactivity and agglomeration of nanoparticles.
- Recycling and disposal: There aren't definitive experimental results about nanoparticles' exposure and their potential toxicity.

The available data about the toxicity of nanomaterials concerns:

- Dose and exposure time effect, which is related to the penetration in cells and depends on the molar concentration of the nanoparticles [62].
- Aggregation and concentration effect, with contradictory points of view. Indeed, some researches assert that the

increase of nanoparticles enhances agglomeration features unable to penetrate the cells, with loss of toxicity.

- Particle size effect: Studies show size-dependent toxicity [63].
- Particle shape effect: Nanoparticles possess different toxicity levels at different aspect ratios [64].
- Surface area effect: Generally the toxicological effect of nanoparticles increases with a decrease in the particle size and an in the surface area.
- Crystal structure effect: Different structures may exhibit different cellular uptake, oxidative mechanisms, and subcellular localizations [65].
- Surface functionalization effect: Different particles' functionalization is related to different oxidation processes [66, 67].
- Pre-exposure effect: Pre-exposure to nanoparticles results in the adaptability of the human body [68].

Protective measures and actions against nanoparticles' risks are available. Mainly they consist in (i) engineering control (use of ventilated/filtered environment and boxes), (ii) administrative controls (appropriate information, clear and exhaustive procedures, good hygiene practices), (iii) personal protective equipment (PPE; respiratory protection devices, dermal protection, and eye protection), and (iv) medical screening and surveillance (proper medical continuous screening).

It is also important to note that nanoparticles are widely used in bioapplications, but evidence about potential adverse health and environmental effects due to protracted exposure to them has not yet been established. However, the environmental impact of such nanomaterials and of their residual waste is expected to increase in the future [69]. Moreover, many factors can influence their ecotoxicity, such as pH, salinity, and organic residues [70].

1.5 Discussion and Conclusion

Nowadays, novel nanoparticles are always available in the market, due to the progress of nanotechnology. There are different types of nanomaterials suitable to be added to cementitious mixtures, providing enhanced properties or new ones. Among other nanofillers, carbon-based ones appear particularly promising for the realization of multifunctional concretes. Also nanotitania and nanosilica are increasingly adopted for their effectiveness and low cost. Issues related to the preparation of composites with nanofillers are associated with the dispersion procedures and the choice of the optimal amount. The homogeneity of nanocomposites is essential to realize feasible materials, and higher additions may determine a deterioration of peculiar properties. A delicate task also concerns the health and environmental effects related to an increasing use of nanoparticles, because in-depth epidemiology studies with univocal and clear results are not completely available.

Acknowledgments

The support of the European Union's Framework Programme for Research and Innovation HORIZON 2020 under Grant Agreement No. 765057 and of the Italian Ministry of Education, University and Research (MIUR) through the funded Project of Relevant National Interest "SMART-BRICK: Novel Strain-Sensing Nanocomposite Clay Brick Enabling Self-Monitoring Masonry Structures" is gratefully acknowledged.

References

1. Horszczaruk, E., Sikora, P., Lukowski, P. (2016). Application of nanomaterials in production of self-sensing concretes: contemporary developments and prospects, *Arch. Civ. Eng.*, **LXII**(3), pp. 61–73.
2. Han, B., Ding, S., Yu, X. (2015). Intrinsic self-sensing concrete and structures: a review, *Measurement*, **59**, pp. 110–128.
3. Kawashima, S., Houa, P., Corr, D. J., Shah, S. P. (2015). Modification of cement-based materials with nanoparticles, *Cem. Concr. Compos.*, **36**, pp. 8–15.
4. Han, B., Yu, X., Ou, J. (2011). Multifunctional and smart carbon nanotube reinforced cement-based materials, in Gopalakrishnan, K., Birgisson, B., Taylor, P., Attoh-Okine, N. O. (eds.) *Nanotechnology in Civil Infrastructure: A Paradigm Shift*, Springer.

5. Cwirzen, A., Habermehl-Cwirzen, K., Nasibulina, L. I., Shandakov, S. D., Nasibulin, A. G., Kauppinen, E. I., Mudimela, P. R., Penttala, V. (2009). CHH cement composite, in *Nanotechnology in Construction 3*, Springer.
6. Bastos, G, Patiño-Barbeito, F., Patiño-Cambeiro, F., Armesto, J. (2016). Nano-inclusions applied in cement-matrix composites: a review, *Materials*, **9**(12), p. 30.
7. Mendes, T., Hotza, D., Repette, W. (2015). Nanoparticles in cement based materials: a review. *Rev. Adv. Master. Sci.*, **40**, pp. 89–96.
8. Yonathan Reches, Y. (2018). Nanoparticles as concrete additives: review and perspectives, *Constr. Build. Mater.*, **175**, pp. 483–495.
9. Konsta-Gdoutos, M. S., Metaxa, Z. S., Shah, S. P. (2010). Highly dispersed carbon nanotube reinforced cement based materials, *Cem. Concr. Res.*, **40**(7), pp. 1052–1059.
10. Collins, F., Lambert, J., Duan, W. H. (2012). The influences of admixtures on the dispersion, workability, and strength of carbon nanotube-OPC paste mixtures, *Cem. Concr. Compos.*, **34**(2), pp. 201–207.
11. Coppola, L., Buoso, A., Corazza, F. (2011). Electrical properties of carbon nanotubes cement composites for monitoring stress conditions in concrete structures, *Appl. Mech. Mater.*, **82**, pp. 118–123.
12. D'Alessandro, A., Ubertini, F., García-Macías, E., Castro-Triguero, R., Downey, A., Laflamme, S., Meoni, A., Materazzi, A. L. (2017). Static and dynamic strain monitoring of reinforced concrete components through embedded carbon nanotube cement-based sensors, *Shock Vib.*, **2017**, 3648403 (11 pp).
13. Han, B., Yu, X., Ou, J. (2011). Multifunctional and smart nanotube reinforced cement-based materials, in Gipalakrishnan, K., Birgisson, B., Taylor P., Attoh-Okine, N. (eds.) *Nanotechnology in Civil Infrastructure: A Paradigm Shift*, Springer, pp. 1–48.
14. Olar, R. (2011). Nanomaterials and nanotechnologies for civil engineering, *Buletinul institutului politehnic din iaşi*, **LIV**(LVIII), 4, pp. 109–118.
15. Cwirzen, A., Cwirzen K., Pentalla, V. (2008). Surface decoration of carbon nanotubes and mechanical properties of cement/carbon nanotube composites, *Adv. Cem. Res.*, **20**(2), pp. 65–73.
16. Konsta-Gdoutos, M. S., Metaxa Z., Shah, S. (2010). Multi-scale mechanical and fracture characteristics and early-age strain capacity of high performance carbon nanotube/cement nanocomposites, *Cem. Concr. Compos.*, **32**(2), pp. 110–115.

17. Nochaiya T., Chaipanich, A. (2011). Behavior of multi-walled carbon nanotubes on the porosity and microstructure of cement-based materials, *Appl. Surf. Sci.*, **257**(6), pp. 1941–1945.
18. Makar, J. M., Chan, G. W. (2009). Growth of cement hydration products on single-walled carbon nanotubes, *J. Am. Ceram. Soc.*, **92**(6), pp. 1303–1310.
19. Yazdani, N., Brown, E. (2016). Carbon nanofibers in cement composites: mechanical reinforcement, in Loh, K., Nagarajaiah, S. (eds.) *Innovative Developments of Advanced Multifunctional Nanocomposites in Civil and Structural Engineering*, Woodhead Publishing, Sawston, UK, pp. 217–246.
20. Metaxa, Z. S., Konsta-Gdoutos, M. S., Shah, S. P. (2010). Mechanical properties and nanostructure of cement-based materials reinforced with carbon nanofibers and Polyvinyl Alcohol (PVA) microfibers, in *Advances in the Material Science of Concrete, Proceedings of the Session at the ACI Spring 2010 Convention*, Chicago, IL, USA, 21–25 March 2010, American Concrete Institute (ACI): Farmington Hills, MI, USA, pp. 115–126.
21. Galao, O., Bañón, L., Baeza, F. J., Carmona, J., Garcés, P. (2016). Highly conductive carbon fiber reinforced concrete for icing prevention and curing, *Materials*, **9**, p. 281.
22. Hanbay, S., lankson, M. A. (2017). Self-sensing damage assessment and image-based surface crack quantification of carbon nanofibre reinforced concrete, *Constr. Build. Mater.*, **134**, pp. 520–529.
23. Chung, D. D. L. (2012). Carbon materials for structural self-sensing, electromagnetic shielding and thermal interfacing, *Carbon*, **50**(9), pp. 3342–3353.
24. Meng W., Khayat K. (2016). Mechanical properties of ultra-high-performance concrete enhanced with graphite nanoplatelets and carbon nanofibers, *Composite Part B*, **107**, pp. 113–122. doi:10.1016/j.compositesb.2016.09.069
25. Le, J.-L., Du, H., Pang, S. D. (2014). Use of 2D graphene nanoplatelets (GNP) in cement composites for structural health evaluation, *Composite Part B*, **67**, pp. 555–563.
26. Wu Y., Yi N., Huang L., Zhang T., Fang S., Chang H., Li N., Oh J., Lee J., Kozlov M. (2015). Three-dimensionally bonded spongy graphene material with super compressive elasticity and near-zero Poisson's ratio, *Nat. Commun.*, **6**, pp. 1–9. doi:10.1038/ncomms7141

27. Chuah, S., Pan, Z., Sanjayan, J. G., Wang, C. M., Duan, W. H. (2014). Nano reinforced cement and concrete composites and new perspective from graphene oxide, *Constr. Build. Mater.*, **73**, pp. 113–124.

28. Graybeal, B. (2011). Ultra-High Performance Concrete. Federal Highway Administration (FHWA), U.S. Department of Transportation; Washington, DC, USA. Publication Number FHWA-HRT-11-038.

29. Masadeh, S. (2015). The effect of added carbon black to concrete mix on corrosion of steel in concrete, *J. Miner. Mater. Charact. Eng.*, **3**, pp. 271–276. doi:10.4236/jmmce.2015.34029

30. Wen, S., Chung, D. D. L. (2007). Partial replacement of carbon fiber by carbon black in multifunctional cement-matrix composites, *Carbon*, **45**, pp. 505–513. doi:10.1016/j.carbon.2006.10.024

31. Xiao, H., Lan, C., Ji, X., Li, H. (2003). Mechanical and sensing properties of structural materials with nanophase materials, *Pac. Sci. Rev.*, **5**, pp. 7–11.

32. Gao, D., Strum, M., Mo, Y. L. (2009). Electrical resistance of carbon-nanofiber concrete, *Smart Mater. Struct.*, **18**, pp. 1–7. doi:10.1088/0964-1726/18/9/095039

33. Monteiro, A., Cachim, P., Costa, P. (2016). Carbon nanoparticles cement-based materials for service life monitoring, *Proceedings of the International RILEM Conference on Materials, Systems and Structures in Civil Engineering*, Lyngby, Denmark, 22–24 August 2016.

34. Paulchamy, B., Arthi, G., Lignesh, B. (2015). A simple approach to step wise synthesis of graphene oxide nanomaterial, *J. Nanomed. Nanotechnol.*, **6**, pp. 1–4.

35. Palermo, V., Kinloch, I., Ligi, S., Pugno, N. (2016). Nanoscale mechanics of graphene and graphene oxide in composites: a scientific and technological perspective, *Adv. Mater.*, **28**, pp. 6232–6238.

36. Babak, F., Abolfazl, H., Alimorad, R., Parviz, G. (2014), Preparation and mechanical properties of graphene oxide: cement nanocomposites, *Sci. World J.*, **2014**, pp. 1–10.

37. Pan, Z., Duan, W., Li, D., Collins, F. (2013). Graphene oxide reinforced cement and concrete. Patent Number: WO2013096990A1, 4 July 2013.

38. Chen, J., Kou, S., Poon, C. (2012). Hydration and properties of nano-TiO_2 blended cement composites, *Cem. Concr. Compos.*, **34**(5), pp. 642–649.

39. Zhang, M., Liu, H. (2011). Pore structure and chloride permeability of concrete containing nano-particles for pavement, *Constr. Build. Mater.*, **25**, pp. 608–616.

40. Senff, L., Hotza, D., Lucas, S., Ferreira, V., Labrincha, J. A. (2012). Effect of nano-SiO_2 and nano-TiO_2 addition on the rheological behavior and the hardened properties of cement mortars, *Mater. Sci. Eng., A*, **532**, pp. 354–361.

41. Oltulu, M., Sahim, R. (2011). Simple and combined effects of nano-SiO_2, nano-Al_2O_3 and nano-Fe_2O_3 powders on compressive strength and capillary permeability of cement mortar containing silica fume, *Mater. Sci. Eng.*, **528**, pp. 7012–7019.

42. Li, H., Xiao, H.-G., Yuan, J., Ou, J. (2004). Microstructure of cement mortar with nanoparticles, *Composite Part B*, **35**(2), pp. 185–189.

43. Land, G., Stephan, D. (2014). Controlling cement hydration with nanoparticles, *Cem. Concr. Compos.*, **57**, pp. 64–67

44. Li, Z., Wang, H., He, S., Lu, Y., Wang, M. (2006). Investigation on the preparation and mechanical properties of the nano-allumina reinforced cement composite, *Mater. Lett.*, **60**(3), pp. 356–359.

45. Broekhuizen, F. A., Broekhuizen, J. C. (2009). Nanotehnologia în industria europeană a construcţiilor – Stadiul actual al tehnologiei (transl. from English), pp. 7–30. www.efbww.org

46. Shah, S., Hou, P., Konsta-Gdoutos, M. (2015). Nano-modification of cementitious material: toward a stronger and durable concrete, *J. Sustainable Cem. Based Mater.*, **5**, pp. 1–22.

47. Mohamed, A. M. (2016). Influence of nano materials on flexural behaviour and compressive strength of concrete, *HBRC J.*, **12**, pp. 212–225.

48. Ji, T. (2005). Preliminary study on the water permeability and microstructure of concrete incorporating nano-SiO_2, *Cem. Concr. Res.*, **35**(10), pp. 1943–1947.

49. Collepardi, M., Olagot, J. J. O., Skarp, U., Troli, R. (2002). Influence of amorphous colloidal silica on the properties of self-compacting concretes, in *Challanges of Concrete Construction*, p. 12.

50. Sanchez, F., Sobolev, K. (2010). Nanotechnology in concrete: a review, *Constr. Build. Mater.*, **24**, pp. 2060–2071.

51. Chithra, S., Senthil Kumar, S. R. R., Chinnaraju, K. (2016). The effect of colloidal nano-silica on workability, mechanical and durability properties of high performance concrete with copper slag as partial fine aggregate, *Constr. Build. Mater.*, **15**, pp. 794–804.

52. Cai, Y., Hou, P., Zhou, Z., Cheng, X. (2016). Effects of nano-$CaCO_3$ on the properties of cement paste: hardening process and shrinkage

at different humidity levels, in *Proceedings of the 5th International Conference on Durability of Concrete Structures*, Shenzhen, China, 30 June–1 July 2016.

53. Liu, X., Chen, L., Liu, A., Wang, X. (2012). Effect of nano-$CaCO_3$ on properties of cement paste, *Energy Procedia*, **16**, pp. 991–996.
54. Sato, T., Beaudoin, J. (2011). Effect of nano-$CaCO_3$ on hydration of cement containing supplementary cementitious materials, *Adv. Cem. Res.*, **23**(1), pp. 33–43.
55. Mutuk, H., Mutuk, T., Gümüş, H., Oktay, B. M. (2016). Shielding behaviors and analysis of mechanical treatment of cements containing nanosized powders, *Acta Phys. Pol.*, **130**, pp. 172–174.
56. Chang, T.-P., Shih, J.-Y., Yang, K., Hsiao, T.-C. (2007). Material properties of Portland cement paste with nano-montmorillonite, *J. Mater. Sci.*, **42**, pp. 7478–7487.
57. Jeevanandam, J., Barhoum, A., Yen, S., Chan, Y. S., Dufresne, A., Danquah, M. K. (2018). Review on nanoparticles and nanostructured materials: history, sources, toxicity and regulations, *Beilstein J. Nanotechnol.*, **9**, pp. 1050–1074.
58. Mathialagan Sumesh, M., Alengaram, U. J., Jumaat, M. Z., Mo, K. H., Alnahhal, M. F. (2017). Incorporation of nano-materials in cement composite and geopolymer based paste and mortar: a review, *Constr. Build. Mater.*, **148**, pp. 62–84.
59. Kennedy, A. J., Hull, M. S., Steevens, J. A., Dontsova, K. M., Chappell, M. A., Gunter, J. C., Weiss, C. A. (2008). Factors influencing the partitioning and toxicity of nanotubes in the aquatic environment, *Jr. Environ. Toxicol. Chem.*, **27**, pp. 1932–1941. doi:10.1897/07-624.1
60. Wang, Y., Li, Y., Pennell, K. D. (2008). Influence of electrolyte species and concentration on the aggregation and transport of fullerene nanoparticles in quartz sands, *Environ. Toxicol. Chem.*, **27**, pp. 1860–1867. doi:10.1897/08-039.1
61. Tervonen, T., Linkov, I., Figueira, J. R., Steevens, J., Chappell, M., Merad, M. (2009). Risk-based classification system of nanomaterials, *J. Nanopart. Res.*, **11**, pp. 757–766. doi:10.1007/s11051-008-9546-1
62. Buzea, C., Pacheco, I. I., Robbie, K. (2007). Nanomaterials and nanoparticles: sources and toxicity, *Biointerphases*, **2**, MR17–MR71. doi:10.1116/1.2815690
63. Ivask, A., Kurvet, I., Kasemets, K., Blinova, I., Aruoja, V., Suppi, S., Vija, H., Käkinen, A., Titma, T., Heinlaan, M. (2014). Size-dependent

toxicity of silver nanoparticles to bacteria, yeast, algae, crustaceans and mammalian cells in vitro, *PLoS One*, **9**, p. e102108. doi:10.1371/journal.pone.0102108

64. Lippmann, M. (1990). Effects of fiber characteristics on lung deposition, retention, and disease, *Environ. Health Perspect.*, **88**, pp. 311–317. doi:10.1289/ehp.9088311

65. Xia, T., Kovochich, M., Brant, J., Hotze, M., Sempf, J., Oberley, T., Sioutas, C., Yeh, J. I., Wiesner, M. R., Nel, A. E. (2006). Comparison of the abilities of ambient and manufactured nanoparticles to induce cellular toxicity according to an oxidative stress paradigm, *Nano Lett.*, **6**, pp. 1794–1807. doi:10.1021/nl061025k

66. Oberdörster, G., Oberdörster, E., Oberdörster, J. (2005). Nanotoxicology: an emerging discipline evolving from studies of ultrafine particles, *J. Environ. Health Perspect.*, **113**, pp. 823–839. doi:10.1289/ehp.7339

67. Sayes, C. M., Fortner, J. D., Guo, W., Lyon, D., Boyd, A. M., Ausman, K. D., Tao, Y. J., Sitharaman, B., Wilson, L. J., Hughes, J. B., West, J. L., Colvin, V. L. (2004). The differential cytotoxicity of water-soluble fullerenes, *Nano Lett.*, **4**, pp. 1881–1887. doi:10.1021/nl0489586

68. Johnston, C. J., Finkelstein, J. N., Mercer, P., Corson, N., Gelein, R., Oberdörster, G. (2000). Pulmonary effects induced by ultrafine PTFE particles, *Toxicol. Appl. Pharmacol.*, **168**, pp. 208–215. doi:10.1006/taap.2000.9037

69. Khana, I., Saeed, K., Khan, I. (2019). Nanoparticles: properties, applications and toxicities, *Arabian J. Chem.*, **12**(7), pp. 908–931.

70. Handy, R. D., von der Kammer, F., Lead, J. R., Hassellöv, M., Owen, R., Crane M. (2008). The ecotoxicology and chemistry of manufactured nanoparticles, *Ecotoxicology*, **17**, pp. 287–314.

Chapter 2

Dispersion Techniques of Nanoinclusions in Cement Matrixes

Matteo Tiecco
Department of Chemistry, Biology and Biotechnology, University of Perugia, Via Elce di Sotto 8, 06124 Perugia, Italy
matteotiecco@gmail.com

2.1 Carbon Nanotubes: Chemical Structure and Properties

Carbon nanotubes (CNTs) are nanoscaled hollow cylinders composed of a hexagonal arrangement of sp^2-hybridized carbon atoms. The CNTs can be interpreted as sheets of graphene rolled up to form cylindrical nanostructures, with diameters ranging from one to tens of nanometers and with lengths up to micrometers [1, 2]. CNTs are generally classified into two main categories: they can be single-walled (SWCNTs, in the case of a single roll of graphene composing them) or multiwalled (MWCNTs, in the case of multiple layers of graphite superimposed) nanotubes [3, 4]. The van der Waals interactions occurring among the sheets determine the merge of the tubes of CNTs into self-aggregating fibers or bundles [5, 6]. In

Nanotechnology in Cement-Based Construction
Edited by Antonella D'Alessandro, Annibale Luigi Materazzi, and Filippo Ubertini

ISBN 978-981-4800-76-1 (Hardcover), 978-0-429-32849-7 (eBook)
www.jennystanford.com

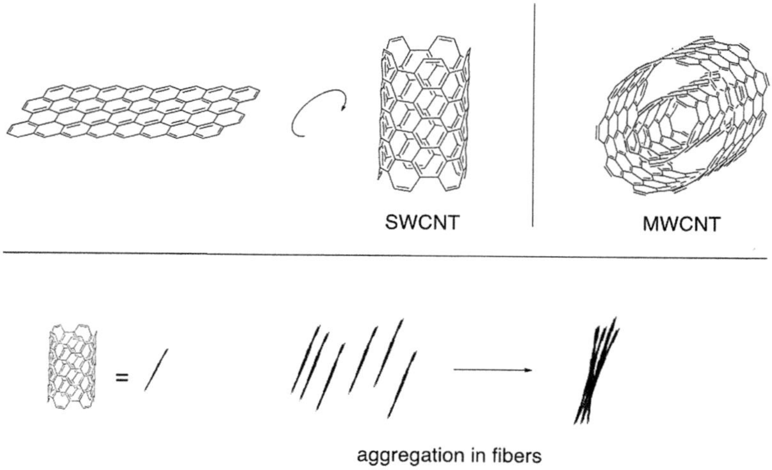

Figure 2.1 Representations of a single-walled carbon nanotube (SWCNT) as a rolled sheet of graphene, a multiwalled carbon nanotube (MWCNT), and the aggregation in bundles/fibers of CNTs.

Fig. 2.1 the structures of SWCNTs, MWCNTs, and a scheme of their aggregations in fibers are reported.

There are many synthetic procedures that can be used to synthesize CNTs; their dimensions, their properties, and the yields of the procedure depend on the method used to realize them, and every method has either advantages or disadvantages. The arc discharge method is one of the most used, thanks to its ease; with this procedure both SWCNTs and MWCNTs can be created, using different catalysts, in a low-pressure inert gas chamber through the arc vaporization of two graphite rods [7, 8]. The disadvantages of this procedure are represented by the by-products obtained and the purifying procedures needed to obtain pure CNTs. This is even more relevant in the case of MWCNT synthesis. The laser ablation method is based on the use of pulsed laser on a graphite target in a high-temperature reactor [9, 10]. This method gives high-purity products, but high energy is required for the procedure.

The applications of CNTs are wide and have been increasing in recent years as they represent unique nanostructured materials. Nanotechnology, nano-electronics, composite materials, and biomedical applications are some relevant examples of scientific fields where both SWCNTs and MWCNTs are finding advantageous roles [4, 11–13]. Moreover, recent studies on the low or absent toxicity of these nanomaterials, as well as their capability of membrane penetration, have opened the way for their application as biological carriers of drugs [14, 15].

Unfortunately, it is well known that both SWCNTs and MWCNTs are practically insoluble in water or in organic solvents; this is due to their chemical structures as well as their high dimensions and to their folding into fibers that lead to even bigger aggregates even harder to solubilize in both polar and nonpolar solvents [5, 6, 16–19]. These facts limit the application of these nanostructured systems if the problem is not faced properly.

In this chapter, the most used and effective techniques used in the literature for the CNT dispersion are treated; the ultrasonication/surfactant method (as well as a description of surfactants structures and properties) will be treated in depth for its high relevance in the field of cement-based sensor realization. The efficacious dispersions of MWCNTs in water is, in fact, a key factor for the inclusion of these nanostructured materials in cement matrixes.

2.2 Dispersion Techniques of Carbon Nanotubes: *Similia Similibus Solvuntur?*

The terms "solubilization" and "dispersion" of CNTs in water or in organic solvents are improperly used in the literature. Both these terms are almost equally reported and used [20]; however, they refer to two different processes: a solution, therefore solubilization, is referred to as a stable solution, while a dispersion is the formation of a colloidal system. These two processes are related also to the dimensions of the solutes: in the case of small dimensions (lower than nanometer units), it is better to refer to a solution; with solutes of higher dimensions (higher than nanometer units), a colloidal dispersion is formed. In the literature of CNT dispersions, the stability times of suspensions are variable, and also the dimensions of CNTs in solution [17]; therefore it is not possible to definitely define them as solutions or dispersions. In this chapter, the term "dispersion" will be considered as the most proper one except in specific cases of solubilization.

There are many efficacious methods reported in the literature for CNT dispersion in different solvents, but they can be simply divided into different categories, depending on the kind of technique used:

- Physical methods (i.e., ultrasonication, plasma treatment, general mechanical methods) are based on the use of physical treatment on CNTs—solvent systems without any chemical modification of the CNTs or without the use of other additives as dispersants.
- Chemical methods (i.e., functionalization of the CNTs' chemical structures, additives) are based on the functionalization of the CNTs' surface by adding hydrophilic/hydrophobic functional groups via covalent bonding to the CNTs or with weak interacting additives such as surfactants.
- Biochemical methods (i.e., use of biochemical molecules as carriers) are based on the use of biological molecules as carriers, in a host–guest manner. They are biovariants of the chemical methods and are relevant in the biomedical application of CNTs.

All these methods were also successfully implemented by using combinations of them: the use of ultrasonication and surfactants is the most interesting method and will be treated in depth in this chapter. Biochemical methods will not be discussed as they represent methods for use in biological systems of CNTs; therefore they are not pertinent to the topic of this chapter and of this book.

2.2.1 Physical Methods for CNT Dispersion

There are many physical methods reported in the literature to facilitate the dispersion of CNTs in water or organic solvents; all of them are based on the disaggregation of the fibers and bundles that the CNTs are able to form in the solvents. The energy applied by the external source is able to form isotropic CNT suspensions that are stable in different times, depending on the technique used. Mechanical mill-based methods result in effective rotating structures that can provoke an effective CNT dispersion in water solutions due to the rotation of disks with fixed pots on them [21]. This method gave interesting results, and it was applied even with the use of low temperatures and with the addition of surfactants [22]. The use of microwaves was found to decrease the density of CNTs in water, leading to the formation of a more isotropic system [23]. The advantage of this method is the easy access to microwave systems for chemical synthesis in chemistry laboratories; this method was successfully employed also in a chemical method to functionalize the CNT surface [24]. Different types of plasma techniques were also successfully used with and without the addition of surfactant molecules [25–27].

2.2.1.1 Ultrasonication physical method

The ultrasonication procedure is based on the application of sound energy to a sample at ultrasonic frequencies (over 20 kHz). In a CNT–solvent system, ultrasonication provokes an exfoliation of the fibers and bundles of CNTs due to a process of cavitation, leading to a dispersion of the CNTs [28, 29]. There are mainly two kinds of ultrasonication that are based on the sonication energy of the procedure and therefore on the instrument used. In Fig. 2.2 the two different procedures are schematically reported: the strong procedure is based on the use of a sonicating probe, while the weak procedure is based on the use of a simple water bath sonicator. Generally, the water bath sonicator is cheaper than the sonicating probe.

To obtain an efficacious dispersion, the power output of the sonicator is not relevant, as well as the sonication times, but is more important the sonication energy. Particular attention must be paid to the ultrasonication procedures, as, at high energies, this treatment can provoke damages on the CNTs and therefore lead to

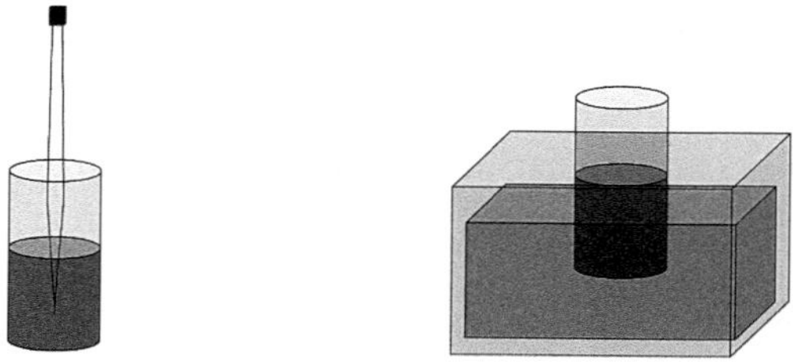

Figure 2.2 Schematic representations of the two ultrasonication methods: strong with a probe immerged in the sample (left) and weak with the sample in a sonicating bath (right).

a loss and/or change of their properties. These damages can be the opening of the CNT rolls or cracks in the CNTs.

There are two cavitation processes that an ultrasonication treatment can provoke on CNTs' bundles or fibers: stable and inertial. The stable process modifies the chemical surface of the CNTs. and the inertial process is a simple exfoliation of the fibers. A proper treatment of the CNT–solvent system (in terms of the sonication energy, times, and the CNTs used) must be applied in order to obtain the desired effects. Many spectroscopic techniques can be used to evaluate the effect of the ultrasonication on CNT suspensions, and particularly eventual damages on the CNT structures. Raman spectroscopy, atomic force microscopy (AFM), scanning electron microscopy (SEM), transmission electron microscopy (TEM), and ultraviolet-visible (UV-Vis) spectroscopy are successfully applied in the literature to evaluate the integrity of samples, and some of these techniques are also nondestructive [30–33].

The ultrasonication method is successfully applied also in combination with dispersing agents such as surfactants [34]; this is one of the cases of mixed physical-chemical techniques. This method has many advantages because the fibers of CNTs can be opened or exfoliated by ultrasonication and the additive can efficaciously interact with the CNT surfaces. This procedure will be better discussed later in this chapter, because it represents one of the most relevant techniques for CNT dispersion.

2.2.2 Chemical Methods for CNT Dispersion

The chemical methods are based on the use of synthetic molecules as additives or on the proper covalent functionalization of the CNTs' surface in order to obtain an effective dispersion of the CNTs in the solvent. These methods are often helped by the use of other physical methods that permit the exfoliation of the CNTs' fibers. The covalent methods are various, and they can be classified in terms of the class of the molecules used for the functionalization.

Some examples of the most commonly used inorganic functionalization methods are:

- Ozone, peroxides, oxides, and strong acids. These are used to functionalize the CNTs' surface to give oxygen derivatives. This functionalization prevents the folding of the CNTs into fibers and increases the dispersion capabilities in different solvents, depending on the hydrophobicity/hydrophilicity of the attached residue [35, 36].
- Decoration of the CNTs with quantum dots. This leads also to efficient use as fluorescent nanoprobes [37].
- Metals, metal nanoparticles, and ions that prevent the folding into fibers into fibers and increase hydrophilicity [38].

Some examples of the use of organic additives/functionalization are represented by:

- Covalent functionalization with alkyl groups that leads to easy dispersion in organic media [39]
- Binding/covalent binding of aromatic molecules that also possess optical properties for "smart functionalization" [40]
- Surfactants

2.2.2.1 Surfactants: structure, properties, and solubilizing capabilities

Surfactants are a class of organic molecules that are formed by two separate and different portions: a hydrophobic one and a hydrophilic one. The smart properties of this class of molecules are due to these peculiar chemical structures [41, 42]. These molecules are also called "amphiphiles." In Fig. 2.3 is reported a

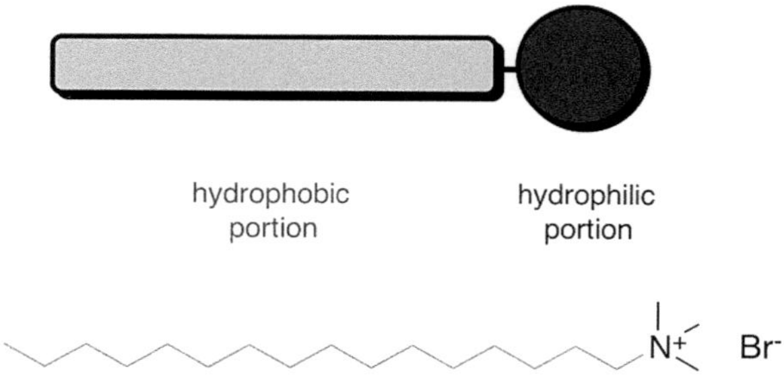

Figure 2.3 (Top) Schematic representation of a surfactant with the hydrophobic portion (tail) and the hydrophilic portion (head). (Bottom) Cetyltrimethylammonium bromide (CTAB) chemical structure.

schematic representation of the chemical structure of a surfactant and an example of one of the most used molecule of this class, cetyltrimethylammonium bromide (CTAB).

The presence in their structures of both polar and nonpolar portions leads to easy solubilization of these molecules both in polar (such as water) and nonpolar solvents. The increase of their concentration in solution leads to a self-aggregation process; this reduces the interactions between the solvent and the differently structured portions, leading to micelle formation [43, 44]. This phenomenon occurs at a concentration that is specific for every surfactant, which is called critical micellar concentration (CMC). This value depends on the structural features of the surfactant, particularly on its hydrophobicity. Micelles are nanoscaled spherical aggregates of molecules of surfactants that are spontaneously formed in order to reduce the contacts between the hydrophobic portions and water or between the hydrophilic portions with apolar solvents. In Fig. 2.4 are reported the structures of a micelle (surfactants in water) and of a reverse micelle (surfactants in apolar/hydrophobic solvents).

The formation of these aggregates is spontaneous because they are energetically favored due to the high reduction of the repulsions between the solvent and the opposite-structured portions. The spherical structures, in fact, can "hide" from the solvent the portions of the molecules that are nonsimilar to it [45]. The increase of the

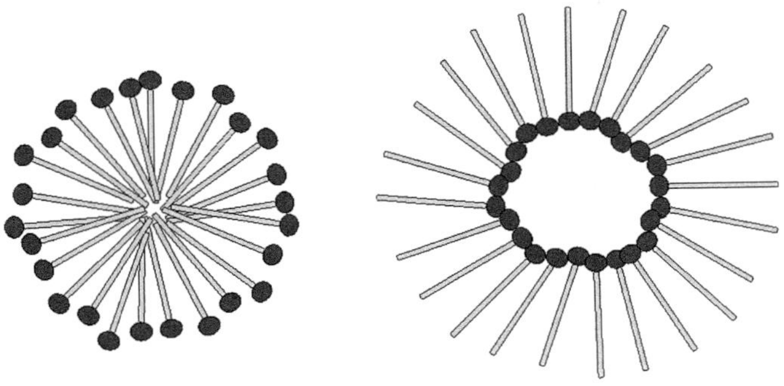

Figure 2.4 Schematic representations of micellar aggregate in water (left) and reverse micellar aggregate in organic/apolar solvents (right).

dimensions of hydrophobic portions in the surfactant molecules decreases the value of the CMC because the process is more favored.

The name "surfactant" derives from a contraction of the terms SURFace ACTing AgeNTS: this is because these molecules have interfacial properties. In water they arrange on the surface, exposing the hydrophobic tail over it and the polar head in water, determining a decrease of surface tension [45].

The amphiphilic properties of these compounds permit to solubilize organic compounds in water and/or polar compounds in organic solvents. This is because the solute can interact with a specific portion of these molecules, while the other portion can interact efficaciously with the solvent, leading to a solubilization effect. This process can be accomplished by the monomers of surfactants or by the micellar aggregates that can be interpreted as vectors for solubilization [46]. The solubilization of greasy or organic materials by surfactants is responsible of the cleansing properties of the commonly used soaps, which can solubilize efficaciously oil material in water [42].

Surfactants can be classified into different types, depending on the structure of the polar head (that can be negatively or positively charged with a counterion or that can be neutral or zwitterionic) or of the number of heads and hydrophobic portions (*gemini* or *twin chain* surfactants). The properties, and therefore the applications, of these molecules can drastically change by small changes in their structures [47].

Due to their structural features, and due to their properties that can be easily modulated by simple chemical changes, the surfactants find a wide range of applications. They can act as gel supramolecular assemblies by the intercalation of organic molecules inside the micellar aggregates [48, 49]; they can act as biocides, or they can act as vector agents of water-insoluble drugs of for DNA, thanks to the structural similarity with the phospholipids of biological membranes [50–52]; they can act as general solubilizing agents to permit synthetic processes in a water solvent [53]; and they permit to obtain novel organic liquids with high solubilizing capabilities [54, 55].

Thanks to these properties, CNTs' dispersion in water is successfully accomplished by surfactants.

2.3 Dispersion of Carbon Nanotubes in Water with Surfactants: *Similia Similibus Solvuntur* (with the Help of Ultrasonication)

Widely used for "oil in water" and "water in oil" dispersions, surfactants are commonly used for CNT dispersions in water, both single-walled and multiwalled CNTs [56, 57].

The interactions between surfactant molecules and CNTs are based on the principle *similia similibus solvuntur*: a Latin phrase that means that similar substances will dissolve similar substances, or "like dissolves like." In the case of surfactants, their two parts (in terms of their different hydrophobicity/hydrophilicity) permit to solubilize/disperse hydrophobic solutes or molecules in polar solvents (such as water) and vice versa (Fig. 2.5).

The interactions between surfactants and CNTs occur at the surfaces of the tubes: CNTs are hydrophobic; therefore the amphiphiles can interact with them with their hydrophobic tails. The polar (and often charged) headgroups of the surfactants therefore permit this CNT–surfactant aggregate to be easily dispersed in water for the favored ion–water interactions [58]. The CNTs therefore

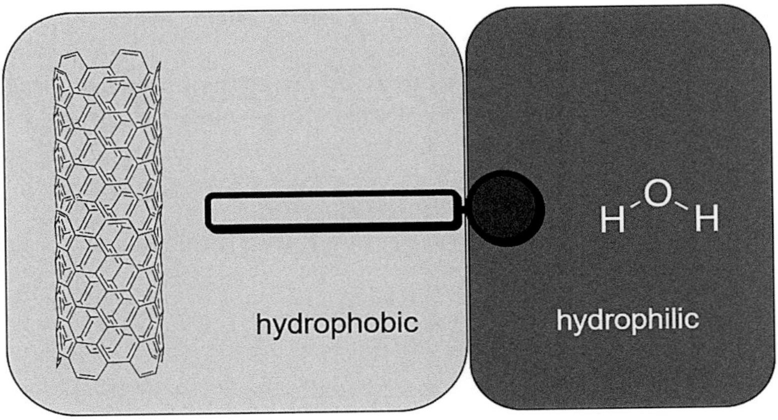

Figure 2.5 Affinity of the two portions of a surfactant with hydrophobic CNTs (with the tail) and with water (with the headgroup).

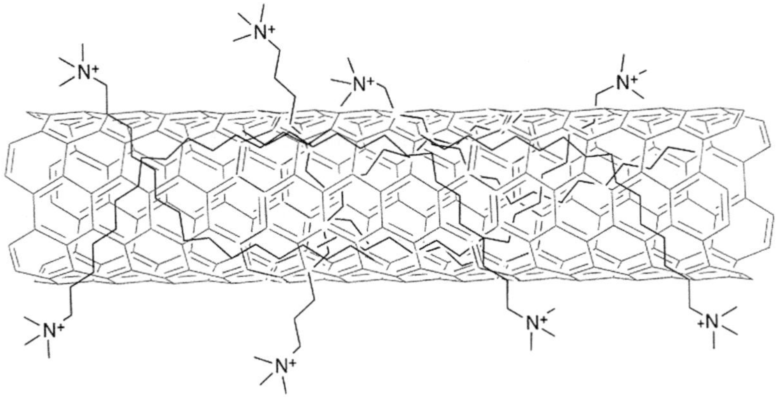

Figure 2.6 Schematic representation of the coating of a CNT's surface with monomers of surfactants.

result as functionalized with polar and/or charged groups on their surfaces. Even if the use of surfactants is a chemical method, these interactions are noncovalent. In Fig. 2.6 is reported a scheme of interaction between CNTs and monomers of surfactants.

The experimental procedures developed in the literature for CNTs' dispersion in water with surfactants are also based on

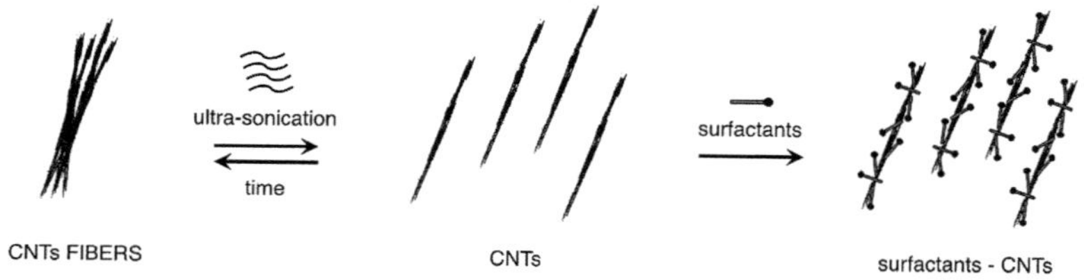

Figure 2.7 Schematic representation of the effect of ultrasonication on CNT bundles (opening of the fibers, reversible after variable times) and of surfactants on CNTs' surfaces (coating, prevention of reaggregation in fibers).

the use of ultrasonication processes; therefore the methods can be considered mixed physical/chemical ones [34, 59, 60]. This dual technique permits an efficacious dispersion of CNTs: first the ultrasonication opens the bundles to give isolated CNTs, and then the surfactants can interact with the surfaces of the dispersed CNTs. The charged surfaces avoid these functionalized nanostructures to spontaneously reaggregate in fibers because of the equal charges' repulsions [61]. Single CNTs can in fact reaggregate because of the van der Waals forces between the tubes that provoke the fiber formation in water. These structures are also smaller than the aggregated structures, so they can be easily dispersed. In Fig. 2.7 the schematic representation of a CNT's dispersion thanks to ultrasonication and surfactants is shown. This procedure is efficaciously used applying ultrasonication on the samples of CNT–surfactants in water.

The surfaces of the CNTs, coated by these charged species, can possess different properties, and this could also help further functionalization of the CNTs, if needed [60, 62].

2.3.1 Optimization of CNT Dispersion with Surfactants

A knowledge of the chemistry of surfactants is needed in order to obtain efficacious and stable CNT dispersions in water and therefore their use in many applications where these dispersions are needed [63]. The first parameter that must be considered is the surfactants' concentration. The surfactants can interact with the surface of CNTs as monomers, as micellar aggregates, or by forming aggregates on

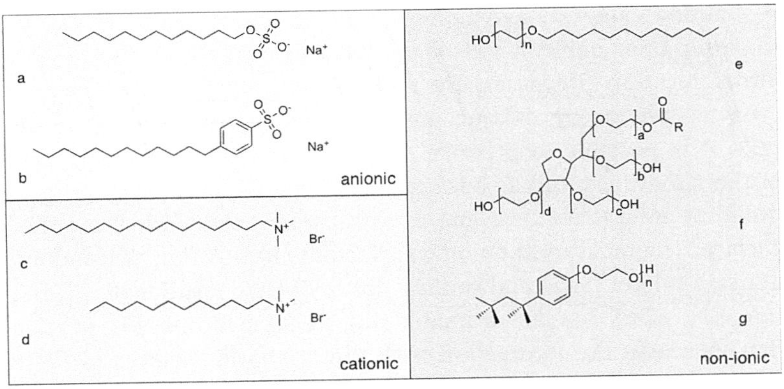

Figure 2.8 Most common commercially available surfactants used in the literature for CNTs' dispersion. Anionic: (a) sodium dodecyl sulfate (SDS) and (b) sodium dodecylbenzenesulfonate (SDBS). Cationic: (c) cetyltrimethylammonium bromide (CTAB) and (d) dodecyltrimethylammonium bromide (DTAB). Non-ionic: (e) Brij, (f) Tween, and (g) Triton X.

their surfaces; these interactions can occur even simultaneously, and they can depend also on the molecular structure of the surfactant used. The concentration of surfactants needed to obtain efficacious surfactant–CNT interactions is generally lower than their CMC. Moreover, the absence of micellar aggregates at concentrations higher than their CMC is evidence of interaction of the monomers with the CNTs: the molecules are involved in the interactions with the tubes' surfaces and cannot form micellar aggregates. An excess of surfactants can also provoke a flocculation of CNTs that favors their precipitation [63]. The temperature also plays an important role in the efficacious dispersions as lower temperatures can obviously decrease the dispersion capabilities of the technique [64].

The most important parameter that must be considered in this topic is represented by the surfactant chemical structure.

2.3.1.1 Commercially available surfactants for CNT dispersions

In the literature many commercially available surfactants are used for the dispersion of CNTs in water [65]. In Fig. 2.8 are reported the structures of the most commonly used amphiphiles in this topic.

Cationic, anionic, and non-ionic amphiphiles are used in this topic for their ability to disperse CNTs in water with the help of ultrasonication processes. By tailoring the surfactant headgroup type or size or by enhancements in the hydrophobicity of the tails, it is possible to provoke different capabilities of dispersion of the nanotubes; this is because the different structures can give different interactions, or more efficacious ones, with the CNTs' surface. The most used anionic surfactants in this field are sodium dodecyl sulfate (SDS) and sodium dodecylbenzenesulfonate (SDBS) [66, 67]. SDBS has an aromatic ring in its hydrophobic portion that increases the interaction with the aromatic surfaces of CNTs and therefore increases the amount of dispersed CNTs [65]. CTAB and dodecyltrimethylammonium bromide (DTAB) are commonly used cationic commercially available surfactants and they were successfully used in this topic [68, 69]. Their solubilizing properties are similar, even though they differ in terms of their hydrophobic tail length. In the class of non-ionic surfactants, the molecular weight of the amphiphile is relevant in order to evaluate the efficacy in CNTs' dispersion [70, 71]. This could be explained for an effect of steric stabilization of "bigger" surfactants on CNTs' surfaces; this avoids the reorganization in fibers of the CNTs [65].

There are also many works reporting the efficacy of mixtures of different amphiphiles and water-soluble polymers for CNTs' dispersions in water [72, 73]. The properties of the mixtures of surfactants (and the properties of these aggregates) are peculiar, and they do not correlate with the properties of the single amphiphiles composing them. In some cases (mixtures of cationic and anionic amphiphiles), charge neutralization phenomena of differently charged amphiphiles could occur, and this changes the amphiphile properties and the properties of the aggregates.

Commercially available amphiphiles are efficaciously used in the literature of this topic; however, due to the low number of commercially available amphiphiles and due to the small changes in their structures, it is not still possible to determine structure–activity relationships of the efficacy of the surfactants in the CNTs' dispersion with the use of commercially available molecules. The available molecules studied are in fact not properly structured in

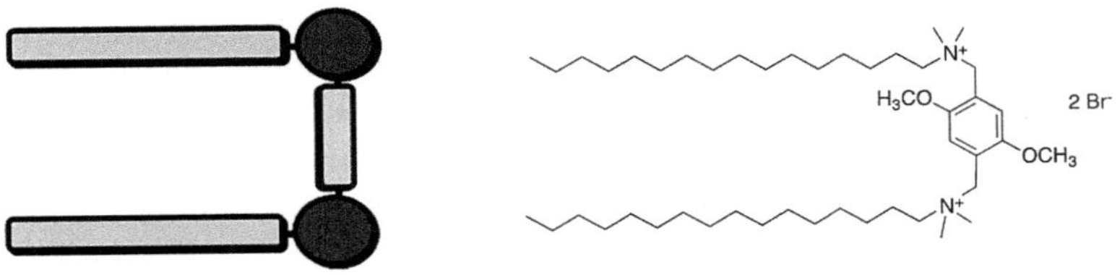

Figure 2.9 Schematic representation of a two-headed and two-tailed *gemini* surfactant linked with a spacer (left). Chemical structure of 2,5-dimethoxy-1,4-bis[N-(n-dodecyl)-*N*,*N*-dimethylammoniomethyl]phenyl-bromide [pXMo(DDA)2Br].

order to evaluate the most effective portions of them involved in the CNTs' dispersion.

2.3.1.2 Increasing CNT dispersion with the use of properly designed surfactants

The use of specifically designed and synthesized surfactants can increase the advantages of the use of these molecules with very simple synthetic procedures. The synthesis of surfactants, in fact, is often very easy and involves few passages from the starting materials [74]. Small changes in the chemical structures of surfactants can lead to dramatic changes in their properties; therefore the design, the realization, and the structure–activity relationships that could be achieved are fundamental for the increase of the effectiveness of these molecules and therefore of the entire process where amphiphiles are involved.

A few papers have been up to now reported in the literature about the use of synthetic surfactants in the dispersion of CNTs in water, but the results are interesting and promising [75]. Many papers are about the use of *gemini* synthetic surfactants: double-headed and double-tailed structures with a linker between the headgroups (Fig. 2.9).

Gemini surfactants have about twice the molecular weight of the correspondent single-chain amphiphiles, but their properties are often more than twice effective than them [47]. The use of a *gemini* surfactant in CNTs dispersions in water has efficacious results, as reported in many different papers in the literature, and this is because of the increase of hydrophobicity of the amphiphiles (as well as the case of non-ionic surfactants) and because of specific interactions occurring between the amphiphile and the CNTs' surface [76]. The aromatic rings often present in these structures (in the spacers between the heads or on the hydrophobic tails) promote the interaction/association of the monomers on the aromatic surfaces of CNTs (*similia similibus solvuntur*).

Here are some relevant examples of the use of *gemini* amphiphiles. Di Crescenzo and co-workers used synthetic *gemini* surfactants with aromatic portions in the spacers and obtained dispersions with a surfactant–CNT ratio four to seven times lower than commercially available CTAB, SDS, and SDBS amphiphiles [77]. Liu and co-workers used imidazolium-based *gemini* surfactants and increased the dispersions of CNTs using 0.1 mg/mL solutions of amphiphiles [78]. Chen and co-workers revealed that only 0.6 wt% of a *gemini* surfactant can lead to 0.5 wt% of multiwalled CNT dispersions [79]. In this field of the dispersion of CNTs in water with surfactants, the use of synthetic amphiphiles could open for the realization of an optimal molecule for the purpose.

References

1. Iijima, S. (1991). Helical microtubules of graphitic carbon, *Nature*, **354**, p. 56.
2. Iijima, S., Ichihashi, T. (1993). Single-shell carbon nanotubes of 1-nm diameter, *Nature*, **363**, pp. 603–605.
3. Peng, X., Wong, S. S. (2009). Functional covalent chemistry of carbon nanotube surfaces, *Adv. Mater.*, **21**, pp. 625–642.
4. Pumera, M., Merkoçi, A., Alegret, S. (2007). Carbon nanotube detectors for microchip CE: comparative study of single-wall and multiwall carbon nanotube, and graphite powder films on glassy carbon, gold, and platinum electrode surfaces, *Electrophoresis*, **28**, pp. 1274–1280.
5. Bandyopadhyaya, R., Nativ-Roth, E., Regev, O., Yerushalmi-Rozen, R. (2002). Stabilization of individual carbon nanotubes in aqueous solutions, *Nano Lett.*, **2**, pp. 25–28.

6. Vigolo, B., Penicaud, A., Coulon, C., Sauder, C., Pailler, R., Journet, C., Bernier, P., Poulin, P. (2000). Macroscopic fibers and ribbons of oriented carbon nanotubes, *Science (80-)*, **290**, pp. 1331–1334.
7. Hutchison, J. L., Kiselev, N. A., Krinichnaya, E. P., Krestinin, A. V., Loutfy, R. O., Morawsky, A. P., Muradyan, V. E., Obraztsova, E. D., Sloan, J., Terekhov, S. V. (2001). Double-walled carbon nanotubes fabricated by a hydrogen arc discharge method, *Carbon N.Y.*, **39**, pp. 761–770.
8. Sugai, T., Yoshida, H., Shimada, T., Okazaki, T., Shinohara, H., Bandow, S. (2003). New synthesis of high-quality double-walled carbon nanotubes by high-temperature pulsed arc discharge, *Nano Lett.*, **3**, pp. 769–773.
9. Sakakibara, Y., Tatsuura, S., Kataura, H., Tokumoto, M., Achiba, Y. (2003). Near-infrared saturable absorption of single-wall carbon nanotubes prepared by laser ablation method, *Jpn. J. Appl. Phys.*, **42**, p. L494.
10. Yu, D. P., Sun, X. S., Lee, C. S., Bello, I., Lee, S. T., Gu, H. D., Leung, K. M., Zhou, G. W., Dong, Z. F., Zhang, Z. (1998). Synthesis of boron nitride nanotubes by means of excimer laser ablation at high temperature, *Appl. Phys. Lett.*, **72**, pp. 1966–1968.
11. Stankovich, S., Dikin, D. A., Dommett, G. H. B., Kohlhaas, K. M., Zimney, E. J., Stach, E. A., Piner, R. D., Nguyen, S. T., Ruoff, R. S. (2006). Graphene-based composite materials, *Nature*, **442**, pp. 282–286.
12. Jiang, K., Li, Q., Fan, S. (2002). Nanotechnology: spinning continuous carbon nanotube yarns, *Nature*, **419**, p. 801.
13. Zhang, Y., Bai, Y., Yan, B. (2010). Functionalized carbon nanotubes for potential medicinal applications, *Drug Discov. Today*, **15**, pp. 428–435.
14. Cellot, G., Cilia, E., Cipollone, S., Rancic, V., Sucapane, A., Giordani, S., Gambazzi, L., Markram, H., Grandolfo, M., Scaini, D. (2009). Carbon nanotubes might improve neuronal performance by favouring electrical shortcuts, *Nat. Nanotechnol.*, **4**, pp. 126–133.
15. Movia, D., Prina-Mello, A., Bazou, D., Volkov, Y., Giordani, S. (2011). Screening the cytotoxicity of single-walled carbon nanotubes using novel 3D tissue-mimetic models, *ACS Nano*, **5**, pp. 9278–9290.
16. Li, C., Chou, T.-W. (2003). Elastic moduli of multi-walled carbon nanotubes and the effect of van der Waals forces, *Compos. Sci. Technol.*, **63**, pp. 1517–1524.
17. Bahr, J. L., Mickelson, E. T., Bronikowski, M. J., Smalley, R. E., Tour, J. M. (2001). Dissolution of small diameter single-wall carbon nanotubes in organic solvents? *Chem. Commun.*, pp. 193–194.

18. Shimoda, H., Oh, S. J., Geng, H. Z., Walker, R. J., Bin Zhang, X., McNeil, L. E., Zhou, O. (2002). Self-assembly of carbon nanotubes, *Adv. Mater.*, **14**, p. 899.
19. Martel, R., Shea, H. R., Avouris, P. (1999). Rings of single-walled carbon nanotubes, *Nature*, **398**, p. 299. http://dx.doi.org/10.1038/18589
20. Premkumar, T., Mezzenga, R., Geckeler, K. E. (2012). Carbon nanotubes in the liquid phase: addressing the issue of dispersion, *Small*, **8**, pp. 1299–1313.
21. Munkhbayar, B., Nine, M. J., Jeoun, J., Bat-Erdene, M., Chung, H., Jeong, H. (2013). Influence of dry and wet ball milling on dispersion characteristics of the multi-walled carbon nanotubes in aqueous solution with and without surfactant, *Powder Technol.*, **234**, pp. 132–140.
22. Lee, J.-H., Rhee, K.-Y. (2010). A study on the dispersion characteristics of carbon nanotubes using cryogenic ball milling process, *J. Korean Soc. Precis. Eng.*, **27**, pp. 49–54.
23. Cho, S.-J., Shrestha, S. P., Lee, S.-B., Boo, J.-H. (2014). Electrical characteristics of carbon nanotubes by plasma and microwave surface treatments, *Bull. Korean Chem. Soc.*, **35**, pp. 905–907.
24. Wang, Y., Iqbal, Z., Mitra, S. (2005). Microwave-induced rapid chemical functionalization of single-walled carbon nanotubes, *Carbon N. Y.*, **43**, pp. 1015–1020.
25. Li, Y., Mann, D., Rolandi, M., Kim, W., Ural, A., Hung, S., Javey, A., Cao, J., Wang, D., Yenilmez, E. (2004). Preferential growth of semiconducting single-walled carbon nanotubes by a plasma enhanced CVD method, *Nano Lett.*, **4**, pp. 317–321.
26. Chen, W., Liu, X., Liu, Y., Kim, H.-I. (2010). Novel synthesis of self-assembled CNT microcapsules by O/W Pickering emulsions, *Mater. Lett.*, **64**, pp. 2589–2592.
27. Choi, Y. C., Shin, Y. M., Lee, Y. H., Lee, B. S., Park, G.-S., Choi, W. B., Lee, N. S., Kim, J. M. (2000). Controlling the diameter, growth rate, and density of vertically aligned carbon nanotubes synthesized by microwave plasma-enhanced chemical vapor deposition, *Appl. Phys. Lett.*, **76**, pp. 2367–2369.
28. Caneba, G. T., Dutta, C., Agrawal, V., Rao, M. (2010). Novel ultrasonic dispersion of carbon nanotubes, *J. Miner. Mater. Charact. Eng.*, **9**, p. 165.
29. Tarasov, V. A., Komkov, M. A., Stepanishchev, N. A., Romanenkov, V. A., Boyarskaya, R. V. (2015). Modification of polyester resin binder by

carbon nanotubes using ultrasonic dispersion, *Polym. Sci. Ser.D*, **8**, pp. 9–16.

30. Osswald, S., Havel, M., Gogotsi, Y. (2007). Monitoring oxidation of multiwalled carbon nanotubes by Raman spectroscopy, *J. Raman Spectrosc.*, **38**, pp. 728–736.
31. Hafner, J. H., Cheung, C.-L., Woolley, A. T., Lieber, C. M. (2001). Structural and functional imaging with carbon nanotube AFM probes, *Prog. Biophys. Mol. Biol.*, **77**, pp. 73–110.
32. Yuzvinsky, T. D., Fennimore, A. M., Mickelson, W., Esquivias, C., Zettl, A. (2005). Precision cutting of nanotubes with a low-energy electron beam, *Appl. Phys. Lett.*, **86**, p. 53109.
33. Grossiord, N., Regev, O., Loos, J., Meuldijk, J., Koning, C. E. (2005). Time-dependent study of the exfoliation process of carbon nanotubes in aqueous dispersions by using UV- visible spectroscopy, *Anal. Chem.*, **77**, pp. 5135–5139.
34. Strano, M. S., Moore, V. C., Miller, M. K., Allen, M. J., Haroz, E. H., Kittrell, C., Hauge, R. H., Smalley, R. E. (2003). The role of surfactant adsorption during ultrasonication in the dispersion of single-walled carbon nanotubes, *J. Nanosci. Nanotechnol.*, **3**, pp. 81–86.
35. Wiggins-Camacho, J. D., Stevenson, K. J. (2011). Mechanistic discussion of the oxygen reduction reaction at nitrogen-doped carbon nanotubes, *J. Phys. Chem. C*, **115**, pp. 20002–20010.
36. Banerjee, S., Hemraj-Benny, T., Wong, S. S. (2005). Covalent surface chemistry of single-walled carbon nanotubes, *Adv. Mater.*, **17**, pp. 17–29.
37. Bottini, M., Cerignoli, F., Dawson, M. I., Magrini, A., Rosato, N., Mustelin, T. (2006). Full-length single-walled carbon nanotubes decorated with streptavidin-conjugated quantum dots as multivalent intracellular fluorescent nanoprobes, *Biomacromolecules*, **7**, pp. 2259–2263.
38. Satishkumar, B. C., Vogl, E. M., Govindaraj, A., Rao, C. N. R. (2003). The decoration of carbon nanotubes by metal nanoparticles, in *Adv. Chem. A Sel.* CNR Rao's Publ., World Scientific, pp. 292–295.
39. Chen, S., Shen, W., Wu, G., Chen, D., Jiang, M. (2005). A new approach to the functionalization of single-walled carbon nanotubes with both alkyl and carboxyl groups, *Chem. Phys. Lett.*, **402**, pp. 312–317.
40. Liu, A., Honma, I., Ichihara, M., Zhou, H. (2006). Poly (acrylic acid)-wrapped multi-walled carbon nanotubes composite solubilization in water: definitive spectroscopic properties, *Nanotechnology*, **17**, p. 2845.

41. Holmberg, K., Jönsson, B., Kronberg, B., Lindman, B. (2002). *Surfactants and Polymers in Aqueous Solution*, Wiley Online Library.
42. Rosen, M. J., Kunjappu, J. T. (2012). *Surfactants and Interfacial Phenomena*, John Wiley & Sons.
43. Lindman, B., Wennerström, H. (1980). Micelles, *Micelles*, pp. 1–83.
44. Hoffmann, H., Ebert, G. (1988). Surfactants, micelles and fascinating phenomena, *Angew. Chemie Int. Ed.*, **27**, pp. 902–912.
45. Ray, A. (1971). Solvophobic interactions and micelle formation in structure forming nonaqueous solvents, *Nature*, **231**, pp. 313–315.
46. Edwards, D. A., Luthy, R. G., Liu, Z. (1991). Solubilization of polycyclic aromatic hydrocarbons in micellar nonionic surfactant solutions, *Environ. Sci. Technol.*, **25**, pp. 127–133.
47. Menger, F. M., Keiper, J. S. (2000). Gemini surfactants, *Angew. Chem. Int. Ed.*, **39**, pp. 1906–1920.
48. Kumar, R., Ketner, A. M., Raghavan, S. R. (2009). Nonaqueous photorheological fluids based on light-responsive reverse wormlike micelles, *Langmuir*, **26**, pp. 5405–5411.
49. Baglioni, P., Braccalenti, E., Carretti, E., Germani, R., Goracci, L., Savelli, G., Tiecco, M. (2009). Surfactant-based photorheological fluids: effect of the surfactant structure, *Langmuir*, **25**, pp. 5467–5475.
50. Shaban, S. M., Saied, A., Tawfik, S. M., Abd-Elaal, A., Aiad, I. (2013). Corrosion inhibition and biocidal effect of some cationic surfactants based on Schiff base, *J. Ind. Eng. Chem.*, **19**, pp. 2004–2009.
51. Farías, T., De Menorval, L.-C., Zajac, J., Rivera, A. (2009). Solubilization of drugs by cationic surfactants micelles: conductivity and 1 H NMR experiments, *Colloids Surf., A*, **345**, pp. 51–57.
52. Corte, L., Tiecco, M., Roscini, L., De Vincenzi, S., Colabella, C., Germani, R., Tascini, C., Cardinali, G. (2015). FTIR metabolomic fingerprint reveals different modes of action exerted by structural variants of *N*-alkyltropinium bromide surfactants on Escherichia coli and Listeria innocua cells, *PLoS One*, **10**(1), p. e0115275.
53. Antonietti, M., Wenz, E., Bronstein, L., Seregina, M. (1995). Synthesis and characterization of noble metal colloids in block copolymer micelles, *Adv. Mater.*, **7**, pp. 1000–1005.
54. Cardellini, F., Germani, R., Cardinali, G., Corte, L., Roscini, L., Spreti, N., Tiecco, M. (2015). Room temperature deep eutectic solvents of (1S)-(+)-10-camphorsulfonic acid and sulfobetaines: hydrogen bond-based

mixtures with low ionicity and structure-dependent toxicity, *RSC Adv.*, **5**, pp. 31772–31786.

55. Bowers, J., Vergara-Gutierrez, M. C., Webster, J. R. P. (2004). Surface ordering of amphiphilic ionic liquids, *Langmuir*, **20**, pp. 309–312.
56. Wang, H., Zhou, W., Ho, D. L., Winey, K. I., Fischer, J. E., Glinka, C. J., Hobbie, E. K. (2004). Dispersing single-walled carbon nanotubes with surfactants: a small angle neutron scattering study, *Nano Lett.*, **4**, pp. 1789–1793.
57. Matarredona, O., Rhoads, H., Li, Z., Harwell, J. H., Balzano, L., Resasco, D. E. (2003). Dispersion of single-walled carbon nanotubes in aqueous solutions of the anionic surfactant NaDDBS, *J. Phys. Chem. B*, **107**, pp. 13357–13367.
58. Vaisman, L., Wagner, H. D., Marom, G. (2006). The role of surfactants in dispersion of carbon nanotubes, *Adv. Colloid Interface Sci.*, **128**, pp. 37–46.
59. Li, J., Ma, P. C., Chow, W. S., To, C. K., Tang, B. Z., Kim, J. (2007). Correlations between percolation threshold, dispersion state, and aspect ratio of carbon nanotubes, *Adv. Funct. Mater.*, **17**, pp. 3207–3215.
60. Ma, P.-C., Siddiqui, N. A., Marom, G., Kim, J.-K. (2010). Dispersion and functionalization of carbon nanotubes for polymer-based nanocomposites: a review, *Composite Part A*, **41**, pp. 1345–1367.
61. Jiang, L., Gao, L., Sun, J. (2003). Production of aqueous colloidal dispersions of carbon nanotubes, *J. Colloid Interface Sci.*, **260**, pp. 89–94.
62. Sahoo, N. G., Cheng, H. K. F., Cai, J., Li, L., Chan, S. H., Zhao, J., Yu, S. (2009). Improvement of mechanical and thermal properties of carbon nanotube composites through nanotube functionalization and processing methods, *Mater. Chem. Phys.*, **117**, pp. 313–320.
63. Rastogi, R., Kaushal, R., Tripathi, S. K., Sharma, A. L., Kaur, I., Bharadwaj, L. M. (2008). Comparative study of carbon nanotube dispersion using surfactants, *J. Colloid Interface Sci.*, **328**, pp. 421–428.
64. Ham, H. T., Choi, Y. S., Chung, I. J. (2005). An explanation of dispersion states of single-walled carbon nanotubes in solvents and aqueous surfactant solutions using solubility parameters, *J. Colloid Interface Sci.*, **286**, pp. 216–223.
65. Moore, V. C., Strano, M. S., Haroz, E. H., Hauge, R. H., Smalley, R. E., Schmidt, J., Talmon, Y. (2003). Individually suspended single-walled carbon nanotubes in various surfactants, *Nano Lett.*, **3**, pp. 1379–1382.

66. Brege, J. J., Gallaway, C., Barron, A. R. (2007). Fluorescence quenching of single-walled carbon nanotubes in SDBS surfactant suspension by metal ions: quenching efficiency as a function of metal and nanotube identity, *J. Phys. Chem. C*, **111**, pp. 17812–17820.
67. Tummala, N. R. (2009). SDS surfactants on carbon nanotubes: aggregate morphology, *ACS Nano*, **3**, pp. 595–602.
68. Bai, Y., Park, I. S., Lee, S. J., Bae, T. S., Watari, F., Uo, M., Lee, M. H. (2011). Aqueous dispersion of surfactant-modified multiwalled carbon nanotubes and their application as an antibacterial agent, *Carbon N. Y.*, **49**, pp. 3663–3671.
69. Hu, C., Xu, Y., Duo, S., Zhang, R., Li, M. (2009). Non-covalent functionalization of carbon nanotubes with surfactants and polymers, *J. Chin. Chem. Soc.*, **56**, pp. 234–239.
70. Chen, R. J., Bangsaruntip, S., Drouvalakis, K. A., Kam, N. W. S., Shim, M., Li, Y., Kim, W., Utz, P. J., Dai, H. (2003). Noncovalent functionalization of carbon nanotubes for highly specific electronic biosensors, *Proc. Natl. Acad. Sci.*, **100**, pp. 4984–4989.
71. Cui, S., Canet, R., Derre, A., Couzi, M., Delhaes, P. (2003). Characterization of multiwall carbon nanotubes and influence of surfactant in the nanocomposite processing, *Carbon N. Y.*, **41**, pp. 797–809.
72. Madni, I., Hwang, C.-Y., Park, S.-D., Choa, Y.-H., Kim, H.-T. (2010). Mixed surfactant system for stable suspension of multiwalled carbon nanotubes, *Colloids Surf., A*, **358**, pp. 101–107.
73. O'Connell, M. J., Boul, P., Ericson, L. M., Huffman, C., Wang, Y., Haroz, E., Kuper, C., Tour, J., Ausman, K. D., Smalley, R. E. (2001). Reversible water-solubilization of single-walled carbon nanotubes by polymer wrapping, *Chem. Phys. Lett.*, **342**, pp. 265–271.
74. Biermann, M., Lange, F., Piorr, R., Ploog, U., Rutzen, H., Schindler, J., Schmid, R. (1987). Synthesis of surfactants, in Falbe, J. (ed.) *Surfactants in Consumer Products: Theory, Technology and Applications*, Springer-Verlag, Berlin, pp. 23–132.
75. Wang, Q., Han, Y., Wang, Y., Qin, Y., Guo, Z.-X. (2008). Effect of surfactant structure on the stability of carbon nanotubes in aqueous solution, *J. Phys. Chem. B*, **112**, pp. 7227–7233.
76. Zhang, S., Lu, F., Zheng, L. (2011). Dispersion of multiwalled carbon nanotubes (MWCNTs) by ionic liquid-based Gemini pyrrolidinium surfactants in aqueous solution, *Colloid Polym. Sci.*, **289**, pp. 1815–1819.

77. Di Crescenzo, A., Germani, R., Del Canto, E., Giordani, S., Savelli, G., Fontana, A. (2011). Effect of surfactant structure on carbon nanotube sidewall adsorption, *Eur. J. Org. Chem.*, **2011**, pp. 5641–5648.

78. Liu, Y., Yu, L., Zhang, S., Yuan, J., Shi, L., Zheng, L. (2010). Dispersion of multiwalled carbon nanotubes by ionic liquid-type Gemini imidazolium surfactants in aqueous solution, *Colloids Surf., A*, **359**, pp. 66–70.

79. Chen, L., Xie, H., Li, Y., Yu, W. (2008). Applications of cationic gemini surfactant in preparing multi-walled carbon nanotube contained nanofluids, *Colloids Surf., A*, **330**, pp. 176–179.

Chapter 3

Use of Styrene Ethylene Butylene Styrene for Accelerated Percolation in Composite Cement–Based Sensors Filled with Carbon Black

Simon Laflamme[a] and Filippo Ubertini[b]

[a] *Department of Civil, Construction, and Environmental Engineering, Iowa State University, USA*
[b] *Department of Civil and Environmental Engineering, University of Perugia, Italy*
laflamme@iastate.edu

3.1 Introduction

Recent advances in nanomaterials and synthetic metals have led to new possibilities in sensor development [3, 16], including new multifunctional materials that enable substantial improvements in the cost-effectiveness of structural health management solutions for geometrically large systems [11, 22]. In particular, sensors fabricated from cementitious-based nanocomposites have been proposed to fabricate self-sensing mortar, concrete, and asphalt [18]. These materials have great potential at improving structural

Nanotechnology in Cement-Based Construction
Edited by Antonella D'Alessandro, Annibale Luigi Materazzi, and Filippo Ubertini

ISBN 978-981-4800-76-1 (Hardcover), 978-0-429-32849-7 (eBook)
www.jennystanford.com

safety and resiliency by providing means to continuously monitor cementitious substrates in a non-intrusive manner. Examples of possible applications include self-sensing road pavements, which would significantly improve the rapidity of road inspections, non-intrusive weigh-in-motion sensors and smart slabs to improve traffic management, self-sensing concrete nuclear containment structures to monitor structural health on the internal layer, and smart mortar to monitor health of historic structures exposed to natural hazards [6].

The sensing principle of cementitious sensors leverages their piezoresistive property, where a change in geometry provokes a measurable change in resistance. It is therefore possible to measure static or dynamic strain of cementitious structures, either through contact (e.g., through electrodes) or noncontact (e.g., eddy current) methods. These sensors are ideally fabricated with a level of conductive particles close to the percolation threshold in order to improve the sensor's sensitivity [4]. Percolation is here defined as the material's phase transition zone indicated by a significant change in its electrical conductivity as a function of the filler content. Several studies have analyzed multifunctional cementitious materials doped with carbon nanoinclusions, including carbon black (CB) [12, 15, 21, 24], carbon fibers [7, 8, 21, 26], and carbon nanotubes (CNTs) [2, 5, 8, 9, 19].

Conductive cementitious materials have been studied and utilized for a few decades, primarily for de-icing of road infrastructures [17, 23, 25]. Nevertheless, the development of multifunctional cementitious materials for sensing purposes is more complex. This is primarily because the nanocomposite must provide a linear and accurately measurable signal with respect to a change in the measured state. There are fundamental challenges that limit the cost-effective fabrication of large volumes of conductive cementitious materials, namely:

- There is a trade-off between the cost of the selected conductive filler and its conductivity. Lower-conductivity nanofillers such as CB require higher loading, which may significantly affect the strength of the nanocomposite. Conversely, higher-conductivity nanofillers such as CNTs require fewer particles but result in a much higher fabrication cost.

- The majority of the proposed conductive cementitious materials necessitate a complex fabrication process to ensure homogeneous dispersion of nanoparticles [19], which impedes scalability. This is a problem particularly associated with CNTs that tend to spontaneously agglomerate in bundles due to van der Waals attraction forces [14].

The authors have previously studied the problem of scalability of existing sensing solutions, and research led to CB-based strategies [13] due to their low cost and ease of dispersion [10]. CB particles are therefore excellent candidates from economical and processing perspectives, although the amount of particle additives required to achieve electrical conductivity is prohibitive (see Wen & Chung [20] and Li et al. [15] for instance). Such a high doping content may have detrimental effects on the materials' strength. Thus, there is a need to develop techniques for assisting the formation of percolating CB networks within cementitious materials, which would decrease the requisite loading.

Various studies have reported the interaction of CB with polymers. For example, [1] showed that the use of styrene-butadiene-styrene to disperse CB particles resulted in a 40% reduction in the percolation threshold. [8] studied the percolation in composites where polyethylene was mixed with polystyrene, and found that CB was selectively distributed on the polyethylene phase and thus led to reduced percolation thresholds compared to the utilization of monophase polymers as composites. Here, we study the utilization of a block copolymer, namely styrene-ethylene/butylene-styrene (SEBS), to accelerate percolation of CB-filled cementitious materials, therefore decreasing the requisite loading of filler for the fabrication of cementitious sensors. SEBS is selected because of previous experience in successfully making conductive paint by adding a low amount of CB particles to SEBS. We termed such sensors "SEBS-CB sensors." The study builds on the results of a preliminary investigation presented in conference proceedings [12]. A more extensive study is presented here to support the hypothesis that SEBS can be used to accelerate percolation of cement paste filled with CB. Also, the fabrication procedure has been updated and validated in order to improve on the dispersion of the SEBS-CB mix.

This chapter is organized as follows. Section 3.2 gives details on materials and fabrication processes. Section 3.3 describes the research methodology. Section 3.4 presents and discusses results. Section 3.5 concludes the chapter.

3.2 SEBS-CB Sensors

3.2.1 Materials

SEBS-type Mediprene was obtained from VTC Elastoteknik AB, Sweden (density = 930 kg/m^3). It is a petroleum-based block copolymer widely used for medical applications because of its purity, softness, elasticity, and strength. CB-type Printex XE-2B (2% ash content and 500 ppm sieve residue 45 μm) was acquired from Orion Engineered Carbons (Kingswood, TX). It is characterized by a high structure (minimum oil absorption 380 cc/100g), which facilitates higher conductivity. Copper meshes to form the electrodes were acquired from McMaster-Carr (Elmhurst, IL). Portland cement type I/II was locally purchased from Ash Grove Cement Company.

All materials were utilized as received, except for CB. CB particles were ball-milled for 24 hours to break down agglomerated particles. Figure 3.1 is a magnified picture of the CB particles before and after ball milling. Ball milling resulted in finer particles of more homogeneous geometrical shapes. The average particle size decreased from approximately 900 μm to approximately 45 μm. Such procedure was used to improve on dispersion of CB.

3.2.2 Sensor Fabrication

Both SEBS-CB and CB-only sensors were fabricated. CB-only samples were used for benchmark purposes. The fabrication process of CB-only sensors is illustrated in Fig. 3.2. The desired levels of CB loadings were weighed and mixed in a high-shear mixer with the

Figure 3.1 CB particles: (a) before ball milling and (b) after 24 hours of ball milling.

required amount of water needed for a 0.45 water/cement (w/c) ratio. CB particles were mixed in water for 5 minutes in a blender. Cement was then added to CB-water, and mixed for 2 minutes in a Hobart mixer at the highest speed. Samples were then cast in molds of dimensions 5.1 × 5.1 × 5.1 cm^3 and vibrated/compacted to eliminate air voids. Two sets of copper mesh electrodes were inserted in each poured sample. The samples were covered with a damp cloth allowed to cure overnight before being demolded and allowed to cure in a curing room at 100% relative humidity at 22°C for 7 days. The samples were then air-cured dry for 4 days to let water evaporate. Figure 3.2, step 6 is a picture of a completed CB-only sample.

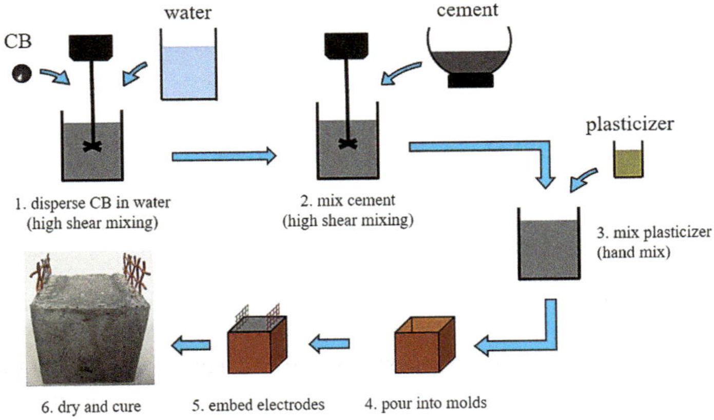

Figure 3.2 Fabrication process of CB-only sensors.

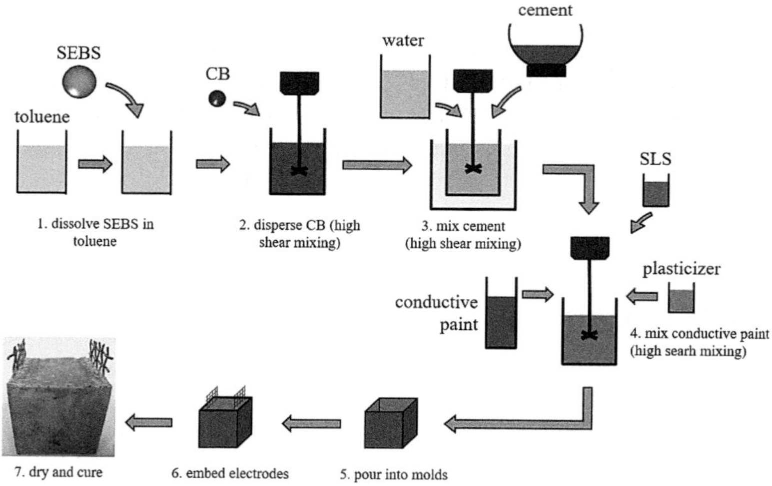

Figure 3.3 Fabrication process of SEBS-CB sensors.

The fabrication process of SEBS-CB sensors is illustrated in Fig. 3.3. SEBS polymer obtained in granular form was first dissolved in toluene in the ratio of 60 g of SEBS per 500 mL of toluene. The amount of SEBS used in the fabrication of SEBS-CB samples was governed by the quantity of CB to be added to ensure that the SEBS solution would not be saturated by CB particles. A mass of 3.41 g of SEBS in solution was used for CB loadings up to 0.71% volume, while a mass of 11.37 g of SEBS was used for CB loadings greater than 0.71% volume. CB was added to the SEBS solution in the desired loading amount before the mixture was mixed at high shear for 3 minutes. A neat cement paste mix was prepared by combining cement and water at a 0.45 w/c ratio and mixing for 2 minutes in a Hobart mixer at the highest speed. After, the SEBS-CB conductive paint was incorporated in the neat cement paste with 0.1 g of sodium lauryl sulfate (SLS) per 131 cm^3 of materials (volume of one sample), a surfactant used to disperse the petroleum-based conductive paint with water, and mixed for additional 2 minutes in the Hobart mixer. The samples were then poured into molds of dimensions $5.1 \times 5.1 \times 5.1$ cm^3 and vibrated/compacted to eliminate air voids. Two sets of copper mesh electrodes were inserted in each poured sample. The samples were covered with a damp cloth and allowed to cure overnight before being demolded and placed in a curing room at 100% relative humidity and 22°C for 48 hours. The samples were air-cured for 4 days to develop a polymer microstructure and let water evaporate. Figure 3.3, step 7 is a picture of a completed SEBS-CB sample.

3.3 Methodology

3.3.1 Mix Proportions

Cementitious mixes with CB-only and SEBS-only samples were prepared. For CB-only samples, a constant w/c ratio of 0.45 was selected, and mixes were designed for constant volumes of 164 cm^3 to produce 25% extra material. The amount of water and cement varied as a function of the quantity of CB added. Table 3.1 lists all different CB-only sample types. Three samples per type were fabricated and tested.

SEBS-CB sample mixes were also designed to keep a constant w/c ratio of 0.45. For samples containing up to 0.71% CB loading, 15 mL of dissolved SEBS (3.41 g) was added to the mix, which corresponds to a concentration of 9% by volume. The quantity of SEBS was doubled for samples containing more than 0.71% to prevent saturation of CB within the SEBS solution. For SEBS-CB

Table 3.1 CB-only samples

Sample type (#)	Water (ml)	CB (g)	Cement (g)	Plasticizer (ml)	w/c ratio	%CB (vol%)
1	293	0	651	6	0.45	0
2	292	1.65	649	6	0.45	0.18
3	291	3.30	647	6	0.45	0.36
4	290	5.00	645	6	0.45	0.54
5	289	6.65	643	6	0.45	0.71
6	286	9.00	630	6	0.45	0.96
7	286	12.0	630	6	0.45	1.25
8	286	15.0	630	6	0.45	1.60

Table 3.2 SEBS-CB samples

Sample type (#)	Water (ml)	CB (g)	SEBS (ml)	Cement (g)	w/c ratio	%CB (vol%)
1	261	0	15	585.9	0.45	0
2	261	1.65	15	585.9	0.45	0.18
3	261	3.30	15	585.9	0.45	0.36
4	261	5.00	15	585.9	0.45	0.54
5	261	6.65	15	585.9	0.45	0.71
6	234	9.00	30	519.8	0.45	0.96

sensors, only loadings up to 0.96% CB were considered because it was found that samples with 1.25% CB loadings and beyond were difficult to fabricate due to the poor workability of the mix, resulting in heterogeneous dispersions. Table 3.2 lists all different SEBS-CB sample types. Three samples per type were fabricated and tested.

3.3.2 Quality Control

The quality of dispersion of CB particles was verified on each sample using thermography. The samples were connected to a 60 V AC current source and placed in an enclosed space. During the electrical loading, a FLIR A35 thermal camera was used to capture thermal images at a rate of approximately one image per second. These images were used to visually inspect the distribution of CB particles that revealed in the form of bright pigments.

The thermal image of an electrically unloaded sample is shown in Fig. 3.4a, where the cementitious block shows as a black cube. Figures 3.4b and 3.4c are typical examples of well-dispersed samples containing high levels of CB particles. Both the CB-only sample with 1.60% CB (Fig. 3.4b) and the SEBS-CB sample with 0.71% CB reveal bright pigments that are well distributed over the observable area. Lastly, Fig. 3.4d shows the example of a badly dispersed sample for which the CB particles sank and agglomerated at the bottom. Samples that were judged as badly dispersed (e.g., Fig. 3.4d) were discarded and refabricated.

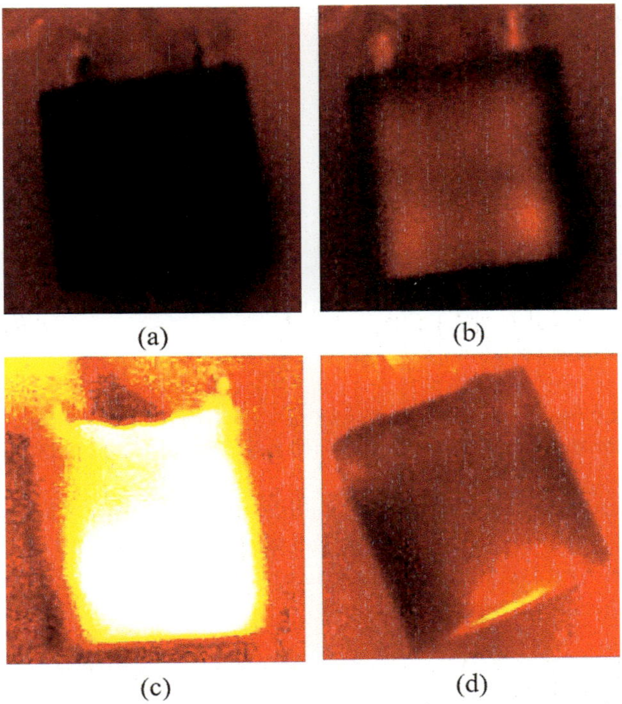

Figure 3.4 Thermal images of (a) electrically unloaded sample (typical), (b) electrically loaded sample (1.6% CB-only), (c) electrically loaded sample (0.71% SEBS-CB), and (d) electrically loaded sample with bad CB dispersion.

3.3.3 Measurements

Tests were conducted to validate the hypothesis that SEBS can be utilized to accelerate percolation of the material and that the resulting conductive cementitious paste can be used as a sensor.

First, the percolation curves were obtained by plotting the resistivity of each sample as a function of CB loadings. Three samples were used for each CB loading, and the resistance was measured using an LCR bridge at 100 kHz.

Second, the strain-sensing capability and linearity of sensors were evaluated by applying strain on each samples and measuring

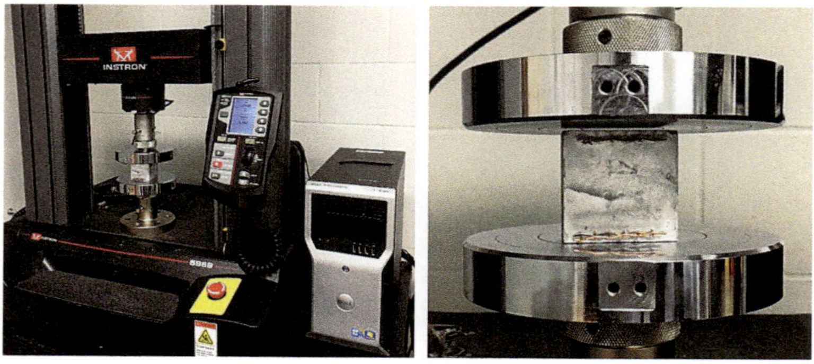

(a) (b)

Figure 3.5 Experimental configuration. (a) Cementitious sensor installed in the universal testing machine and (b) zoom on the cementitious sensors.

the change in resistance. The samples were placed in a universal testing machine of type Instron (50 kN capacity), loaded in steps of 40 $\mu\epsilon$ at a rate of 0.0005 in/second up to 400 $\mu\epsilon$, and then unloaded using the same steps and rate.

3.3.4 Electromechanical Model

To develop the electromechanical model, consider a cementitious sensor with two embedded electrodes of embedded width b and height h, spaced by a distance L, subjected to a uniaxial strain (perpendicular to the electrodes) provoking a change in the electrode distance ΔL. The sensor can be modeled as a resistor of resistance R

$$R = \rho \frac{L}{A} \tag{3.1}$$

where ρ is the resistivity of the material and $A = b \cdot w$ the cross section area of the embedded electrodes. Assuming small strain, taking the finite difference of Eq. 3.1 yields

$$\begin{aligned} \frac{\Delta R}{R} &= \frac{\Delta \rho}{\rho} + \frac{\Delta L}{L} - \frac{\Delta A}{A} \\ &= \frac{\Delta \rho}{\rho} + (1 + 2\nu)\varepsilon \end{aligned} \tag{3.2}$$

with

$$\varepsilon = \frac{\Delta L}{L} \tag{3.3}$$

where ν is the Poisson's ratio of the material and ε the axial strain. Using Eq. 3.2, an expression for the gauge factor λ can be obtained:

$$\begin{aligned} \lambda &= \frac{\frac{\Delta R}{R}}{\varepsilon} \\ &= (1 + 2\nu) + \frac{\frac{\Delta \rho}{\rho}}{\varepsilon} \end{aligned} \tag{3.4}$$

The first term in Eq. 3.4 represents the change in resistance due to the change in the sensor's geometry, while the second term represents the piezoresistive effect.

3.4 Results and Discussion

This section presents and discusses results on the percolation study, effect of SEBS loading, and strain sensitivity of the cementitious sensors.

3.4.1 Percolation Thresholds

Figure 3.6 shows a plot of the resistivity of the sensors as a function of CB loading for both CB-only and SEBS-CB samples. It can be observed that the percolation threshold for CB-only samples is located between 0.71% and 1.25%, while the percolation threshold appears to be located between 0.36% and 0.71% for SEBS-CB samples. This initial comparison of thresholds shows that the utilization of SEBS results in reducing the percolation threshold by approximately 50%. However, samples with SEBS initially have a higher value of resistivity because of the insulating nature of the polymer, and SEBS-CB samples after percolation have a value of resistivity that compares with that of CB-only samples before percolation.

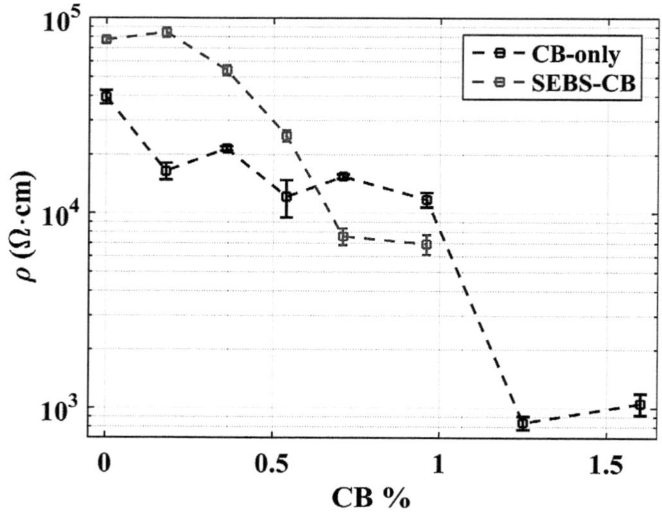

Figure 3.6 Resistivity versus CB% for CB-only and SEBS-CB samples.

3.4.2 Strain Sensitivity

The relative changes in the sensors' resistance as a function of strain was studied for samples located around the percolation threshold found in the percolation study. Results are plotted in Figs. 3.7 (CB-only samples) and 3.8 (SEBS-CB samples). The slope of the linear fit is the gauge factor λ (or sensitivity) of the sensor to strain. The gauge factor for each sample type is listed in Table 3.3. Results for CB-only samples (Fig. 3.7) show a higher gauge factor for CB-only samples at 0.96% CB, while results for SEBS-CB samples exhibit a higher gauge factor at 0.54%CB loading. Thus, the inclusion of SEBS

Table 3.3 Experimental gauge factors λ

CB%	SEBS-CB	CB-only
0.18%	25.8	–
0.36%	30.8	–
0.54%	38	47.3
0.71%	17.7	82.5
0.96%	–	178
1.25%	–	17

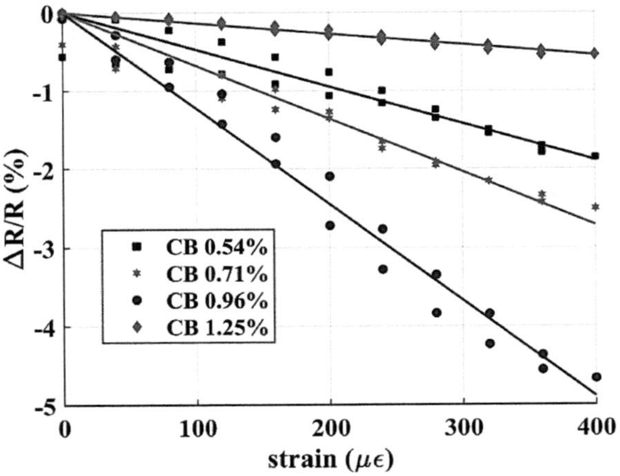

Figure 3.7 Percentage change in resistance vs. CB loading for CB-only samples.

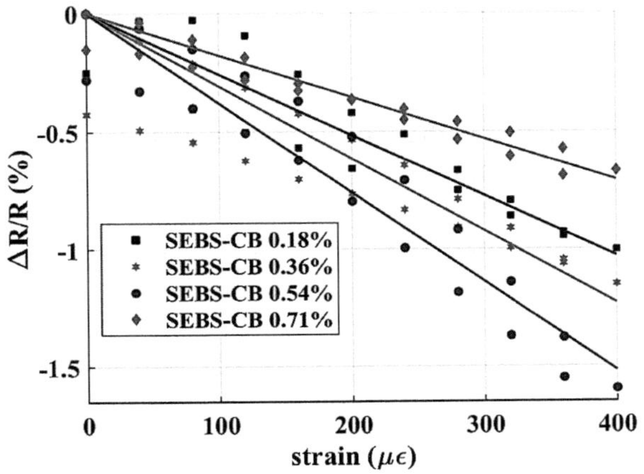

Figure 3.8 Percentage change in resistance vs. CB loading for CB with SEBS samples.

in the cementitious matrix allows to reach the optimal sensitivity with a lower CB loading, but this optimal gauge factor is significantly lower than one obtained through CB-only samples. A comparison of the optimal gauge factors shows that CB-only samples yield a gauge factor ($\lambda = 178$) that is 4.7 times higher.

Also, a closer inspection of the results shows that the CB-only samples exhibit a good linearity, except for the 0.96% loading at low levels of strain, which nonlinearity could be attributed to the piezoresistive effect being more important at low strain levels around the percolation threshold. The SEBS-CB samples show higher nonlinearities. While the piezoresistive effect may also explain the nonlinearity at low strain levels, it is possible that the capacitance of the material is also affected by strain. Lastly, the presence of the polymer may create a significant hysteresis, which can also be observed in Fig. 3.8. The modeling of the nonlinearities in the SEBS-CB samples is left to future work.

While the SEBS-CB samples underperform the CB-only samples in terms of gauge factor and linearity, results from this section demonstrate that the inclusion of SEBS in the cementitious mix results in reducing the location of the percolation threshold.

3.5 Conclusion

This chapter investigated the use of SEBS to accelerate percolation of cementitious composites filled with CB particles. While CB is an inexpensive conductive filler, it is less conductive compared to the more expensive counterparts such as CNTs. It results that a larger amount of CB particles is necessary to reach the percolation threshold, which provides a sensor with high level of sensitivity. However, using a large amount of conductive fillers may result in a significant loss in the material's mechanical properties such as strength. The motivation of the research was to create optimally sensitive cementitious sensors with a smaller amount of conductive particles, therefore maintaining good mechanical properties.

Conductive cement paste sensors were fabricated with SEBS mixed with various concentrations of CB particles (SEBS-CB). The resistivity of each sample was measured for each concentration, which was used to find the percolation threshold. Results were compared with conductive cement paste sensors fabricated without SEBS (CB-only). It was found that the utilization of SEBS reduced the percolation threshold by approximately 50%. However, percolated SEBS-CB samples yielded a resistivity similar to the CB-only samples before percolation, attributed to the insulating nature of SEBS.

The sensitivity or gauge factor of specimens around the percolation threshold was investigated by measuring the change in electrical resistance as a function of applied strain. Results show that a concentration of 0.54% of CB resulted in an optimal gauge factor of 38 for SEBS-CB samples, while this concentration was 0.86% for CB-only samples, yet yielding a gauge factor of 178. This result confirmed that SEBS could be used to accelerate the percolation of the material. However, the optimal gauge factor of SEBS-CB sensors is significantly smaller than the optimal gauge factor of CB-only sensors. In addition, an investigation of the linearity of the sensors showed that SEBS-CB samples were significantly nonlinear compared to CB-only samples. Several factors were listed to explain the nonlinearity, including the piezoresistive effect at low levels of strain, the nonnegligible participation of the material's capacitance, and the hysteresis generated by the use of a polymer.

This study demonstrated the promise of SEBS for fabricating cementitious sensors with smaller amounts of CB particles. While the overall electromechanical properties of SEBS-CB samples were found inferior to those of CB-only samples, the utilization of smaller amount of CB could lead to significantly enhanced mechanical properties. Future work includes the study of such mechanical properties, as well as alternative fabrication procedures of SEBS-CB samples (e.g., concentration of SEBS, use of different surfactant, etc.) to improve on the electromechanical properties.

Acknowledgments

The authors acknowledge Dr. Joseph Schaefer from the Department of Aerospace Engineering at Iowa State University for his assistance with the electromechanical tests. The authors are also grateful to Dr. Steve Holland from Center for Non-Destructive Evaluation at Iowa State University for his help with thermography.

References

1. (a) Laflamme, S., Eisenmann, D., Wang, K., Ubertini, F., Pinto, I., DeMoss, A. (2018). Smart concrete for enhanced nondestructive evaluation, *Mater. Eval.*, **76**(10), pp. 1395–1404. (b) Al-Saleh, M. H., Sundararaj, U. (2008). An innovative method to reduce percolation threshold of carbon black filled immiscible polymer blends, *Composites Part A*, **39**(2), pp. 284–293.
2. Arshak, K., Morris, D., Arshak, A., Korostynska, O. (2008). Sensitivity of polyvinyl butyral/carbon-black sensors to pressure, *Thin Solid Films*, **516**(10), pp. 3298–3304.
3. Chiacchiarelli, L., Rallini, M., Monti, M., Puglia, D., Kenny, J., Torre, L. (2013). The role of irreversible and reversible phenomena in the piezoresistive behavior of graphene epoxy nanocomposites applied to structural health monitoring, *Compos. Sci. Technol.*, **80**, pp. 73–79.
4. Cochrane, C., Koncar, V., Lewandowski, M., Dufour, C. (2007). Design and development of a flexible strain sensor for textile structures based on a conductive polymer composite, *Sensors*, **7**(4), pp. 473–492.
5. D'Alessandro, A., Rallini, M., Ubertini, F., Materazzi, A., Kenny, J., Laflamme, S. (2015). A comparative study between carbon nanotubes and carbon nanofibers as nanoinclusions in self-sensing concrete, in *2015 IEEE 15th International Conference on Nanotechnology (IEEE-NANO)*, Rome, pp. 698–701.
6. D'Alessandro, A., Ubertini, F., Laflamme, S., Materazzi, A. L. (2016). Towards smart concrete for smart cities: recent results and future application of strain-sensing nanocomposites, *J. Smart Cities*, **1**, p. 1.
7. Galao, O., Baeza, F., Zornoza, E., Garcés, P. (2014). Strain and damage sensing properties on multifunctional cement composites with cnf admixture, *Cem. Concr. Compos.*, **46**, pp. 90–98.
8. Gubbels, F., Jérôme, R., Teyssie, P., Vanlathem, E., Deltour, R., Calderone, A., Parente, V., Brédas, J.-L. (1994). Selective localization of carbon black in immiscible polymer blends: a useful tool to design electrical conductive composites, *Macromolecules*, **27**(7), pp. 1972–1974.
9. Han, B., Yu, X., Kwon, E. (2009). A self-sensing carbon nanotube/cement composite for traffic monitoring, *Nanotechnology*, **20**(44), p. 445501.
10. Huang, J.-C. (2002). Carbon black filled conducting polymers and polymer blends, *Adv. Polym. Technol.*, **21**(4), pp. 299–313.

11. Laflamme, S., Cao, L., Chatzi, E., Ubertini, F. (2015). Damage detection and localization from dense network of strain sensors, *Shock Vib.*, **2016**, p. 2562949.
12. Laflamme, S., Pinto, I., Elkashef, M., Wang, K., Cochran, E. W., Ubertini, F. (2015). Conductive paint-filled cement paste sensor for accelerated percolation, *Proc. SPIE*, **9437**, p. 943722.
13. Laflamme, S., Saleem, H. S., Vasan, B. K., Geiger, R. L., Chen, D., Kessler, M. R., Rajan, K. (2013). Soft elastomeric capacitor network for strain sensing over large surfaces, *IEEE/ASME Trans. Mechatron.*, **18**(6), pp. 1647–1654.
14. Lourie, O., Cox, D., Wagner, H. (1998). Buckling and collapse of embedded carbon nanotubes, *Phys. Rev. Lett.*, **81**(8), p. 1638.
15. McCarter, W., Starrs, G., Chrisp, T. (2000). Electrical conductivity, diffusion, and permeability of portland cement-based mortars, *Cem. Concr. Res.*, **30**(9), pp. 1395–1400.
16. Monti, M., Natali, M., Petrucci, R., Kenny, J., Torre, L. (2011). Impact damage sensing in glass fiber reinforced composites based on carbon nanotubes by electrical resistance measurements, *J. Appl. Polym. Sci.*, **122**, pp. 2829–2836.
17. Tumidajski, P., Xie, P., Arnott, M., Beaudoin, J. (2003). Overlay current in a conductive concrete snow melting system, *Cem. Concr. Res.*, **33**(11), pp. 1807–1809.
18. Ubertini, F., Laflamme, S., D'Alessandro, A. (2016). Smart cement paste with carbon nanotubes, in Loh, K. J., Nagarajaiah, S. (eds.) *Innovative Developments of Advanced Multifunctional Nanocomposites in Civil and Structural Engineering*, Woodhead Publishing, pp. 97–120.
19. Ubertini, F., Materazzi, A. L., D'Alessandro, A., Laflamme, S. (2014). Natural frequencies identification of a reinforced concrete beam using carbon nanotube cement-based sensors, *Eng. Struct.*, **60**, pp. 265–275.
20. Wen, S., Chung, D. (1999). Piezoresistivity in continuous carbon fiber cement-matrix composite, *Cem. Concr. Res.*, **29**(3), pp. 445–449.
21. Wen, S., Chung, D. (2007). Partial replacement of carbon fiber by carbon black in multifunctional cement–matrix composites, *Carbon*, **45**(3), pp. 505–513.
22. Wu, J., Song, C., Saleem, H. S., Downey, A., Laflamme, S. (2015). Network of flexible capacitive strain gauges for the reconstruction of surface strain, *Meas. Sci. Technol.*, **26**(5), p. 055103.

23. Wu, S., Mo, L., Shui, Z., Chen, Z. (2005). Investigation of the conductivity of asphalt concrete containing conductive fillers, *Carbon*, **43**(7), pp. 1358–1363.

24. Xiao, H., Li, H., Ou, J. (2010). Modeling of piezoresistivity of carbon black filled cement-based composites under multi-axial strain, *Sens. Actuators, A*, **160**(1), pp. 87–93.

25. Yehia, S. A., Tuan, C. Y. (2000). Thin conductive concrete overlay for bridge deck deicing and anti-icing, *Transp. Res. Rec.*, **1698**(1), pp. 45–53.

26. Zhang, W., Dehghani-Sanij, A. A., Blackburn, R. S. (2007). Carbon based conductive polymer composites, *J. Mater. Sci.*, **42**(10), pp. 3408–3418.

Chapter 4

Advancements in Silica Aerogel–Based Mortars

António Soares, Inês Flores-Colen, and Jorge de Brito
CERIS, Department of Civil Engineering, Architecture and Georresources, Instituto Superior Técnico, Universidade de Lisboa, Portugal
ortiz.soares@gmail.com; ines.flores.colen@tecnico.ulisboa.pt; jb@civil.ist.utl.pt

Thermal mortars are thermal insulation solutions for the opaque envelope of buildings. However, the solutions with greater weight in the market still possess thermal conductivity values higher than those obtained by current thermal insulation materials, which reach at least 0.04 W/m · K. One way of overcoming this threshold is through the incorporation of superinsulating materials, such as silica aerogel.

Silica aerogel is a nanostructured material with high porosity (75% to 98%) and pores of small sizes (between 10 and 100 nm), which allows it to have thermal conductivities between 0.012 and 0.021 W/m · K, lower than the air thermal conductivity at ambient pressure (0.026 W/m · K). It is produced through a sol-gel process, and it comes as monoliths or small-size (grain) aerogels with different characteristics regarding mechanical strength, thermal behavior, density, degree of hydrophobicity, and opacity.

Nanotechnology in Cement-Based Construction
Edited by Antonella D'Alessandro, Annibale Luigi Materazzi, and Filippo Ubertini

ISBN 978-981-4800-76-1 (Hardcover), 978-0-429-32849-7 (eBook)
www.jennystanford.com

There are already applications of aerogel in insulation solutions in buildings, from thermal insulation of glazed siding to aerogel applied in blankets. The incorporation of silica aerogel has only recently begun to be exploited, and there are still only a few mortars with aerogel incorporation in the European market and some patented mortars. Mortars with aerogel incorporation may reach thermal conductivity values lower than 0.03 W/m · K. However, most studies on these mortars are focused mostly on their thermal behavior, and there is still lack of information on the remaining relevant properties to the overall performance of a thermal mortar.

This chapter presents a summary of the studies on thermal mortars with aerogel and some of the main difficulties concerning the incorporation of this superinsulating material in mortars.

4.1 Introduction

There has been a long-standing concern about thermal comfort inside buildings, with the first thermal insulation panels emerging in the 19th century through the processing of organic materials (Bozsaky, 2010). On the other hand, the growth of the world population, associated with an increase in the consumption habits of society, has made human impact increasingly significant on the environment (Wenzel et al., 1997; Milutienė et al., 2012). But in the recent past scientists around the world have recognized that climate change and global warming are the result of human action (Gupta, 2016), and therefore it is important to take measures to reduce energy consumption and gaseous pollutant emission. In this sense, with the Kyoto Protocol, which Portugal signed, a set of policies and measures has been created to counter this tendency (United Nations, 1998; Parliament et al., 2003). Since buildings embody 40% of the total energy consumption in the European Union (EU), there is a significant potential for reducing energy consumption by improving the performance of buildings (EU, 2010). This improvement can be achieved through the use of thermal insulation systems, focusing mainly on reducing the thermal conductivity of materials and on the thermal transmission of the elements of the building envelope (Jelle et al., 2010).

One of the ways to improve the thermal performance of buildings is to minimize losses and thermal gains through the opaque envelope (Gonçalves and Graça, 2004), and this can be achieved through the use of thermal insulation systems such as mortars with thermal performance improvement, which have the same procedure application as those traditionally applied on buildings (Leopolder, 2010, cited by Barbero-Barrera et al., 2014).

For a mortar to be classified as thermal, it needs to comply with certain levels of performance present in the EN 998-1 (CEN, 2010) standard, as shown in Table 4.1.

The thermal conductivity coefficient (λ), expressed in $W/m \cdot K$ or $W/m \cdot {}^{\circ}C$, characterizes thermally homogeneous products by the amount of heat (expressed in W per unit area [m^2]) that crosses a unit thickness (m), when a unit temperature difference (1°C or 1 K) is established between two flat and parallel faces (Pina dos Santos and Matias, 2006).

To achieve thermal conductivity values compatible with the requirements of EN 998-1 (2010), the incorporation of lightweight aggregates has been seen as a way of reducing the bulk density of mortars and consequently their thermal conductivity.

In Fig. 4.1 several scientific publications and technical data sheets are presented, where it is possible to observe a trend of reduction of the thermal conductivity of different products with the reduction of bulk density.

It is also possible to observe that the variation of the thermal conductivity with the bulk density is not linear, reaching a level from which it is difficult to reduce the thermal conductivity at the expense of the density, and the same tendency can be observed in commercial mortars (Barbero-Barrera et al., 2014). It is also observed that industrial mortars reach values of thermal conductivity (indicated in datasheets) lower than those obtained in research papers. However, the lower values of thermal conductivity presented are approximately 0.04 $W/m \cdot K$ for mortars with bulk density less than 250 kg/m^3.

Nevertheless, regardless of the density of the material, the thermal conductivity of the gas (air) remains constant (~0.026 $W/m \cdot K$), which causes a barrier (lower limit of thermal conductivity) of 0.03 $W/m \cdot K$ to the performance of current thermal insulation materials

Table 4.1 Characteristics and requirements for a thermal mortar, in accordance with EN 998-1 (data from CEN, 2010)

	Properties with requirements to fulfill				**Properties with declared values**			
Properties	**Compressive strength (N/mm^2)**	**Water absorption due to capillary ($kg/m^2 \cdot min^{0.5}$)**	**Water vapor permeability coefficient**	**Thermal conductivity coefficient ($W/m \cdot K$)**	**Dry bulk density (kg/m^3)**	**Adhesion (N/mm^2)**	**Reaction to fire**	**Durability**
Classes/limits	0.4 a 2.5 (CS I) 1.5 a 5.0 (CS II)	$Cc \leq 0.4$ (W1)	$\mu \leq 15$	$\lambda \leq 0.1$ (T1) $\lambda \leq 0.2$ (T2)	–	$\geq$ declared value and rupture type	–	–
Testing method	EN 1015-11 (CEN, 1999b)	EN 1015-18 (CEN, 2002b)	EN 1015-19 (CEN, 2004)	EN 1745 (CEN, 2002a)	EN 1015-10 (CEN, 1999a)	EN 1015-12 (CEN, 2000)	EN 13501-1 (CEN, 2007)	–

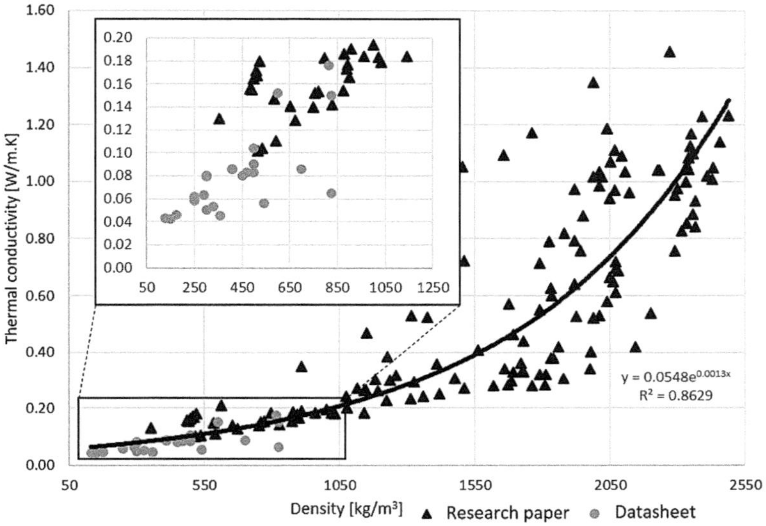

Figure 4.1 Thermal conductivity versus mortar's density. Sources: (Fu and Chung, 1997; Xu and Chung, 2000b; Xu and Chung, 2000a; Demirboga, 2003; Uysal et al., 2004; Demirboga et al., 2007; Benazzouk et al., 2008; Azichem, 2009; Ashour et al., 2010; Sengul et al., 2011; FassaBortolo, 2011; Panesar and Shindman, 2012; Gutiérrez-González et al., 2012b; Gutiérrez-González et al., 2012a; Taoukil et al., 2012; Barreca and Fichera, 2013; Herrero et al., 2013; Thermowall, 2013; Barbero-Barrera et al., 2014; Tecresa, 2015; ARDEX, 2016; EDILTECO, 2016; Secil, 2016; Weber, 2016; RÖFIX, 2016; Chiraema, 2017; Diasen, 2017; CVR, 2017; LATERLITE, 2017; Kimia, 2017).

(Simmler et al., 2005, cited by Cuce et al., 2014), so it is necessary to use innovative materials to exceed this limit.

4.1.1 Nanomaterials

Nanotechnology allows the manipulation of materials with sizes and precision between 0.1 and 100 nm. At the nanoscale, surface effects are more important than bulk properties, which can lead to materials with enhanced properties (Drexler, 1981, cited by Pacheco-Torgal and Jalali, 2011; Holister, 2002; Zhu et al., 2004; SCENIHR, 2010).

According to the International Organization for Standardization (ISO, 2010, cited by Potočnik, 2011), a "nanomaterial" describes any material with external sizes to the nanoscale or with the internal structure or surface structure at the nanoscale. Nanomaterials, according to their unconfined sizes at the nanoscale, can still be classified into 0D (nanoparticles), 1D (nanotubes, nanofilaments, and nanofibers), 2D (nanofilms and nanocoatings), and 3D (monoliths) (Gonçalves, 2012).

However, what makes these sizes so interesting is the high ratio between surface area and volume. Thus, there is a greater exposure of atoms/molecules of the material to the surrounding environment and a greater amount of possible connections available to the surface, which makes the particle chemically more active. For example, in a particle with the approximate diameter of 10 nm, about 20% to 25% of the atoms are exposed to the surrounding medium, whereas in a particle with 3 nm this percentage increases from 45% to 60% (Sengupta and Sarkar, 2015).

There are two types of approaches to nanomaterial synthesis: top down or bottom up. With the top-down approach, the size of the structure is reduced to the nanoscale, while retaining the original properties. Thus, nanomaterials are obtained by methods such as high-energy milling, mechanochemical processing, pickling, electric blasting, ultrasonic agitation, cathodic spraying, and laser ablation. Immediately after the process of obtaining the nanoparticles, they are highly reactive and can easily form agglomerates. Thus, these methods have to be carried out under vacuum or in an inert atmosphere, because if the procedure is performed in the presence of some reactive gas, additional reactions may occur (Taylor, 2002; Nass et al., 2004; Zhu et al., 2004; Al-Jabri and Shoukry, 2014).

In turn, the bottom-up approach involves the manipulation of atoms and molecules with synthesis of materials to create nanostructures, through chemical reactions, chemical processes based on the transformation of solutions, such as the sol-gel process, vapor deposition by chemical process, plasma or flame spray synthesis, and atomic or molecular condensation. Of these methods, the sol-gel process is highlighted, which differs from the rest because it occurs at low temperatures, which makes it a versatile and cost-effective process (Taylor, 2002; Nass et al., 2004; Zhu et al., 2004; Al-Jabri and Shoukry, 2014).

Nanomaterials can improve some characteristics of the construction products (like strength and durability) and add new functions such as photocatalytic (self-cleaning, pollution reduction, and antimicrobial ability), antifogging, and self-sensing abilities (Jayapalan et al., 2013). Therefore, the introduction of nanomaterials in conventional construction products can lead to construction materials with excellent properties and make buildings with a more efficient performance (Li et al., 2004; Senff et al., 2013).

Even though nanotechnology is not a new science or technology, only in the past two decades has there been an increase in research in this field (Ge and Gao, 2008). Recently some studies have emerged regarding the incorporation of nanomaterials in mortars, leading to mortars with enhanced and innovative properties to be applied on building facades (nanorenders).

Due to their nanosize, nanomaterials generally affect the strength, shrinkage, and durability properties of Portland cement paste and help to improve the quality and longevity of structures (Ltifi et al., 2011; Hanus and Harris, 2013). With nanomaterials it is also possible to prevent the initiation of microcracks through the control of nanocracks with nanosized fibers (Konsta-Gdoutos et al., 2010a), to enhance the thermal performance (Stahl et al., 2012), and to give photocatalytic and self-cleaning properties to the renders (Jayapalan et al., 2013).

To achieve these improvements, good dispersion of the nanoparticles in a cement mortar matrix is important, which is difficult to reach due to their high surface energy and strong interparticle forces, which can lead to poor dispersion, resulting in weak zones or potential areas with concentrated stresses (Guskos et al., 2010; Nazari and Riahi, 2011; Givi et al., 2011, cited by Hanus and Harris, 2013). The possible risks that nanomaterials could pose on health are also a concern, among them being lung inflammation or cancer, and because of that, they should be handled like a material with unknown toxicity (Grassian et al., 2007; Hallock et al., 2009, cited by Pacheco-Torgal and Jalali, 2011), especially powder nanomaterials (nanoparticles).

Multiwalled carbon nanotubes (CNTs) have a tensile strength and Young's modulus that are hundreds and tens of times, respectively, those of steel. Their incorporation in small amounts, associated with a higher aspect ratio, leads to an effective arrest of the nanocracks (Konsta-Gdoutos et al., 2010a; 2010b; Hanus and Harris, 2013). Multiwalled CNTs can also improve the durability of the cement matrix because they reduce the fine pores, which results in a decrease of the capillary stresses. Therefore, it is possible to obtain a mortar with good properties for protection against chemical attack (Konsta-Gdoutos et al., 2010b).

However, it is very important to get a uniform dispersion of CNTs in the cement matrix (Xie et al., 2005), which is difficult to accomplish because of their poor interaction with cement, resulting from their intrinsically hydrophobic nature and because of the tightly bound formation of agglomerates or bundles resulting from their production (Grobert, 2007, cited by Konsta-Gdoutos et al., 2010a; Sáez de Ibarra et al., 2006, cited by Hanus and Harris, 2013).

Several studies have emerged regarding the incorporation of nanosilica (nano-SiO_2), with good results in improving the compressive and splitting tensile strengths of cement-based materials (Shekari and Razzaghi, 2011) and reducing the permeability of cement mortar (Sadrmomtazi et al., 2009). However, the adverse effect on cement-based materials' workability due to the high specific surface area of nano-SiO_2 and the dispersion of these nanoparticles can cause lower strength of mortars in later ages (Kawashima et al., 2013). To counteract the difficulties in achieving a good distribution of nano-SiO_2, some authors suggest a low percentage of particles (1–5 wt%), while others propose contents of 10 wt% by weight of cement (Senff et al., 2009, cited by Ltifi et al., 2011).

Nano–titanium dioxide (nano-TiO_2), with photocatalytic and self-cleaning properties, can enable mortars with new functionalities that contribute to the reduction of air pollutants, like biocide properties and self-cleaning effects, leading to the maintenance of aesthetic characteristics, such as color, and decrease the environmental impact of construction materials (Cárdenas et al., 2012; Hanus and Harris, 2013; Jayapalan et al., 2013). Once these particles need light stimulation for catalytic action, nano-TiO_2 particles are normally applied on external renders or paints (Senff et al., 2013). However, some studies (Hanus and Harris, 2013) found that the self-cleaning and photocatalytic effects of TiO_2 may form harmful side products, which indicates the need for more research on their side effects. The use of fine particles may also has a negative effect on the rheological properties, which results in higher water demand and shorter setting time (Hüsken et al., 2009; Chen et al., 2012; Senff et al., 2012, cited by Senff et al., 2013).

Summing up, with the incorporation of nanomaterials in mortars it is possible to improve several characteristics, as shown in Table 4.2. In general, an increase of mortars' compacity was observed due to the nanosize of the particles that fill their voids. However, in some cases, a decrease of mortars' workability was noticed due to the high specific surface area of the nanomaterials, and some difficulties in achieving good dispersion of nanoparticles in the mortar matrix were reported.

Although there are materials that do not have a nanoscale size that allows them to react or fill certain spaces, they can present a nanostructure that gives the material exceptional properties, as is the case of silica aerogel. The advantages and disadvantages of this nanomaterial are also found in Table 4.2, with a more detailed description of its influence on mortars in the following section.

To sum up, an environmental concern has been noticed with the use of nanomaterials in mortars. Nevertheless, despite their high initial impact, nanomaterials can contribute positively in the long term to more sustainable buildings. However, there is still a lack of research on their health impact. Nowadays, despite the efforts to reduce the production cost of some nanomaterials, their production cost is still high and there is a considerable lack of information regarding the real costs of renders with nanomaterials. Therefore, it is important to research the effects of nanomaterials in mortars to identify the aspects to be improved and, in the future, achieve products to apply in renders with advanced properties at a low or competitive cost.

4.2 Silica-Based Aerogel

Silica aerogel is a nanostructured material, produced by the sol-gel process, consisting of a large number of pores of small size (Aspen, 2014; citados por Thapliyal and Singh, 2014; Kyushu, 2014).

Table 4.2 Effect of various nanomaterials on mortars (data from Soares et al., 2014a)

Material	%	Workability and mixing	Flexural strength	Compressive strength	Physical properties	Durability	Cost	Environmental impact
CNT	(–)	Decrease	Increase	Increase	Reduction of capillary stresses	Increase	*e e e*	n.s.
Nano-SiO_2	(–)	Decrease	Increase	Increase	Reduction of water permeability	n.s.	n.s.	n.s.
Silica aerogel	(+)	n.s.	n.s.	n.s.	Reduction of thermal conductivity	n.s.	*e e e*	Initial increase Final decrease
Nano-TiO_2	(–)	Decrease	n.s.	Initial increase Decrease at 28 days	Photocatalytic and self-cleaning properties (new properties)	n.s.	*e e*	Initial increase Final decrease

%, percentage of incorporation; (–), low; (+) high; *e e e* , very expensive; *e e*, expensive; n.s., not specified.

The structure of the aerogel is composed of small spherical silica agglomerates (SiO_2 particles), generally 2 to 5 nm in size, which are connected together in chain, resulting in a "pearl necklace," forming a porous space grid, in which the pores have an average size of 20 to 40 nm, generally ranging between 10 and 100 nm.

Thus, aerogels with different particle and pore sizes can be obtained, as well as different porosity values, which in general range from 75% to 98% (Rubin and Lampert, 1983; Zeng et al., 1994; Pajonk, 1998; Schmidt and Schwertfeger, 1998; Wagh et al., 1999; Akimov, 2003; Ilharco et al., 2007; Soleimani Dorcheh and Abbasi, 2008).

These characteristics confer unique properties to the aerogel, as a very low density, and in some cases can reach values very close to the density of air (1.2 kg/m^3). This is the case of a silica aerogel synthesized by Dr. Steven Jones of NASA's Jet Propulsion Laboratory, which in 2002 was considered by the Guinness World Records book as the lightest solid material in the world with density 3 kg/m^3 (NASA, 2002; Akimov, 2003). This milestone was already surpassed in 2004 by another silica aerogel with density 1.9 kg/m^3 (S and TR, 2003) and recently (2012) by graphene aerogel (0.18 kg/m^3) (Mecklenburg et al., 2012), with a graphene aerogel with density 0.16 kg/m^3 being the current holder of the Guinness World Record for the lightest solid in the world (since 2013) (Guinness, 2013).

Although aerogel is classified as a nanomaterial due to its nanostructure, it is composed of a continuous pore network, which allows obtaining aerogels of small size, such as aerogel in granules (Fig. 4.2a) or monoliths of considerable size (Fig. 4.2b) (Hüsing and Schubert, 1998; Pajonk, 1998; Bhagat et al., 2007; Zhu et al., 2007; Roy and Hossain, 2008).

The high thermal insulation capacity of aerogel can be justified by its high total porosity and small particle size that make up the structure, resulting in a solid structure occupying a small part of the aerogel. In addition, it has an open structure that has a high amount of loose ends, that is, dead ends for heat transfer by conduction through the structure, resulting in a path with high tortuosity and inefficiency for the heat transfer through the solid part of the aerogel (Zeng et al., 1994; Akimov, 2003; Wei et al., 2011; Baetens et al., 2011; Thapliyal and Singh, 2014).

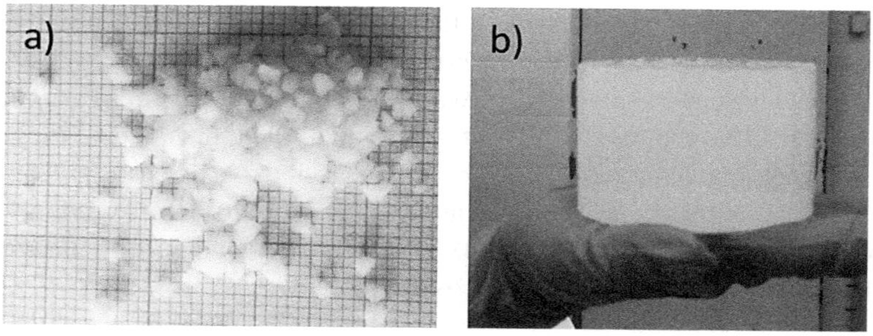

Figure 4.2 Example of silica-based aerogels: (a) grain and (b) monolith.

The heat transfer component associated with the solid structure depends on the density of the aerogel and can be evaluated using vacuum. For example, the thermal conductivity component associated only with the solid structure has values in the order of 0.01 W/m · K for aerogels with density 270 kg/m^3 and can reach 0.001 W/m · K for aerogels with density 80 kg/m^3 (Heinemann et al., 1996; Hüsing and Schubert, 1998; Zhu et al., 2007; Baetens et al., 2011).

Because the average pore size (about 20 nm) is less than the space required for air to move (about 70 nm) at ambient temperature and pressure, the movement of the gas molecules within the silica aerogel porous network is strongly suppressed, thus reducing gas conduction (Wei et al., 2011), resulting in a marked decrease of thermal conductivity in pores from 1 μm to 10 nm (Jelle et al., 2010).

Thus, it is possible to assume that the air molecules become entrapped in the porous structure, whereby the heat transfer by air convection inside the pores is negligible. Therefore, the contribution of the air inside the aerogel to the thermal conductivity is 0.005 to 0.01 W/m · K, considerably lower than the air conductivity without air convection (0.026 W/m · K) (Rubin and Lampert, 1983; Kuhn et al., 1995; Zhu et al., 2007).

Silica aerogel can absorb radiation with wavelengths greater than 5 μm (infrared radiation), so it is efficient to block the radiation emitted by walls and windows at an environment temperature and can be considered a translucent thermal insulator, which efficiently conducts sunlight but blocks infrared radiation.

However, at wavelengths between 3 μm and 5 μm, its absorption is reduced, which results in a deterioration of the thermal insulation capacity with the temperature increase (Fricke and Tillotson, 1997; Fricke, 1988; Ebert, 2011). The radiation conduction component has values between 0.0015 and 0.003 W/m · K for aerogels with density 100 kg/m^3, at a temperature of 26.85°C (Hümmer et al., 1993).

In this way, the aerogel is presented as a material with high potential for thermal insulation. In fact, this material, synthesized for the first time in 1931 by Steven S. Kistler, has considerably reduced thermal conductivities ranging from 0.012 to 0.021 W/m · K. Even the highest values of thermal conductivity recorded in silica aerogels are lower than the thermal conductivity of air at ambient pressure (0.026 W/m · K) (Hüsing and Schubert, 1998; Schmidt and Schwertfeger, 1998; Zhu et al., 2007; Baetens et al., 2011; Buratti and Moretti, 2012; Jelle et al., 2014).

Despite their remarkable properties of thermal insulation and good acoustic properties, the characteristics that make aerogels good insulators are also responsible for making aerogel a weak material, which is a disadvantage for its use in building insulation materials (Parmenter and Milstein, 1998; Wei et al., 2011).

Silica aerogels are cellular solid with a "pearl necklace"-type skeleton net. The fragility of silica aerogel is attributed to the "necks" (weak points of this structure that are in the regions between particles) that form the structural network of the aerogel (Woignier and Phalippou, 1988; Ma et al., 2000; Tan et al., 2001; Zhang et al., 2004; She and Ohji, 2002, cited by Lu et al., 2011). It is possible to improve the strength of the aerogel monoliths by increasing the width of these link points between particles (Leventis et al., 2002).

Another reason for the aerogel's fragility is that the solid structure presents a disordered morphology, which causes an irregular distribution of pores with zones of greater porosity and others of dense agglomeration of particles (with little defined connections), which results in a low-organized material with density gradients (Obrey et al., 2011).

Because of the fragility of the aerogels and the complexity involved in handling the preparation of specimens for mechanical testing, it is difficult to carry out tests of pure traction, torsion, or multiaxial stresses, so usually only compression and bending tests are carried out (Woignier and Phalippou, 1988, cited by Lu et al., 2011).

The lower values of the compressive strength of aerogels of greater density may also be related to the "neck" region. In fact, unlike low-porosity materials in which the materials are treated as a whole and the connection between the particles of the solid is not taken into account, in materials of high porosity the way the particles are linked together is an important parameter (Woignier and Phalippou, 1988).

The most common process for preparing silica aerogels is the sol-gel process, which includes three main steps: hydrolysis reactions, condensation (gelling) reactions, and aging and drying (Brinker and Scherer, 1990). The sol-gel process consists of obtaining a solid 3D network of an oxide, by chemical reaction in solution, at low temperatures. To this end, starting from suitable precursors, a colloid (sol) is produced that can generate, by polymerization reactions, a rigid 3D structure (gel) from which it is possible to obtain ceramic materials.

A "sol" may be defined as a stable dispersion of colloidal particles or polymers in a solvent, wherein "colloidal" represents a dispersion of suspended solids of such a small size (1 to 1000 nm) that the gravitational forces are negligible and the interactions may be called short-range forces, such as van der Waals forces and surface charges, with no agglomeration or sedimentation occurring. In turn, a "gel" consists of a continuous 3D network involving the liquid phase (Brinker and Scherer, 1990; Ghosh, 2015).

Hydrolysis of precursors (generally metal alkoxides) and condensation between hydrolyzed oligomers are catalyzed, in a co-solvent, with controlled stirring and temperature. The synthesis and process parameters determine the aerogel structure, affecting its properties and applications (Menaa et al., 2010; Pierre and Rigacci, 2011). For example, adequate aging of the wet gel, during which the condensation reactions proceed, strengthening the network, leads to aerogels with higher mechanical strength (Ma et al., 2000).

Thus, the sol-gel process can be described as a two-step process to form a structural network (hydrolysis and condensation reactions). In the first stage of the reaction, polymers or rings are established, subsequently resulting in colloids defining the "sol." In the second step, due to condensation reactions, the polymer network grows with the union of the molecules, establishing macromolecules that extend throughout the solution, obtaining the transition from "sol" to "gel" (De Souza et al., 2011; Pandey and Mishra, 2011).

In this manner, the sol-gel process can be performed with different precursors and/or solvents and with different drying methods, which allows obtaining materials of different sizes (small grains to monoliths of considerable sizes) and different characteristics with regard to mechanical resistance, thermal behavior, density, degree of hydrophobicity, and opacity (Júlio and Ilharco, 2017).

Since silica aerogel was discovered, many potential applications have been described (Schmidt and Schwertfeger, 1998) and can be used from pharmacy or agriculture (as carrier material), electronics, kinetic energy absorbers (to collect dust particles suspended in the atmosphere), and thermal insulation (mostly for cryogenic ambient conditions or high temperature) (Tsou, 1995; Schmidt and Schwertfeger, 1998; Pajonk, 2003; Jones, 2006).

With regard to the application of silica aerogel in construction, there has been a high interest in the thermal insulation of glazed siding of buildings in order to reduce energy costs with domestic and industrial heating (Schmidt and Schwertfeger, 1998; Pajonk, 2003). Aerogels are also applied in blankets for insulation of the opaque building envelope (Baetens et al., 2011; Ibrahim et al., 2015). Their high thermal insulation and good acoustic behavior make them a potential material to be used in buildings (Schmidt and Schwertfeger, 1998; Acharya et al., 2013).

4.3 Aerogel-Based Mortars

The incorporation of silica aerogel in coating materials has been studied only recently, with the first research paper on the subject emerging in 2012 (Stahl et al., 2012).

Since mortars are produced using water, it is important to understand the behavior of aerogels in the presence of mixing

Figure 4.3 Collapsed particles of hydrophilic aerogel in mortar under (a) a binocular microscope and (b) scanning electronic microscopy (SEM).

water. These may exhibit different degrees of hydrophobicity, from hydrophilic to superhydrophobic, resulting in different behaviors of the mortars at the fresh and hardened states (Soares et al., 2014b; Júlio et al., 2016b; Júlio and Ilharco, 2017; Soares et al., 2018).

Hydrophilic aerogels have high structural sensitivity to the presence of water. When incorporated in mortars, internal fractures appear in the aerogel that lead to its collapse, with poor adhesion between the aerogel and the remaining mortar paste, resulting in a mortar with poor cohesion at the hardened state (Fig. 4.3). This property makes hydrophilic aerogels unsuitable for aerogel incorporation into mortars, especially for high incorporation rates.

On the other hand, hydrophobic aerogels are presented as a possible solution of insulating aggregates for incorporation in high-performance thermal mortars, because they do not allow the entrance of water into their interior and consequent collapse.

Stahl et al. (2012) prepared a hydrophobic mineral cement-free mortar of high thermal performance (0.025 $W/m \cdot K$) and a density of 200 kg/m^3, incorporating an aerogel in 60% to 90% of the total volume, and some additions to improve the workability, which are not identified. However, the study does not refer to mechanical strength, water behavior, or other fundamental properties that allow the complete performance characterization of the mortars.

Buratti et al. (2014) presented a study on the thermal and acoustic performance of a mineral-based render with incorporation of an aerogel, between 80% and 99% of the total volume, with thermal conductivities between 0.014 and 0.05 W/m · K and densities between 125 and 300 kg/m^3. The renders produced by the authors showed an increase in the sound absorption coefficient in relation to the reference material (laminated gypsum plaster).

Ibrahim et al. (2014; 2015) studied the hygrothermal behavior of a patented mortar based on mineral and/or hydraulic binder, an insulating filler composed of hydrophobic silica xerogel powder or grains. The studied mortar had a low thermal conductivity and density (0.0268 W/m · K and 156 kg/m^3, respectively), good water behavior with a coefficient of resistance to water vapor of 4.25% ±6%, and a coefficient of water absorption of 0.184% ±15% (kg/m^2 s$^{1/2}$) in which they evaluated the change in thermal conductivity with relative humidity. The same authors found that the coefficient of thermal conductivity changes only for relative humidity values higher than 60% and also presents a good thermal insulation behavior (~0.041 W/m · K) for relative humidity higher than 90%. However, no indication is given regarding the mechanical behavior of this mortar.

Ng et al. (2015) carried out a study of cement-based mortars incorporating aerogel at between 20% and 80% of the total volume, where they analyzed the thermal and mechanical performance of the mortars with an increase in the aerogel content. A mortar with aerogel in the amount of 80% of the total volume reached the thermal conductivity value of 0.31 W/m · K and a density of 690 kg/m^3, with negligible resistance values (very close to zero). The same authors also studied the aerogel interface with the binder matrix, where they found a gap (spaces) between the two materials. These aspects were also analyzed by Gao et al. (2014) for concrete, which shows the difficulty of incorporating hydrophobic aerogels into mortars or concrete, due to problems of connection at the interface between the aerogel and the cement matrix.

Júlio et al. (2016b) bypassed this situation through the incorporation of surfactants, resulting in a better connection between the hydrophobic aerogel and the remaining mortar mix. The same authors indicate that the content of surfactants should be adjusted to the hydrophobicity of the aerogel, since to a certain extent its incorporation is positive for the connection between the aerogel and the remaining matrix, but overcoming a limit, gaps (spaces)

Figure 4.4 Aerogel-based mortar: (a) production, (b) application on a brick substrate, (c) application on a wall, and (d) the final aspect.

occur again around the aerogel. Thus, with correct adjustment of the content of surfactants as a function of the hydrophobicity of aerogel, it is possible to produce mortars with a good workability for application to the spoon (Fig. 4.4).

4.4 Performance of Aerogel-Based Mortars

There are already a few mortars with aerogel on the market; one of them is FIXIT222 with claimed thermal conductivity (λ_{10}) of 0.0261 W/m · K (EMPA, 2016). Several studies have been developed with this mortar (Koebel et al., 2012; Ibrahim et al., 2014; Wakili et al., 2015). This mortar should have a reinforced finishing, similar to the ETICS-type system, and these studies are mainly focused on thermal conductivity.

Table 4.3 synthesizes the relevant properties for a thermal mortar according to EN 998-1 (CEN, 2010), which are available for aerogel-based mortars from various authors, as well as industrial mortars and patents (presented in the previous sections).

As shown in this table, the values of thermal conductivity are between 0.014 and 0.076 W/m · K, clearly proving the potential of these materials as aggregates. However, the lack of information on

Table 4.3 Properties of aerogel-based thermal renders

Composition		Properties with declared values				Properties with requirements to fulfill				Reference
% vol.	**Binder**	ρ **(kg/m³)**	f_u **(N/mm²)**	**RF**	***D***	C_S **(N/mm²)**	W **(kg/m²s⁰·⁵)**	μ	λ **(W/m · K)**	
60 a 90	Mineral binder	200	–	–	–	–	–	–	0.025	(Stahl et al., 2012)
96 a 99	Calcium hydroxide	115 to 125	–	–	–	–	–	–	0.014 to 0.016	(Buratti et al., 2014)
80	Portland cement, silica fume	<100	–	–	–	negligible	–	–	0.31	(Ng et al., 2015)
–	Cement	–	–	–	–	34.77	–	–	0.076	(Khamidi et al., 2014)
–	Mineral binder and/or organic hydraulic binder	156	–	–	–	–	0.184 ± 15%	4.25 ± 6%	0.027 ± 3%	(Ibrahim et al., 2014)
–	Hydraulic lime NHL 5, calcium hydroxide, white cement	220	–	–	–	–	–	4 to 5	0.0261	(EMPA, 2016)
–	Cement	144 to 318	–	–	–	0.13 a >0.5	–	–	0.015 to 0.066	(Doshi et al., 2011)
–	Lime, gypsum, cement	450 to 630	–	–	–	–	–	–	0.15 to 0.25	(Frank et al., 2000)
–	Alumina cement, Portland cement, lime, calcium sulfate	–	–	–	–	0.08	–	–	0.034	(Achard et al., 2011)

(Continued)

Table 4.3 (*Continued*)

Composition		Properties with declared values				Properties with requirements to fulfill				Reference
% vol.	Binder	ρ (kg/m^3)	f_u (N/mm^2)	RF	D	C_S (N/mm^2)	W (kg/m^{2}s$^{0.5}$)	μ	λ (W/m · K)	
–	–	153	–	–	–	0.43	–	–	0.042	(Permanyer and Sanz, 2014)
12.25 a 49.5	Lime putty	–	–	–	–	<0.3[a]	–	5.24 to 8.08	0.050 to 0.350	(Westgate et al., 2018)
25 a 95	Lime plaster	345 to 630	–	–	–	–	–	–	0.036 to 0.114	(Nosrati and Berardi, 2017)
25 a 90	Lime plaster	199 to 789	–	–	–	–	–	5.93 to 9.86	0.027 to 0.113	(Nosrati and Berardi, 2018)
34.5 to 59.2	Gypsum	200 to 730	–	–	–	<0.01 to 0.82	–	–	0.028 to 0.096	(Sanz-Pont et al., 2016)
48	Portland cement	758	–	–	–	1.63	2.2	19	0.097	(Júlio et al., 2016a)
48 a 80	Portland cement, fly ash	652 to 718	–	–	–	0.92 a 1.03	1.3 a 1.8	16	0.084 to 0.085	
80	Portland cement, fly ash	418	–	–	–	0.47	0.48	14	0.066	(Soares et al., 2018)
–	Cement, lime	135 to 140	–	–	–	–	–	–	0.029 to 0.041	(Gomes et al., 2018)

% vol., volume incorporation of aerogel; ρ, dry bulk density; f_u, adhesion; RF, reaction to fire; D, durability; C_S, compressive strength; W, water absorption due to capillary; μ, water vapor permeability coefficient; λ, thermal conductivity coefficient.

[a] Approximate value (obtained from a graph).

mortar composition, namely the percentage of aerogel incorporation, and on properties such as mechanical behavior and/or water behavior is noticeable in most products.

In the works referred to in Table 4.3, it can be observed that mortars with lower thermal conductivity (reaching values less than 0.03 W/m · K) have lime or gypsum binders in their composition and generally volume of the aerogel (when indicated) greater than 90 vol%. Nevertheless, these mortars present low mechanical resistance with values between 0.08 and 0.82 N/mm^2. On the other hand, mortars with a cement binder can reach values of compressive strength higher than 1 N/mm^2. Furthermore, information on the composition of aerogel-based thermal renders is scarce and only in some cases the amount (by volume) and the nature of the incorporated aerogel are indicated. The lack of information on the properties of the renders is also visible: in some cases, only the dry bulk density and/or the compressive strength are mentioned. Moreover, only in recently emerging studies aerogel-based renders are discussed in detail, in terms of both structure and other properties relevant to the overall performance of a render.

4.5 Conclusions

Silica aerogel is a nanostructured material with high total porosity and small pore size. It presents unique characteristics of thermal insulation with thermal conductivity values between 0.012 and 0.021 W/m · K, below the conductivity of air at ambient pressure (0.026 W/m · K), making aerogel incorporation a viable option to obtain high-thermal-insulating mortars. However, silica aerogel presents low mechanical strength, which limits its application in some cases.

Only in the recent past have studies analyzed the incorporation of silica aerogel in cement-based materials, even though there is already in the market a few aerogel-based renders, which must be applied with a protective layer. However, most of the existing publications focus only the thermal performance of aerogel-based renders and do not approach other important properties for global performance of renders, such as mechanical strength and water behavior. Sometimes, there is also a lack of information about mortar composition (binder, other aggregates, and admixtures), water/binder ratio, aerogel type (e.g., grain size, hydrophobicity degree), and aerogel amount. This lack of information makes it difficult to understand the performance of these mortars and the interaction between the different constituents, with only a few studies explaining possible relations, at the microscopic and molecular levels, between aerogel characteristics and mortar performance.

Thus, the incorporation of silica aerogel in superinsulating mortars has already reached thermal conductivity values lower than 0.03 W/m · K (less than current solutions of ETICS) but needs further research in order to develop multifunctional thermal mortars, fulfilling the EN 998 requirements.

Acknowledgments

Acknowledgments by the authors to the financial support of the Foundation for Science and Technology (FCT): projects PTDC/ECM/118262/2010, NANORENDER (2012-2015); PhD Grant SFRH/BD/97182/2013 from FCT, POCI-01-0247-FEDER-017417 from COMPETE 2020, and CERIS research center from Instituto Superior Técnico, University of Lisbon.

References

Achard, P., Rigacci, A., Echantillac, T., Bellet, A., Aulagnier, M., Daubresse, A. (2011). Insulating silica xerogel plaster, WO2011083174 A1.

Acharya, A., Joshi, D., Gokhale, V. (2013). AEROGEL: a promising building material for sustainable buildings, *Chem. Process Eng. Res.*, **9**, pp. 1–7.

Akimov, Y. K. (2003). Fields of application of aerogels: review, *Instrum. Exp. Tech.*, **46**(3), pp. 287–299.

Al-Jabri, K., Shoukry, H. (2014). Use of nano-structured waste materials for improving mechanical, physical and structural properties of cement mortar, *Constr. Build. Mater.*, **73**, pp. 636–644.

ARDEX (2016). ISOLTECO, Available at: http://www.ardex.es

Ashour, T., Wieland, H., Georg, H., Bockisch, F.-J., Wu, W. (2010). The influence of natural reinforcement fibres on insulation values of earth plaster for straw bale buildings, *Mater. Des.*, **31**(10), pp. 4676–4685.

Aspen (2014). http://www.aerogel.com/

Azichem (2009). Sanawarme, Available at: http://img.edilportale.com

Baetens, R., Jelle, B. P., Gustavsen, A. (2011). Aerogel insulation for building applications: a state-of-the-art review, *Energy Build.*, **43**(4), pp. 761–769.

Barbero-Barrera, M. M., García-Santos, A., Neila-González, F. J. (2014). Thermal conductivity of lime mortars and calcined diatoms. Parameters influencing their performance and comparison with the traditional lime and mortars containing crushed marble used as renders, *Energy Build.*, **76**, pp. 422–428.

Barreca, F., Fichera, C. R. (2013). Use of olive stone as an additive in cement lime mortar to improve thermal insulation, *Energy Build.*, **62**, pp. 507–513.

Benazzouk, A., Douzane, O., Mezreb, K., Laidoudi, B., Quéneudec, M. (2008). Thermal conductivity of cement composites containing rubber waste particles: experimental study and modelling, *Constr. Build. Mater.*, **22**(4), pp. 573–579.

Bhagat, S. D., Kim, Y. H., Moon, M. J., Ahn, Y. S., Yeo, J. G. (2007). A cost-effective and fast synthesis of nanoporous SiO_2 aerogel powders using water-glass via ambient pressure drying route, *Solid State Sci.*, **9**(7), pp. 628–635.

Bozsaky, D. (2010). The historical development of thermal insulation materials, *Period. Polytech. Archit.*, **41**(2), pp. 49–56.

Buratti, C., Moretti, E. (2012). Experimental performance evaluation of aerogel glazing systems, *Appl. Energy*, **97**, pp. 430–437.

Buratti, C., Moretti, E., Belloni, E., Agosti, F. (2014). Development of innovative aerogel based plasters: preliminary thermal and acoustic performance evaluation, *Sustainability*, **6**(9), pp. 5839–5852.

Cárdenas, C., Tobón, J. I., García, C., Vila, J. (2012). Functionalized building materials: photocatalytic abatement of NOx by cement pastes blended with TiO_2 nanoparticles, *Constr. Build. Mater.*, **36**, pp. 820–825.

CEN (1999a). Methods of test for mortar for masonry. Part 10: Determination of dry bulk density of hardened mortar, *EN 1015-10*. Comité Européen de Normalisation, Brussels.

CEN (1999b). Methods of test for mortar for masonry. Part 11: Determination of flexural and compressive strength of hardened mortar, *EN 1015-11*. Comité Européen de Normalisation, Brussels.

CEN (2000). Methods of test for mortar for masonry. Part 12: Determination of adhesive strength of hardened rendering and plastering mortars on substrates, *EN 1015-12*. Comité Européen de Normalisation, Brussels.

CEN (2002a). Masonry and masonry products: methods for determining design thermal values, *EN 1745*. Comité Européen de Normalisation, Brussels.

CEN (2002b). Methods of test for mortar for masonry. Part 18: Determination of water absorption coeffcient due to capillary action of hardened mortar, *EN 1015-18*. Comité Européen de Normalisation, Brussels.

CEN (2004). Methods of test for masonry. Part 19: Determination of water vapour permeability of hardened rendering and plastering mortars, *EN 1015-19*. Comité Européen de Normalisation, Brussels.

CEN (2007). Fire classification of construction products and building elements. Part 1: Classification using data from reaction to fire tests, *EN 13501-1*. Comité Européen de Normalisation, Brussels.

CEN (2010). Specification for mortar for masonry. Part 1: Rendering and plastering mortar, *EN 998-1*. Comité Européen de Normalisation, Brussels.

Chen, J., Kou, S. C., Poon, C. S. (2012). Hydration and properties of nano-TiO_2 blended cement composites, *Cem. Concr. Compos.*, **34**(5), pp. 642–649.

Chiraema (2017). Chiraema, Available at: http://www.chiraema.it

Cuce, E., Cuce, P. M., Wood, C. J., Riffat, S. B. (2014). Toward aerogel based thermal superinsulation in buildings: a comprehensive review, *Renew. Sust. Energ. Rev.*, **34**, pp. 273–299.

CVR (2017). Termopor, Available at: http://www.cvr.it

Demirboga, R. (2003). Influence of mineral admixtures on thermal conductivity and compressive strength of mortar, *Energy Build.*, **35**(2), pp. 189–192.

Demirboga, R., Türkmen, İ., Karakoç, M. B. (2007). Thermo-mechanical properties of concrete containing high-volume mineral admixtures, *Build. Environ.*, **42**(1), pp. 349–354.

Diasen (2017). Diathonite, Available at: http://l.diasen.com

Doshi, D., Miller, T., Chase, J., Norwood, C. (2011). Aerogel composites and methods for making and using them, WO2011066209 A2.

Drexler, K. (1981). Molecular engineering: an approach to the development of general capabilities for molecuar manipulation, *Proc. Natl. Acad. Sci. U.S.A.*, **78**(9), pp. 5275–5278.

Ebert, H.-P. (2011). Thermal properties of aerogels, in Aegerter, M., Leventis, N., Koebel, M. (eds.) *Aerogels Handbook*, Springer, London, pp. 537–564.

EDILTECO (2016). EDILTECO, Available at: http://www.edilteco.it

EMPA (2016). Technical information sheet Fixit 222, Available at: http://www.fixit.ch/aerogel/pdf/upload/1463124179-TM Fixit 222 EN.pdf

EU (2010). Directiva 2010/31/UE, *Official Journal of the European Union*, pp. 13–35.

European Parliament Council of the European Union (2003). The European Parliament and the Council of the EU (4 September), pp. 9–19.

FassaBortolo (2011). KT 48 Reboco Termo-isolante, Available at: http://www.fassabortolo.pt

Frank, H., Zimmermann, G., Stuhler, N. (2000). Composite material containing aerogel, process for manufacturing the same and the use thereof, US006080475A.

Fricke, J. (1988). Aerogels, *Sci. Am.*, **258**, pp. 92–97.

Fricke, J., Tillotson, T. (1997). Aerogels: production, characterization, and applications, *Thin Solid Films*, **297**(1–2), pp. 212–223.

Fu, X., Chung, D. D. L. (1997). Effects of silica fume, latex, methylcellulose, and carbon fibers on the thermal conductivity and specific heat of cement paste, *Cem. Concr. Res.*, **27**(12), pp. 1799–1804.

Gao, T., Jelle, B. P., Gustavsen, A., Jacobsen, S. (2014). Aerogel-incorporated concrete: an experimental study, *Constr. Build. Mater.*, **52**, pp. 130–136.

Ge, Z., Gao, Z. (2008). Applications of nanotechnology and nanomaterials in construction, in *First Inter. Confer. Construc. Develop. Countries*, pp. 235–240.

Givi, A., Rashid, S., Aziz, F., Salleh, M. (2011). Investigations on the development of the permeability properties of binary blended concrete with nano-SiO_2 particles, *J. Compos. Mater.*, **45**(19), pp. 1931–1938.

Gomes, M. G., Flores-Colen, I., da Silva, F., Pedroso, M. (2018). Thermal conductivity measurement of thermal insulating mortars with EPS and silica aerogel by steady-state and transient methods, *Constr. Build. Mater.*, **172**, pp. 696–705.

Gonçalves, H., Graça, J. M. (2004). *Bioclimatic Concepts for Buildings in Portugal (in Portuguese)*. INETI, Lisboa.

Gonçalves, M. (2012). Nanomaterials in *Science and Engineering of Construction Materials (in Portuguese).* IST Press, Lisboa, pp. 725–771.

Grassian, V. H., O'Shaughnessy, P. T., Adamcakova-Dodd, A., Pettibone, J. M., Thorne, P. S. (2007). Inhalation exposure study of Titanium dioxide nanoparticles with a primary particle size of 2 to 5 nm, *Environ. Health Perspect.*, **115**(3), pp. 397–402.

Grobert, N. (2007). Carbon nanotubes: becoming clean, *Mater. Today*, **10**(1–2), pp. 28–35.

Guinness (2013). Least dense solid, Available at: http://www. guinnessworldrecords. com/world-records/least-dense-solid

Gupta, A. (2016). Climate change and Kyoto Protocol: an overview, in *Handbook of Environmental and Sustainable Finance*, pp. 3–23.

Guskos, N., Zolnierkiewicz, G., Typek, J., Blyszko, J., Kiernozycki, W., Narkiewicz, U. (2010). Ferromagnetic resonance and compressive strength study of cement mortars containing carbon encapsulated nickel and iron nanoparticles, *Rev. Adv. Mater. Sci.*, **23**(1), pp. 113–117.

Gutiérrez-González, S., Gadea, J., Rodríguez, A., Blanco-Varela, M. T., Calderón, V. (2012a). Compatibility between gypsum and polyamide powder waste to produce lightweight plaster with enhanced thermal properties, *Constr. Build. Mater.*, **34**, pp. 179–185.

Gutiérrez-González, S., Gadea, J., Rodríguez, A., Junco, C., Calderón, V. (2012b). Lightweight plaster materials with enhanced thermal properties made with polyurethane foam wastes, *Constr. Build. Mater.*, **28**(1), pp. 653–658.

Hallock, M. F., Greenley, P., DiBerardinis, L., Kallin, D. (2009). Potential risks of nanomaterials and how to safely handle materials of uncertain toxicity, *J. Chem. Health Saf.*, **16**(1), pp. 16–23.

Hanus, M. J., Harris, A. T. (2013). Nanotechnology innovations for the construction industry, *Prog. Mater. Sci.*, **58**(7), pp. 1056–1102.

Heinemann, U., Caps, R., Fricke, J. (1996). Radiation-conduction interaction: an investigation on silica aerogels, *Int. J. Heat Mass Transfer*, **39**(10), pp. 2115–2130.

Herrero, S., Mayor, P., Hernández-Olivares, F. (2013). Influence of proportion and particle size gradation of rubber from end-of-life tires on mechanical, thermal and acoustic properties of plaster-rubber mortars, *Mater. Des.*, **47**, pp. 633–642.

Holister, P. (2002). Nanotech: the tiny revolution. CMP Científica, Available at: http://www.nanotech-now.com/CMP-reports/NOR_White_Paper-July2002.pdf

Hümmer, E., Rettelbach, T., Lu, X., Fricke, J. (1993). Opacified silica aerogel powder insulation, *Thermochim. Acta*, **218**, pp. 269–276.

Hüsing, N., Schubert, U. (1998). Aerogels - airy materials: chemistry, structure, and properties, *Angew. Chem. Int. Ed.*, **37**(1/2), pp. 22–45.

Hüsken, G., Hunger, M., Brouwers, H. J. H. (2009). Experimental study of photocatalytic concrete products for air purification, *Build. Environ.*, **44**(12), pp. 2463–2474.

Ibrahim, M., Biwole, P. H., Achard, P., Wurtz, E. (2015). Aerogel-based materials for improving the building envelope's thermal behavior: a brief review with a focus on a new aerogel-based rendering, in *Energy Sustainability Through Green Energy*. Springer India (Green Energy and Technology), New Delhi, pp. 163–188.

Ibrahim, M., Wurtz, E., Biwole, P. H., Achard, P., Sallee, H. (2014). Hygrothermal performance of exterior walls covered with aerogel-based insulating rendering, *Energy Build.*, **84**, pp. 241–251.

Ilharco, L. M., Fidalgo, A., Farinha, J. P., Martinho, J. M., Rosa, M. E. (2007). Nanostructured silica/polymer subcritical aerogels, *J. Mater. Chem.*, **17**(21), pp. 2195–2198.

ISO (2010). Nanotechnologies: Vocabulary; Part 1: core terms. *ISO/TS 80004-1*. Geneva: International Organization for Standardization.

Jayapalan, A. R., Lee, B. Y., Kurtis, K. E. (2013). Can nanotechnology be "green"? Comparing efficacy of nano and microparticles in cementitious materials, *Cem. Concr. Compos.*, **36**(1), pp. 16–24.

Jelle, B. P., Gao, T., Ingunn, L., Sandberg, C., Tilset, B. G., Grandcolas, M., Gustavsen, A. (2014). Thermal superinsulation for building applications - from concepts to experimental investigations, **1**(2), pp. 43–50.

Jelle, B. P., Gustavsen, A., Baetens, R. (2010). The path to the high performance thermal building insulation materials and solutions of tomorrow, *J. Build. Phys.*, **34**(2), pp. 99–123.

Jones, S. (2006). Aerogel: space exploration applications, *J. Sol-Gel Sci. Technol.*, **40**(2–3), pp. 351–357.

Júlio, M. de F., Ilharco, L. M. (2017). Ambient pressure hybrid silica monoliths with hexamethyldisilazane: from vitreous hydrophilic xerogels to superhydrophobic aerogels, *ACS Omega*, **2**(8), pp. 5060–5070.

Júlio, M. de F., Soares, A., Ilharco, L. M., Flores-Colen, I., de Brito, J. (2016a). Aerogel-based renders with lightweight aggregates: correlation between molecular/pore structure and performance, *Constr. Build. Mater.*, **124**, pp. 485–495.

Júlio, M. de F., Soares, A., Ilharco, L. M., Flores-Colen, I., de Brito, J. (2016b). Silica-based aerogels as aggregates for cement-based thermal renders, *Cem. Concr. Compos.*, **72**, pp. 309–318.

Kawashima, S., Hou, P., Corr, D. J., Shah, S. P. (2013). Modification of cement-based materials with nanoparticles, *Cem. Concr. Compos.*, **36**(1), pp. 8–15.

Khamidi, M. F., Glover, C., Farhan, S. A., Puad, N. H. A., Nuruddin, M. F. (2014). Effect of silica aerogel on the thermal conductivity of cement paste for the construction of concrete buildings in sustainable cities, *WIT Trans. Built Environ.*, **137**, pp. 665–674.

Kimia (2017). Tectoria TH1, Available at: http://www.kimia.it

Koebel, M., Rigacci, A., Achard, P. (2012). Aerogel-based thermal superinsulation: an overview, *J. Sol-Gel Sci. Technol.*, **63**(3), pp. 315–339.

Konsta-Gdoutos, M. S., Metaxa, Z. S., Shah, S. P. (2010a). Highly dispersed carbon nanotube reinforced cement based materials, *Cem. Concr. Res.*, **40**(7), pp. 1052–1059.

Konsta-Gdoutos, M. S., Metaxa, Z. S., Shah, S. P. (2010b). Multi-scale mechanical and fracture characteristics and early-age strain capacity of high performance carbon nanotube/cement nanocomposites, *Cem. Concr. Compos.*, **32**(2), pp. 110–115.

Kuhn, J., Gleissner, T., Arduini-Schuster, M. C., Korder, S., Fricke, J. (1995). Integration of mineral powders into SiO_2 aerogels, *J. Non-Cryst. Solids*, **186**, pp. 291–295.

Kyushu (2014). http://www.chem-eng.kyushu-u.ac.jp/e/research.html

LATERLITE (2017). Termointonaco laterlite, Available at: http://www.edilio.it

Leopolder, F. (2010). The global drymix mortar industry (Part 1), *ZKG Int.*, **63**(4), pp. 32–45.

Leventis, N., Sotiriou-Leventis, C., Zhang, G., Rawashdeh, A. M. M. (2002). Nanoengineering strong silica aerogels, *Nano Lett.*, **2**(9), pp. 957–960.

Li, H., Xiao, H. G., Yuan, J., Ou, J. (2004). Microstructure of cement mortar with nano-particles, *Composites Part B*, **35**(2), pp. 185–189.

Ltifi, M., Guefrech, A., Mounanga, P., Khelidj, A. (2011). Experimental study of the effect of addition of nano-silica on the behaviour of cement mortars, *Procedia Eng.*, **10**, pp. 900–905.

Lu, H., Luo, H., Leventis, N. (2011). Mechanical characterization of aerogels, in Aegerter, M., Leventis, N., Koebel, M. (eds.) *Aerogels Handbook*, Springer, London, pp. 499–535.

Ma, H.-S., Roberts, A. P., Prévost, J.-H., Jullien, R., Scherer, G. W. (2000). Mechanical structure–property relationship of aerogels, *J. Non-Cryst. Solids*, **277**(2–3), pp. 127–141.

Mecklenburg, M., Schuchardt, A., Mishra, Y. K., Kaps, S., Adelung, R., Lotnyk, A., Kienle, L., Schulte, K. (2012). Aerographite: ultra lightweight, flexible nanowall, carbon microtube material with outstanding mechanical performance, *Adv. Mater.*, **24**(26), pp. 3486–3490.

Menaa, B., Menaa, F., Aiolfi-Guimarães, C., Sharts, O. (2010). Silica-based nanoporous sol-gel glasses: from Bioencapsulation to protein folding studies, *Int. J. Nanotechnol.*, **7**(1), pp. 1–45.

Milutienė, E., Staniškis, J. K., Kručius, A., Auguliene, V., Ardickas, D. (2012). Increase in buildings sustainability by using renewable materials and energy, *Clean Technol. Environ. Policy*, **14**(6), pp. 1075–1084.

NASA (2002). http://www.jpl.nasa.gov/releases/2002/release_2002_108.html

Nass, R., Campbell, R., Dellwo, U., Schuster, F., Tenegal, F., Kallio, M., Lintunen, P., Salatra, O., Remškar, M., Zumer, M., Hoet, P., Brüske-Hohlfeld, I., Lipscomb, S., Luther, W., Malanowski, N., Zweck, A. (2004). *Industrial Application of Nanomaterials: Chances and Risks - Technology Analysis*, Luther, W. (ed.), Future Technologies Division of VDI Technologiezentrum GmbH.

Nazari, A., Riahi, S. (2011). The effects of SiO_2 nanoparticles on physical and mechanical properties of high strength compacting concrete, *Composites Part B*, **42**(3), pp. 570–578.

Ng, S., Sandberg, L. I. C., Jelle, B. P. (2015). Insulating and strength properties of an aerogel-incorporated mortar based an UHPC formulations, *Key Eng. Mater.*, **629–630**, pp. 43–48.

Nosrati, R., Berardi, U. (2017). Long-term performance of aerogel-enhanced materials, *Energy Procedia*, **132**, pp. 303–308.

Nosrati, R. H., Berardi, U. (2018). Hygrothermal characteristics of aerogel-enhanced insulating materials under different humidity and temperature conditions, *Energy Build.*, **158**, pp. 698–711.

Obrey, K. A. D., Wilson, K. V., Loy, D. A. (2011). Enhancing mechanical properties of silica aerogels, *J. Non-Cryst. Solids*, **357**(19–20), pp. 3435–3441.

Pacheco-Torgal, F., Jalali, S. (2011). Nanotechnology: advantages and drawbacks in the field of construction and building materials, *Constr. Build. Mater.*, **25**(2), pp. 582–590.

Pajonk, G. M. (1998). Transparent silica aerogels, *J. Non-Cryst. Solids*, **225**, pp. 307–314.

Pajonk, G. M. (2003). Some applications of silica aerogels, *Colloid Polym. Sci.*, **281**(7), pp. 637–651.

Pandey, S., Mishra, S. B. (2011). Sol–gel derived organic–inorganic hybrid materials: synthesis, characterizations and applications, *J. Sol-Gel Sci. Technol.*, **59**(1), pp. 73–94.

Panesar, D. K., Shindman, B. (2012). The mechanical, transport and thermal properties of mortar and concrete containing waste cork, *Cem. Concr. Compos.*, **34**(9), pp. 982–992.

Parmenter, K. E., Milstein, F. (1998). Mechanical properties of silica aerogels, *J. Non-Cryst. Solids*, **223**(3), pp. 179–189.

Permanyer, C., Sanz, D. (2014). Composition de mortier isolant, WO2014162097 A1.

Pierre, A., Rigacci, A. (2011). SiO_2 aerogels, in Aegerter, M., Leventis, N., Koebel, M. (eds.) *Aerogels Handbook*, Springer, London, pp. 21–45.

Pina dos Santos, C., Matias, L. (2006). Thermal coefficients for buildings envelope. Technical Data Building, *ITE 50* (in Portuguese). LNEC, Lisbon.

Potočnik, J. (2011). Commission recommendation of 18 October 2011 on the definition of nanomaterial. *Official Journal of the European Union*, L 275/38-L 275/40.

RÖFIX (2016). RÖFIX, Available at: http://www.roefix.it

Roy, S., Hossain, A. (2008). Modeling of stiffness, strength, and structure-property relationship in crosslinked silica aerogel, in *Multiscale Modeling and Simulation of Composite Materials and Structures*, pp. 463–494.

Rubin, M., Lampert, C. M. (1983). Transparent silica aerogels for window insulation, *Sol. Energy Mater.*, **7**(4), pp. 393–400.

Sadrmomtazi, A., Fasihi, A., Balalaei, F., Haghi, A. (2009). Investigation of mechanical and physical properties of mortars containing silica fume and nano-SiO_2, in *The Third International Conference on Concrete and Development*. Building and Housing Research Center, Tehran, pp. 1153–1161.

Sáez de Ibarra, Y., Gaitero, J. J., Erkizia, E., Campillo, I. (2006). Atomic force microscopy and nanoindentation of cement pastes with nanotube dispersions, *Phys. Status Solidi A*, **203**(6), pp. 1076–1081.

Sanz-Pont, D., Sanz-Arauz, D., Bedoya-Frutos, C., Flatt, R., López-Andrés, S. (2016). Anhydrite/aerogel composites for thermal insulation, *Mater. Struct.*, **49**(8), pp. 3647–3661.

SCENIHR (2010). Scientific basis for the definition of the term "nano-material", Available at: http://ec.europa.eu/health/scientific_committees/emerging/docs/scenihr_o_032.pdf

Schmidt, M., Schwertfeger, F. (1998). Applications for silica aerogel products, *J. Non-Cryst. Solids*, **225**, pp. 364–368.

Secil (2016). Secil, Available at: http://www.secilargamassas.pt

Senff, L., Hotza, D., Lucas, S., Ferreira, V. M., Labrincha, J. A. (2012). Effect of nano-SiO_2 and nano-TiO_2 addition on the rheological behavior and the hardened properties of cement mortars, *Mater. Sci. Eng., A*, **532**, pp. 354–361.

Senff, L., Labrincha, J. A., Ferreira, V. M., Hotza, D., Repette, W. L. (2009). Effect of nano-silica on rheology and fresh properties of cement pastes and mortars, *Constr. Build. Mater.*, **23**(7), pp. 2487–2491.

Senff, L., Tobaldi, D. M., Lucas, S., Hotza, D., Ferreira, V. M., Labrincha, J. A. (2013). Formulation of mortars with nano-SiO_2 and nano-TiO_2 for degradation of pollutants in buildings, *Composites Part B*, **44**(1), pp. 40–47.

Sengul, O., Azizi, S., Karaosmanoglu, F., Tasdemir, M. A. (2011). Effect of expanded perlite on the mechanical properties and thermal conductivity of lightweight concrete, *Energy Build.*, **43**(2–3), pp. 671–676.

Sengupta, A., Sarkar, C. K. (eds.) (2015). *Introduction to Nano: Basics to Nanoscience and Nanotechnology*, Springer-Verlag Berlin Heidelberg.

She, J., Ohji, T. (2002). Porous mullite ceramics with high strength, *J. Mater. Sci. Lett.*, **21**(23), pp. 1833–1834.

Shekari, A. H., Razzaghi, M. S. (2011). Influence of nano particles on durability and mechanical properties of high performance concrete, *Procedia Eng.*, **14**, pp. 3036–3041.

Simmler, H., Brunner, S., Heinemann, U., Schwab, H., Kumaran, K., Mukhopadhyaya, P., Erb, M. (2005). Vacuum insulation panels study on VIP-components and panels for service life prediction of VIP in building applications (subtask A). EMPA; ZAE-Bayern; NRC-IRC; CSTB; Fraunhofer IVV; TU Delft Technical University of Delft; Dr.Eicher+Pauli AG.

Soares, A., Fátima, M. De; Flores-Colen, I., Ilharco, L. M., Brito, J. De (2018). EN 998-1 performance requirements for thermal aerogel-based renders, *Constr. Build. Mater.*, **179**, pp. 453–460.

Soares, A., Flores-Colen, I., de Brito, J. (2014a). Nanorenders on building facades: technical, economic and environmental performance, in *XIII International Conference on Durability of Building Materials and Components - XIII DBMC*. RILEM Publications, pp. 483–490.

Soares, A., Júlio, M., Flores-Colen, I., Ilharco, L., de Brito, J., Gaspar Martinho, J. (2014b). Water-resistance of mortars with lightweight aggregates, *Key Eng. Mater.*, **634**, pp. 46–53.

Soleimani Dorcheh, A., Abbasi, M. H. (2008). Silica aerogel; synthesis, properties and characterization, *J. Mater. Process. Technol.*, **199**(1–3), pp. 10–26.

De Souza, K. C., Mohallem, N. D. S., De Sousa, E. M. B. (2011). NanocompóSitos magnéticos: potencialidades de aplicações em biomedicina, *Quim. Nova*, **34**(10), pp. 1692–1703.

Stahl, T., Brunner, S., Zimmermann, M., Ghazi Wakili, K. (2012). Thermo-hygric properties of a newly developed aerogel based insulation rendering for both exterior and interior applications, *Energy Build.*, **44**, pp. 114–117.

Tan, C., Fung, B., Newman, J., Vu, C. (2001). Organic aerogels with very high impact strength, *Adv. Mater.*, **13**(9), pp. 644–646.

Taoukil, D., El bouardi, A., Ajzoul, T., Ezbakhe, H. (2012). Effect of the incorporation of wood wool on thermo physical proprieties of sand mortars, *KSCE J. Civ. Eng.*, **16**(6), pp. 1003–1010.

Taylor, J. (2002). New dimensions for manufacturing: a UK strategy for nanotechnology. UK Advisory Group on Nanotechnology Applications.

Tecresa (2015). Mortero tecwool, Available at: https://mercortecresa.com/morteros-tecwool

Thapliyal, P. C., Singh, K. (2014). Aerogels as promising thermal insulating materials: an overview, *J. Mater.*, **2014**, p. 10.

Thermowall (2013). THW, Available at: www.thermowall.pt

Tsou, P. (1995). Silica aerogel captures cosmic dust intact, *J. Non. Cryst. Solids*, **186**, pp. 415–427.

United Nations. (1998). Kyoto Protocol to the United Nations framework, *Rev. Eur. Community Int. Environ. Law*, **7**, pp. 214–217.

Uysal, H., Demirboğa, R., Şahin, R., Gül, R. (2004). The effects of different cement dosages, slumps, and pumice aggregate ratios on the thermal conductivity and density of concrete, *Cem. Concr. Res.*, **34**(5), pp. 845–848.

Wagh, P. B., Begag, R., Pajonk, G. M., Rao, A. V., Haranath, D. (1999). Comparison of some physical properties of silica aerogel monoliths synthesized by different precursors, *Mater. Chem. Phys.*, **57**(3), pp. 214–218.

Wakili, K. G., Stahl, T., Heiduk, E., Schuss, M., Vonbank, R., Pont, U., Sustr, C., Wolosiuk, D., Mahdavi, A. (2015). High performance aerogel containing plaster for historic buildings with structured façades, *Energy Procedia*, **78**, pp. 949–954.

Weber (2016). Aislone, Available at: https://www.weber.com.pt

Wei, G., Liu, Y., Zhang, X., Yu, F., Du, X. (2011). Thermal conductivities study on silica aerogel and its composite insulation materials, *Int. J. Heat Mass Transfer*, **54**(11–12), pp. 2355–2366.

Wenzel, H., Hauschild, M., Alting, L. (1997). *Environmental Assessment of Products: Volume 1 Methodology, Tools and Case Studies in Product Development.* Springer US.

Westgate, P., Paine, K., Ball, R. J. (2018). Physical and mechanical properties of plasters incorporating aerogel granules and polypropylene monofilament fibres, *Constr. Build. Mater.*, **158**, pp. 472–480.

Woignier, T., Phalippou, J. (1988). Mechanical strength of silica aerogels, *J. Non-Cryst. Solids*, **100**(1–3), pp. 404–408.

Xie, X., Mai, Y., Zhou, X. (2005). Dispersion and alignment of carbon nanotubes in polymer matrix: a review, *Mater. Sci. Eng., R*, **49**(4), pp. 89–112.

Xu, Y., Chung, D. D. L. (2000a). Cement of high specific heat and high thermal conductivity, obtained by using silane and silica fume as admixtures, *Cem. Concr. Res.*, **30**(7), pp. 1175–1178.

Xu, Y., Chung, D. D. L. (2000b). Effect of sand addition on the specific heat and thermal conductivity of cement, *Cem. Concr. Res.*, **30**(1), pp. 59–61.

Zeng, S. Q., Hunt, A. J., Cao, W., Greif, R. (1994). Pore size distribution and apparent gas thermal conductivity of silica aerogel, *J. Heat Transfer*, **116**(3), p. 756.

Zhang, G., Dass, A., Rawashdeh, A.-M. M., Thomas, J., Counsil, J. A., Sotiriou-Leventis, C., Fabrizio, E. F., Ilhan, F., Vassilaras, P., Scheiman, D. A., McCorkle, L., Palczer, A., Johnston, J. C., Meador, M. A., Leventis, N. (2004). Isocyanate-crosslinked silica aerogel monoliths: preparation and characterization, *J. Non-Cryst. Solids*, **350**, pp. 152–164.

Zhu, Q., Li, Y., Qiu, Z. (2007). Research progress on aerogels as transparent insulation materials, in *International Conference on Power Engineering*. Springer Berlin Heidelberg, Hangzhou, pp. 1117–1121.

Zhu, W., Bartos, P. J. M., Porro, A. (2004). Application of nanotechnology in construction: summary of a state-of-the-art report, *Mater. Struct.*, **37**(9), pp. 649–658.

Chapter 5

Multifunctional Cement-Based Carbon Nanocomposites

Liqing Zhang,[a] Siqi Ding,[b] Sufen Dong,[c] Xun Yu,[d] and Baoguo Han[e]

[a]*School of Civil Engineering and Architecture, East China Jiaotong University, Nanchang 330013, China*
[b]*Department of Civil and Environmental Engineering, The Hong Kong Polytechnic University, Hung Hom, Kowloon, Hong Kong*
[c]*School of Material Science and Engineering, Dalian University of Technology, Dalian 116024, China*
[d]*Department of Mechanical Engineering, New York Institute of Technology, NY 11568, USA*
[e]*School of Civil Engineering, Dalian University of Technology, Dalian 116024, China*
hithanbaoguo@163.com; hanbaoguo@dlut.edu.cn

5.1 Introduction

Cement-based materials are the most consumable building materials. They greatly affect the civil infrastructures' quality. To improve the safety of civil infrastructures, enhancing the mechanical behaviors and sensing behaviors of cement-based materials is becoming more and more important [1]. In addition, there are still some functions that are important for civil infrastructures. For example,

Nanotechnology in Cement-Based Construction
Edited by Antonella D'Alessandro, Annibale Luigi Materazzi, and Filippo Ubertini

ISBN 978-981-4800-76-1 (Hardcover), 978-0-429-32849-7 (eBook)
www.jennystanford.com

Table 5.1 Properties of NCMs [9, 14, 15]

Properties		NGPs	CNTs	CNFs	NCB
Elastic modules/TPa		1 (in-plane)	0.3–1	0.4–0.6	–
Strength/GPa		10–20	10–500	2.7–7.0	–
Electrical resistivity/ μΩ·cm		50 (in-plane)	5–50	55	0.22 Ω·cm*
Dimensions	Diameter	1–20 μm	0.75–30 nm	50–200 nm	10–400 nm
	Thickness (*T*)/ Length (*L*)	*T*: ~30 nm	*L*: 0.1–50 μm	*L*: 50–100 μm	–
Surface area/m^2/g		~2630	>400	~200	~1056*
Aspect ratio		50–300	~1000	100–500	–

*From Ref. [15].

damping behavior is vital for civil infrastructures in earthquake-prone regions [2]. Good electromagnetic shielding/absorbing behaviors are necessary for military structures, electronic equipment, and human health [3, 4]. In cold regions, self-heating behavior is useful for improving the safety of humans and protecting the environment [5–7]. As for all the civil infrastructures, durability is important for saving resources, saving manpower, and reducing construction waste. All of these needs motivate us to develop a new generation of cement-based materials with multifunctional behaviors.

The recently developed carbon nanomaterials (CNMs) provide an efficient way to develop multifunctional cement-based materials [8–15]. CNMs include nanographite platelets (NGPs), carbon nanotubes (CNTs), carbon nanofibers (CNFs), and nanocarbon black (NCB). The representative properties of CNMs are listed in Table 5.1, and scanning electron microscopy (SEM) photos of CNMs are shown in Fig. 4.1. Structurally, NGPs are, in general, layered and planar. The numbers of layers are from a few to several and the sum thickness of layers is in the nanoscale. In addition, graphene oxide (GO) is a single layer of NGPs. CNTs have a hollow cylindrical nanostructure, which can be visualized as the rolling of a graphite sheet [8]. According to the number of rolled layers of graphite, CNTs can be categorized into

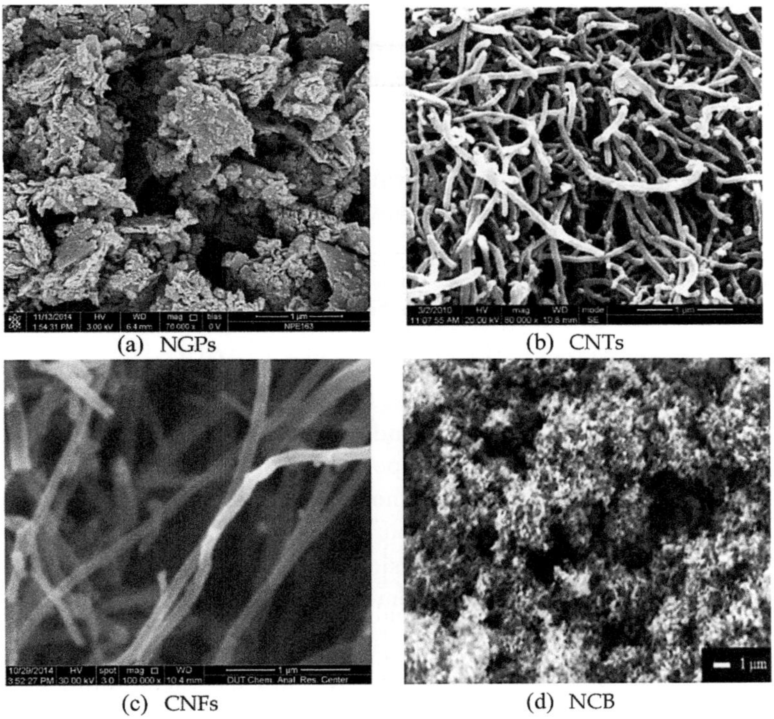

Figure 5.1 SEM photos of (a) NGPs, (b) CNTs, (c) CNFs, and (d) NCB.

single-walled carbon nanotubes (SWCNTs) and multiwalled carbon nanotubes (MWCNTs). CNFs can be thought of as regularly stacked, truncated conical or planar layers along the filament length. NCB is a form of amorphous carbon material. In general, CNMs have good mechanical and electrical properties, chemical durability, and a large surface area. CNMs possess mechanical, electrical, and magnetic properties that are significantly improved compared to those at the micro- and macroscales. This is due to the relatively larger surface area at the nanolevel, which can affect their strength and electrical properties. Adding CNMs into cement-based materials can make cement-based carbon nanocomposites multifunctional, with enhanced mechanical behaviors, sensing behavior, and damping behavior [9–13].

This chapter aims to provide a review of multifunctional cement-based carbon nanocomposites, which includes their design and manufacture, mechanical behaviors, electrical behavior, sensing behavior, damping behavior, electromagnetic shielding/absorbing behaviors, self-heating behavior, and durability. The future development and challenges of multifunctional cement-based carbon nanocomposites have also been discussed.

5.2 Design and Manufacture of Multifunctional Cement-Based Carbon Nanocomposites

Strong van der Waals forces cause the agglomeration of nanoparticles, so the key issue of design and manufacture of multifunctional cement-based carbon nanocomposites is dispersing CNMs. The traditional CNM dispersion methods include physical dispersion, chemical dispersion, and a combination of these two methods [16, 17]. However, traditional dispersion can just disperse CNMs in a cement matrix at some level. Some researchers have put forward new dispersion methods that include in situ–growing CNMs on cement/mineral admixture particles [18] (Fig. 5.2), spraying CNM-based films on aggregates [19, 20] (Fig. 5.3), and electrostatic self-assembly of CNMs [21] (Fig. 5.4). These new dispersion methods would effectively propel multifunctional cement-based carbon nanocomposites from lab to large-scale application.

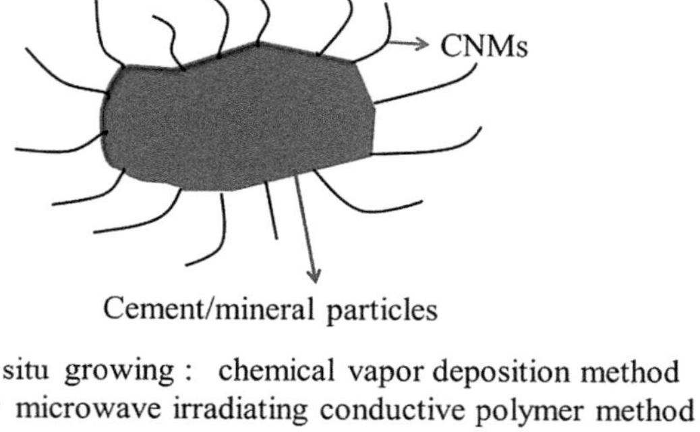

Figure 5.2 In situ growing CNMs.

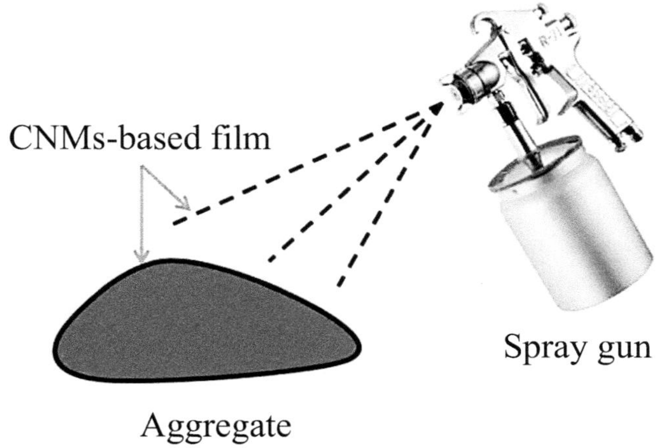

Figure 5.3 Spray CNM-based thin film on aggregates.

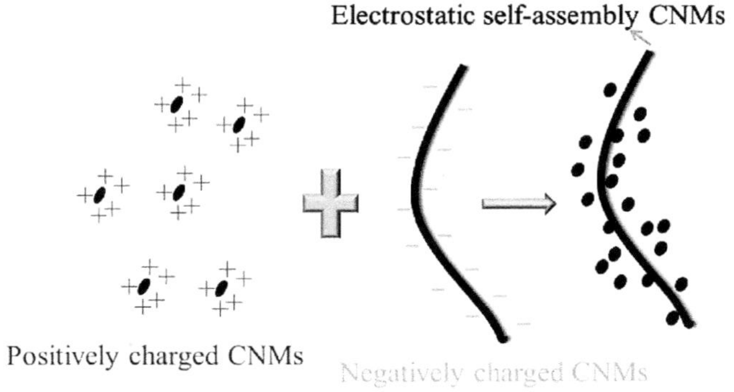

Figure 5.4 Electrostatic self-assembly of CNMs.

5.3 Behaviors of Multifunctional Cement-Based Carbon Nanocomposites

5.3.1 Mechanical Behaviors

Cement-based composites play the role of structural material in civil engineering. Therefore, mechanical behaviors are important and fundamental for cement-based composites. Cement-based composites are heterogeneous materials, which include gas phase, liquid phase, and solid phase. Due to the heterogeneous property, cement-based composites are suitable for the compressive zone but not for the tensile region in structural elements. The tensile strength of cement-based composites is only about 1/10 of the compressive strength. In general, cement-based composites are very brittle and low in tensile strength and strain. So, buildings and structures made of cement-based composites can fail without obvious deformation under tension stresses. Steel reinforcements were first used to improve the mechanical behavior of cement-based composites. And then fibers at the micrometer scale, such as steel fibers, carbon fibers (CFs), and basalt fibers, were employed as reinforcing materials, which can prevent cracks from propagating. However, fibers at the micrometer scale cannot prevent cracking in cement-based composites. In the early 20th century, nanomaterials were introduced into cement-based composites. Nanomaterials, especially CNMs, can improve the mechanical properties of cement-based composites at the nanoscale.

Till now, much effort has been devoted into the research on enhancing the mechanical behaviors of cement-based composites by adding CNMs [22–32]. In general, a little amount of CNMs can make big improvements in the mechanical behaviors of cement-based composites, which include compressive strength, flexural strength, the Young's modulus, hardness, the critical opening displacement, fracture toughness, and ductility. The best performance enhancement of cement-based carbon nanocomposites includes a 300% increase in compressive strength [33], a 34.28% increase in tensile strength [34], a 269% increase in flexural strength [35], a 270% increase in fracture toughness [36], a 14% increase in fracture energy [37], over 600% improvement in Vickers's hardness at the early ages [38], a 2200% increase in deflection [24], a 130% increase in ductility [35], over 430% improvement in resilience [23] and a 227% increase in the Young's modulus [39].

However, the reinforcing mechanism of CNMs in cement-based composites is not fully understood. The recognized and possible mechanisms are as follows. CNMs can make cement-based composites denser and hydration products well distributed due to the filling effect [15, 40–42] and the nucleation effect [38, 43]. As fiber materials, CNTs and CNFs also can improve the mechanical properties by the fiber's pull-out, crack bridging, pinning effect, crack deflection, fiber debonding, and fiber breaking [16]. CNMs all have such a big surface area that they can adsorb some water. This will reduce the actual water/cement ratio compared to the designed water/cement ratio, which may also contribute to inner curing and final mechanical behaviors.

In addition, it also can be seen that the research results, even for the same CNMs, are different. The reasons may be as follows. First, the reinforced matrix is different: it includes cement paste, cement mortar, and concrete. With the same content of CNMs, the reinforced effect decreases when the reinforced matrix changes from cement paste to concrete. Second, the water/cement ratio can affect the bond between CNMs and the cement matrix, and the distribution of CNMs. Third, the types of CNMs also have an effect on reinforcement results. For instance, CNTs have better mechanical properties than CNFs due to a larger length–diameter ratio. Fourth, the content of CNMs also has an effect on the mechanical properties. In general, the mechanical properties of cement-based carbon nanocomposites first increase and then decrease with increasing content of CNMs. Fifth, the dispersion issue is a key problem in using CNMs in cement-based composites. Homogeneously dispersed CNMs can greatly improve the mechanical behavior of cement-based composites. Although many studies have been conducted on uniform-dispersing CNMs, the dispersion issue cannot be solved now.

Cement-based carbon nanocomposites can help form high-strength concrete and high-performance concrete. In addition, they also can improve the early strength of concrete. However, there are still many issues that are needed to be studies, such as the underlying reinforcing mechanism, the relationship between microstructures and macroscopical performance, and constitutive models.

5.3.2 Electrically Conductive Behavior

Dry cement-based composites are slightly conductive in nature. CNMs have excellent electrical conductive properties, as shown in Table 5.1. Cement-based composites change from slightly conductive

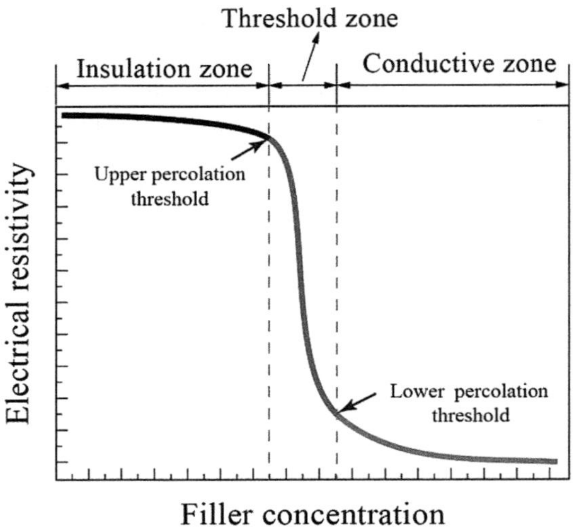

Figure 5.5 Change of electrical resistivity along with filler concentration [47].

to semiconductor or conductor with increasing content of CNMs [44–46]. Cement-based carbon nanocomposites show typical features of percolation phenomena, as shown in Fig. 5.5 [47]. The curve can be divided into three zones: insulation zone, threshold zone, and conductive zone. In the insulation zone, electrical resistivity slowly decreases with increasing content of CNMs. In the threshold zone, resistivity varies dramatically. This zone is very important for the sensing properties of cement based carbon nanocomposites. In the conductive zone, cement-based carbon nanocomposites become conductive materials whose resistivity also changes slightly with increasing CNMs.

CNMs can sharply reduce the electrical resistivity of cement-based carbon nanocomposites, as proved by many studies [10, 15, 44–46]. Singh et al. found that differences in the conductivity values of pristine cement are of the order of 10^{-7} compared to composites with the addition of 15 wt% of MWCNTs [1]. Han et al. found that the electrical resistivity reduction of cement mortar with 1.52 vol% of a CNT/NCB composite filler is 99.9% [21].

The conductivity of cement-based carbon nanocomposites is mainly affected by the following factors. First, the conductive properties of CNMs have direct effects. Second, the shapes of CNMs also wield a lot of influence. Compared to granulated NCB and lamellar NGPs, CNTs and CNFs can more easily reach the threshold zone due to a high length–diameter ratio [48]. Third, the distribution and dispersion of CNMs contribute to reaching the threshold zone.

5.3.3 Sensing Behavior

Sensing behavior in the cement-based composites field refers to the capacity of sensing strain, stress, damage, temperature, etc., which are important parameters to monitor the "health" of civil infrastructures [46, 49]. Sensing behavior of cement-based composites can be achieved by embedding, attaching, or remoting sensors. Traditional sensors, such as optical fiber sensors, electrical resistance strain gauges, and piezoelectric ceramics, are limited by the high cost, poor durability, and compatibility with cement-based composites.

The recently developed cement-based carbon nanocomposites are conductive and piezoresistive and can be used as sensors. They have high sensitivity, good durability, and compatibility. Cement-based carbon nanocomposites can be placed into infrastructures in the form of integral type, embedded type, and sandwich type. They can sense smoke [50], chloride penetration [51], stress/strain [10, 12, 13, 40, 44–46, 52–62], and damage [63]. Among the sensing behaviors, sensing stress and strain arouse great interest. Usually, the change of stress or strain causes a change in the electrical properties of cement-based carbon nanocomposites. Therefore, getting electrical information helps know stress or strain.

Generally, various interpretations are accepted by researchers of cement-based carbon nanocomposites. First, the "pull out" and "pull in" effect may be responsible for the change of electrical resistivity. CNT and CNF cement-based composites may exist the "pull out" and "pull in" effects, which are similar to those of CF cement-based composites. "Pull out" will increase the electrical resistivity when the sensor is in a tension state. In contrast, "pull in" will decrease the electrical resistivity, which indicates the sensor in a compressive state. Second, the probability of the tunnel effect among CNMs is different when CNMs in a cement matrix are in different stress states. For instance, the probability of the tunnel effect among CNMs is high when they are in a compressive state. From the macroscopic point of view, electrical resistivity of cement-based carbon nanocomposites decreases.

Self-sensing behaviors of cement-based carbon nanocomposites are influenced by external factors and interior factors. The external factors include temperature, humidity, loading amplitudes, and loading rates. The interior factors include types of added fillers, the content of CNMs, the dispersion state of CNMs, the water binder ratio, types of matrices, and the curing age. Also, hybrid CNMs and uniform dispersion of CNMs are two important features in self-sensing cement-based carbon nanocomposites.

5.3.4 Damping Behavior

The capacity of changing vibration energy into other forms of energy is called damping behavior, and it is very important for the safety, comfort, reliability, and durability of bridges and industrial and civilian infrastructures. However, conventional cement-based composites, as the main materials used in infrastructures, have low damping behavior under frequent vibration loads. Cement-based composites improve their damping behavior by adding functional fillers such as CNMs, silica fume (SF), polymer latex, and methylcellulose (MC) [64, 65]. Among CNMs, CNTs and NGPs have been especially introduced into cement-based materials to improve the damping behavior.

Luo and Duan carried out a serial of studies on the damping behavior of cement-based composites with CNTs [66–68]. They found that the critical damping ratio of concrete with 2% of CNTs is 1.6 times that of plain concrete [66, 68]. Duan and Luo also studied the vibration damping behavior of cement-based composites with SF (15 wt%), MC (0.4 wt%), and CNTs (0, 0.1 wt%, 0.2 wt%, and 0.5 wt%). The more the amount of CNTs, the higher the damping ratio. The maximum increment of damping ratio can reach up to 20.55%

[67]. Muthusamy et al. studied the vibration damping capacity of cement-based composites with 8 vol% of NGPs and found that it had high values of both loss tangent (0.81) and loss modulus (7.5 GPa) under flexure at 0.2 Hz [69].

The known improvement mechanisms of damping behavior of cement-based carbon nanocomposites are the following. First, low interfacial shear strength between CNMs and the matrix and high interfacial area lead to an increase in damping behavior. Second, the microcrack bridging and interfacial "stick slip" capacity of CNTs in cement matrix contribute to balanced enhancement on structural damping behavior. Third, CNM networks are microscopic in scale, with a close network that is important for achieving high damping [69].

5.3.5 Electromagnetic Shielding/Absorbing Behaviors

With the rapid development of electrical and electronic devices, such as automobile communication systems, computers, and mobile phones, electromagnetic interference (EMI) shielding/absorbing has attracted wide attention. Proper EMI protection would not only ensure good working of related electronic devices but also minimize radiative damage to the human body [70–72]. However, conventional cement-based composites, commonly used in engineering construction, have poor EMI-protective properties. It is possible to shield or absorb some electromagnetic waves by adding conductive fillers. Till now, metals and carbon materials are the main shielding materials. Compared to metal materials, CNMs have intrinsic advantages, such as low weight, strong resistance against corrosion, and tunable electric conductivity [73].

Some studies on EMI shielding/absorbing of cement-based carbon nanocomposites are available in the literature. Nam et al. investigated the EMI shielding effectiveness (SE) of cement-based composites with CNTs and found the most effective shielding performance at a frequency range from 0.1 to 18 GHz was attained with 1.5 wt% of CNTs [74]. In addition, Nam et al. also studied the EMI SE of concrete containing both silica fume and CNTs. When 0.6 wt% of CNTs and 20 wt% of silica fume are added to concrete, the greatest EMI SE is obtained at 0.94 GHz, 1.56 GHz, and 2.46 GHz in a frequency range of 45 MHz–18 GHz [75]. Singh et al. found that concrete with 15wt% of CNTs produces an SE (it is dominated by absorption) more than 27 dB in the range of 8.2–12.4 GHz [31]. Singh et al. studied the EMI shielding properties of GO-ferrofluid-cement-based composites and found that the total SE (SET) of the GO-ferrofluid-cement-based composites is 46 dB, which is much higher than that of the pristine cement-based composites (4 dB) [3]. Dai et al. explored the absorbing electromagnetic waves of NCB cement-based composites (CBCC) and found that CBCC exhibited good performance of absorbing electromagnetic waves in the frequency range from 8 to 26.5 GHz. The minimum reflectivity of cement mortar with 2.5 wt% of NCB reached 20.30 dB in the frequency of 20.6 GHz [15].

5.3.6 Self-Heating Behavior

Self-heating cement-based composites, also called electrothermal cement-based composites, are able to generate heat by applying a voltage to themselves based on Joule's law. The design of self-heating cement-based composites mainly should consider the following factors. First, the electrical resistivity of cement-based composites should be in a suitable range. The suitable electrical resistivity of self-heating cement-based composites is about 10 Ω·cm [76]. Second, the electrical resistivity of self-heating cement-based composites should be stable. Studies have indicated that the resistivity of conductive cement-based composites tends to be stable when the content of conductive fillers is near or more than the percolation threshold [4, 77–79]. In addition, the stability of electrical resistivity is also related to the type of function filler. The resistivity of self-heating concrete with steel fibers increased by about 60 times after 1 year [76]. It was because the steel fibers formed a passive film in a high-alkaline environment of cement-based composites. Furthermore, CNMs have good chemical durability. Therefore, they are good candidates for making self-heating cement-based composites. Third, cement-based composites are structural materials. Therefore, the functional filler should have no or a slightly bad effect on the mechanical properties of cement-based composites. CNMs, uniformly dispersed, improve mechanical properties.

Many studies have studied self-heating cement-based carbon nanocomposites. For example, Sun et al. built a room using NCB mortar slabs as the electrical floor heating element and found that an electrical power of about 123.8 W/m^2 resulted in an indoor temperature rise of 10°C within 330 min [79]. Yao et al. employed 0.5 wt% of CNTs to improve the positive thermoelectric power of the CFs in cementitious composites, with an increase of 260% in the positive thermoelectric power (from 8.8 μV/°C to 22.6 μV/°C) [80]. Kim et al. found that heat generation capability of CNT-embedded cementitious composites improved by an increase in the amount of CNTs, and composites having CNTs less than 0.6 wt% were appropriate as heating elements [81]. In addition, Galaoa et al. [82] prepared cement pastes with the addition of different CNF dosages (from 0% to 5% by mass of cement) and found that temperatures up to 138°C were registered during shotcrete-5% CNF cement paste tests (showing initial 10°C/min heating rates).

Self-heating cement-based carbon nanocomposites need to be further investigated on long-term stability, temperature control, economic analysis, energy consumption, and safety.

5.3.7 Durability

Durability of cement-based composites is the ability to keep good performance and integrity when subjected to an external environment, such as a corrosive gas or liquid, or temperature. First, CNMs have good durability, even in a high-alkaline cement matrix, due to their stable chemical properties and excellent mechanical properties. Second, CNMs in cement-based composites can promote hydration and make hydration products well distributed by the nucleation effect. In addition, CNMs also can fill the pores in a cement matrix. All of these make cement-based carbon nanocomposites much denser compared to plain cement-based composites. Hence, CNMs can improve the durability of cement-based composites.

Studies mainly focus on abrasion resistance [24], leaching property, moisture sorptivity, acid resistance [83, 84], and transport properties [85]. For instance, Peyvandi et al. found that addition

0.145 vol% of CNFs and 3.55 vol% of steel fibers to ultrahigh-performance concrete can improve the abrasion resistance by 1700% [24]. In another research, Peyvandi et al. observed that 0.05 vol% of NGPs can significantly improve the moisture transport performance and acid resistance of cement-based composites [83]. Brown et al. found that the mass loss and flexural strength loss of cement-based composites with 0.2% of CNFs were less than those of plain cement paste after exposure to ammonium nitrate solutions for 125 days [84]. Du et al. found that water penetration depth, chloride diffusion, and migration coefficients of concrete with 1.5% of GNP were reduced by 80%, 80%, and 37%, respectively [85].

In addition, CNT-CF/cement-based materials' overlay can act as auxiliary anodes to improve cathodic protection of steel in cement-based composites structures [86].

5.4 Conclusions

Compared to conventional cement-based composites, properly designed multifunctional cement-based carbon nanocomposites can be applied to optimize the safety, longevity, and performance of infrastructures and reduce the life cycle costs, resource consumption, and environment pollution. Multifunctional cement-based carbon nanocomposites, as an innovative technology in the field of construction materials, inject novel vitality for construction materials despite challenges. The development of multifunctional cement-based carbon nanocomposites will promote their application to a broader prospect, producing enormous socioeconomic effects.

Although many studies have been conducted on the behavior of multifunctional cement-based carbon nanocomposites over the past decade, there are still many challenges needed to be addressed. Future works may include (i) learning more about the relationship between the nanoscale behavior and macroproperties, namely finding the nanogenomic code and the nanobehavior blueprint of multifunction cement-based carbon nanocomposites; (ii) developing a bottom-up design method of fabricating large-scale multifunctional cement-based carbon nanocomposites; (iii) studying the working ability and rheological behaviors of multifunction cement-based carbon nanocomposites; (iv) expanding new function behaviors, such as thermally conductive behaviors, abrasive resistance, and impact resistance; and (v) applying multifunctional cement-based carbon nanocomposites in infrastructures.

Acknowledgments

The authors thank funding support from the National Science Foundation of China (51578110), the Department of Education of Jiangxi Province (GJJ180299), the China Postdoctoral Science Fundation (2019M651116), and the Fundamental Research Funds for the Central Universities in China (DUT18GJ203).

References

1. Singh, A. P., Gupta, B. K., Mishra, M., Chandra, A., Mathur, R. B., Dhawan, S. K. (2013). Multiwalled carbon nanotube/cement composites with exceptional electromagnetic interference shielding properties, *Carbon*, **56**, pp. 86–96.
2. Khitab, C. A., Anwar, W. (2016). Nano-scale behavior and nano-scale modification of cement and concrete materials, in Zhang, L., Ding, S., Sun, S., Han, B., Yu, X., Ou, J. (eds.) *Advanced Research on Nanotechnologe for Civil Engineering Applications*, IGI Global, USA, pp. 28–79.
3. Singh, A. P., Mishra, M., Chandra, A., Dhawan, S. K. (2011). Graphene oxide/ferrofluid/cement composites for electromagnetic interference shielding application, *Nanotechnology*, **22**, p. 465701.
4. Gomis, J., Galao, O., Gomis, V., Zornoza, E., Garcés, P. (2015). Self-heating and deicing conductive cement. Experimental study and modeling, *Constr. Build. Mater.*, **75**, pp. 442–449.
5. Loh, K., Nagarajaiah S. (2015). Nano carbon materials filled cementitious composites: fabrication, properties and application, in Han, B., Sun, S., Ding, S., Zhang, L., Dong, S., Yu, X. (eds.) *Innovative Developments of Advanced Multifunctional Nanocomposites in Civil and Structural Engineering*, Elsevier, USA, pp. 153–181.
6. Zhang, K., Han, B., Yu, X. (2011). Nickel particle based electrical resistance heating cementitious composites, *Cold Reg. Sci. Technol.*, **69**, pp. 64–69.

7. Sanchez, F., Sobolev, K. (2010). Nanotechnology in concrete: a review, *Constr. Build. Mater.*, **24**, pp. 2060–2071.
8. Sindu, B. S., Sasmal, S., Gopinath, S. (2014). A multi-scale approach for evaluating the mechanical characteristics of carbon nanotube incorporated cementitious composites, *Constr. Build. Mater.*, **50**, pp. 317–327.
9. Han, B., Yu, X., Ou, J. (2014). *Self-Sensing Concrete in Smart Structures*, Elsevier, USA.
10. Han, B., Zhang, K., Yu, X., Kwon, E., Ou, J. (2012). Electrical characteristics and pressure-sensitive response measurements of carboxyl MWNT/cement composites, *Cem. Concr. Compos.*, **34**, pp. 794–800.
11. Han, B., Zhang, K., Yu, X., Kwon, E., Ou, J. (2012). Fabrication of piezoresistive CNT/CNF cementitious composites with superplasticizer as dispersant, *J. Mater. Civ. Eng.*, **24**, pp. 658–665.
12. Ubertini, F., Materazzi, A. L., D'Alessandro, A., Laflamme, S. (2014). Natural frequencies identification of a reinforced concrete beam using carbon nanotube cement-based sensors, *Eng. Struct.*, **60**, pp. 265–275.
13. Materazzi, A. L., Ubertini, F., D'Alessandro, A. (2013). Carbon nanotube cement-based transducers for dynamic sensing of strain, *Cem. Concr. Compos.*, **37**, pp. 2–11.
14. Han, B., Wang, Y., Sun, S., Yu, X., Ou, J. (2014). Nanotip-induced ultrahigh pressure-sensitive composites: principles, properties and applications, *Composites Part A*, **59**, pp. 105–114.
15. Dai, Y., Sun, M., Liu, C., Li, Z. (2010). Electromagnetic wave absorbing characteristics of carbon black cement-based composites, *Cem. Concr. Compos.*, **32**, pp. 508–513.
16. Gopalakrishnan, K., Birgisson, B., Taylor, P., Attoh-Okine, N. O. (2011). Multifunctional and smart carbon nanotube reinforced cement-based materials, in Han, B., Yu, X., Ou, J. (eds.) *Nanotechnology in Civil Infrastructure*, Springer, Germany, pp. 1–47.
17. Raki, L., Beaudoin, J. J., Alizadeh, R., Makar, J. M., Sato, T. (2010). Cement and concrete nanoscience and nanotechnology, *Materials*, **3**, pp. 918–942.
18. Sun, S., Yu, X., Han, B., Ou, J. (2013). In situ growth of carbon nanotubes/carbon nanofibers on cement/mineral admixture particles: a review, *Constr. Build. Mater.*, **49**, pp. 835–840.
19. Gupta, S., Gonzalez, J. G., Loh, K. J. (2016). Self-sensing concrete enabled by nano-engineered cement-aggregate interfaces, *Struct. Heal. Monit.*, pp. 1–15.

20. Mortensen, L. P., Ryu, D. H., Zhao, Y. J., Loh, K. J. (2013). Rapid assembly of multifunctional thin film sensors for wind turbine blade monitoring, *Key Eng. Mater.*, pp. 515–522.

21. Han, B., Zhang, L., Sun, S., Yu, X., Dong, X., Wu, T., Ou, J. (2015). Electrostatic self-assembled carbon nanotube/nano carbon black composite fillers reinforced cement-based materials with multifunctionality, *Composites Part A*, **79**, pp. 103–115.

22. Lv, S., Ma, Y., Qiu, C., Sun, T., Liu, J., Zhou, Q. (2013). Effect of graphene oxide nanosheets of microstructure and mechanical properties of cement composites, *Constr. Build. Mater.*, **49**, pp. 121–127.

23. Metaxa, Z. S., Konsta-Gdoutos, M. S., Shah, S. P. (2013). Carbon nanofiber cementitious composites: effect of debulking procedure on dispersion and reinforcing efficiency, *Cem. Concr. Compos.*, **36**, pp. 25–32.

24. Peyvandi, A., Sbia, L. A., Soroushian, P., Sobolev, K. (2013). Effect of the cementitious paste density on the performance efficiency of carbon nanofiber in concrete nanocomposite, *Constr. Build. Mater.*, **48**, pp. 265–269.

25. Yazdanbakhsh, A., Grasley, Z., Tyson, B., Abu Al-Rub, R. (2012). Challenges and benefits of utilizing carbon nanofilaments in cementitious materials, *J. Nanomater.*, **2012**, pp. 1–8.

26. Galao, O., Zornoza, E., Baeza, F. J., Bernabeu, A., Garcés, P. (2012). Effect of carbon nanofiber addition in the mechanical properties and durability of cementitious materials, *Materiales de Construcción*, **62**, pp. 343–357.

27. Tyson, B. M., Abu Al-Rub, R. K., Yazdanbakhsh, A., Grasley, Z. (2011). Carbon nanotubes and carbon nanofibers for enhancing the mechanical properties of nanocomposite cementitious materials, *J. Mater. Civ. Eng.*, **23**, pp. 1028–1035.

28. Metaxa, Z. S., Konsta-Gdoutos, M. S., Shah, S. P. (2010). Carbon nanofiber-reinforced cement-based materials, *J. Transp. Res. Board*, **2142**, pp. 114–118.

29. Cwirzen, A., Habermehl-Cwirzen, K., Shandakov, D., Nasibulina, L. I., Nasibulin, A. G., Mudimela, P. R., Penttala, V., Kauppinen, E. I. (2009). Properties of high yield synthesised carbon nano fibres/Portland cement composite, *Adv. Cem. Res.*, **21**, pp. 141–146.

30. Musso, S., Tulliani, J., Ferro, G., Tagliaferro, A. (2009). Influence of carbon nanotubes structure on the mechanical behavior of cement composites, *Compos. Sci. Technol.*, **69**, pp. 1985–1990.

31. Luo, J., Duan, Z., Li, H. (2009). The influence of surfactants on the processing of multi-walled carbon nanotubes in reinforced cement matrix composites, *Phys. Status Solidi A*, **206**, pp. 2783–2790.
32. Habermehl-Cwirzen, K., Penttala, V., Cwirzen, A. (2008). Surface decoration of carbon nanotubes and mechanical properties of cement/carbon nanotube composites, *Adv. Cem. Res.*, **20**, pp. 65–73.
33. Nasibulina, L. I., Anoshkin, I. V., Semencha, A. V., Tolochko, O. V., Malm, J. E., Karppinen, M. J., Nasibulin, A. G., Kauppinen, E. I. (2012). Carbon nanofiber/clinker hybrid material as a highly efficient modificator of mortar mechanical properties, *Mater. Phys. Mech.*, **13**, pp. 77–84.
34. Ludvig, P., Ladeira, L. O., Calixto, J. M., Gaspar, I., Melo, V. S. (2009). In-situ synthesis of multiwall carbon nanotubes on Portland cement clinker, *Proc. 11th International Conference on Advanced Materials*, Brazil.
35. Abu Al-Rub, R. K., Ashour, A. I., Tyson, B. M. (2012). On the aspect ratio effect of multi-walled carbon nanotube reinforcements on the mechanical properties of cementitious nanocomposites, *Constr. Build. Mater.*, **35**, pp. 647–655.
36. Tyson, B. M., Abu Al-Rub, R. K., Yazdanbakhsh, A., Grasley, Z. (2011). Carbon nanotubes and carbon nanofibers for enhancing the mechanical properties of nanocomposite cementitious materials, *J. Mater. Civ. Eng.*, **23**, pp. 1028–1035.
37. Hlavacek, P., Smilauer, V., Padevet, P., Nasibulina, L., Nasibulin, A. G. (2011). Cement grains with surface-shyntetized carbon nanofibres: mechanical properties and nanostructure, *Proc. 3rd International Conference Nanocon*, Czech, pp. 75–80.
38. Makar, J. M., Margeson, J. C., Luh, J. (2005). Carbon nanotube/cement composites-early results and potential applications, *Proc. 3rd International Conference on Construction Materials: Performance, Innovation and Structural*, Canada, pp. 1–10.
39. Sáez De Ibarra, Y., Gaitero, J. J., Erkizia, E., Campillo, I. (2006). Atomic force microscopy and nanoindentation of cement pastes with nanotube dispersions, *Phys. Status Solidi A*, **203**, pp. 1076–1081.
40. Li, G. Y., Wang, P. M., Zhao, X. (2007). Pressure-sensitive properties and microstructure of carbon nanotube reinforced cement composites, *Cem. Concr. Compos.*, **29**, pp. 377–382.
41. Nochaiya, T., Chaipanich, A. (2011). Behavior of multi-walled carbon nanotubes on the porosity and microstructure of cement-based materials, *Appl. Surf. Sci.*, **257**, pp. 1941–1945.

42. Konsta-Gdoutos, M. S., Metaxa, Z. S., Shah, S. P. (2010). Multi-scale mechanical and fracture characteristics and early-age strain capacity of high performance carbon nanotube/cement nanocomposites, *Cem. Concr. Compos.*, **32**, pp. 110–115.

43. Kerienė, J., Kligys, M., Laukaitis, A., Yakovlev, G., Špokauskas, A., Aleknevičius, M. (2013). The influence of multi-walled carbon nanotubes additive on properties of non-autoclaved and autoclaved aerated concretes, *Constr. Build. Mater.*, **49**, pp. 527–535.

44. Du, H., Quek, S. T., Pang, S. D. (2013). Smart multifunctional cement mortar containing graphite nanoplatelet, *Proc. SPIE Smart Structures and Materials+ Nondestructive Evaluation and Health Monitoring*, USA, pp. 869238-869238-10.

45. Gao, D., Sturm, M., Mo, Y. L. (2009). Electrical resistance of carbon-nanofiber concrete, *Smart Mater. Struct.*, **18**, p. 95039.

46. Han, B., Sun, S., Ding, S., Zhang, L., Yu, X., Ou. J. (2015). Review of nanocarbon-engineered multifunctional cementitious composites, *Composites Part A*, **70**, pp. 69–81.

47. Stauffer, D., Aharony, A. (1994). *Introduction to Percolation Theory*, CRC Press, USA.

48. Wen, S., Chung, D. D. L. (2006). Partial replacement of carbon fiber by carbon black in multifunctional cement–matrix composites, *Carbon*, **45**, pp. 505–513.

49. Dong, S., Han, B., Ou, J., Li, Z., Han, L., Yu, X. (2016). Electrically conductive behaviors and mechanisms of short-cut super-fine stainless wire reinforced reactive powder concrete, *Cem. Concr. Compos.*, **72**, pp. 48–65.

50. Shukla, P., Bhatia, V., Gaur, V., Basniwal, R. K., Singh, B. K., Jain, V. K. (2012). Multiwalled carbon nanotubes reinforced Portland cement composites for smoke detection, *Solid State Phenom.*, pp. 21–24.

51. Kim, H. (2015). Chloride penetration monitoring in reinforced concrete structure using carbon nanotube/cement composite, *Constr. Build. Mater.*, **96**, pp. 29–36.

52. Sixuan, H. (2012). Multifunctional graphite nanoplatelets (GNP) reinforced cementitious composites. Dissertation for the Degree of Master of Engineering, National University of Singapore, Singapore.

53. Galao, O., Baeza, F. J., Zornoza, E., Garcés, P. (2014). Strain and damage sensing properties on multifunctional cement composites with CNF admixture, *Cem. Concr. Compos.*, **46**, pp. 90–98.

54. Baeza, F., Galao, O., Zornoza, E., Garcés, P. (2013). Multifunctional cement composites strain and damage sensors applied on reinforced concrete (RC) structural elements, *Materials*, **6**, pp. 841–855.
55. Howser, R. N., Dhonde, H. B., Mo, Y. L. (2011). Self-sensing of carbon nanofiber concrete columns subjected to reversed cyclic loading, *Smart Mater. Struct.*, **20**, p. 85031.
56. Kim, H. K., Nam, I. W., Lee, H. K. (2014). Enhanced effect of carbon nanotube on mechanical and electrical properties of cement composites by incorporation of silica fume, *Compos. Struct.*, **107**, pp. 60–69.
57. Han, B., Yu, X., Kwon, E., Ou, J. (2011). Effects of CNT concentration level and water/cement ratio on the piezoresistivity of CNT/cement composites, *J. Compos. Mater.*, **46**, pp. 19–25.
58. Han, B., Yu, X., Zhang, K., Kwon, E., Ou, J. (2011). Sensing properties of CNT-filled cement-based stress sensors, *J. Civ. Struct. Health Moint.*, **1**, pp. 17–24.
59. Han, B., Yu, X., Ou, J. (2010). Effect of water content on the piezoresistivity of MWNT/cement composites, *J. Mater. Sci.*, **45**, pp. 3714–3719.
60. Yu, X., Kwon, E. (2009). A carbon nanotube/cement composite with piezoresistive properties, *Smart Mater. Struct.*, **18**, p. 55010.
61. Han, B., Yu, X., Kwon, E. (2009). A self-sensing carbon nanotube/cement composite for traffic monitoring, *Nanotechnology*, **20**, p. 445501.
62. Han, B., Ding, S., Yu, X. (2015). Intrinsic self-sensing concrete and structures: a review, *Measurement*, **59**, pp. 110–128.
63. Gupta, S., Gonzalez, J., Loh, K. J. (2015). Damage detection using smart concrete engineered with nanocomposite cement-aggregate interfaces, *Proc. 10th International Workshop on Structural Health Monitoring*, USA, pp. 3033–3041.
64. Orak, S. (2000). Investigation of vibration damping on polymer concrete with polyester resin, *Cem. Concr. Res.*, **30**, pp. 171–174.
65. Kuang, T., Chang, L., Chen, F., Sheng, Y., Fu, D., Peng, X. (2016). Facile preparation of lightweight high-strength biodegradable polymer/multi-walled carbon nanotubes nanocomposite foams for electromagnetic interference shielding, *Carbon*, **105**, pp. 305–313.
66. Luo, J., Duan, Z., Xian, G., Li, Q., Zhao, T. (2013). Damping performances of carbon nanotube reinforced cement composite, *Mech. Adv. Mater. Struct.*, **22**, pp. 224–232.

67. Duan, Z., Luo, J. (2007). Effect of multi-walled carbon nanotubes on the vibration-reduction behavior of cement, *Proc. SPIE, International Conference on Smart Materials and Nanotechnology in Engineering*, China, pp. 64230R-64230R-6.

68. Lou, J. (2009). Fabrication and functional properties of multi-walled carbon nanotube/cement composites. Dissertation for the Doctoral degree, Harbin Institute of Technology, Harbin, China.

69. Muthusamy, S., Wang, S., Chung, D. D. L. (2010). Unprecedented vibration damping with high values of loss modulus and loss tangent, exhibited by cement–matrix graphite network composite, *Carbon*, **48**, pp. 1457–1464.

70. Cao, M., Song, W., Hou, Z., Wen, B., Yuan, J. (2010). The effects of temperature and frequency on the dielectric properties, electromagnetic interference shielding and microwave-absorption of short carbon fiber/silica composites, *Carbon*, **48**, pp. 788–796.

71. Wang, C., Li, K., Li, H., Guo, L., Jiao, G. (2008). Influence of CVI treatment of carbon fibers on the electromagnetic interference of CFRC composites, *Cem. Concr. Compos.*, **30**, pp. 478–485.

72. Li, N., Huang, Y., Du, F., He, X., Lin, X., Gao, H., Ma, Y., Li, F., Chen, Y., Eklund, P. C. (2006). Electromagnetic interference (EMI) shielding of single-walled carbon nanotube epoxy composites, *Nano Lett.*, **6**, pp. 1141–1145.

73. Kuang, T., Chang, L., Chen, F., Sheng, Y., Fu, D., Peng, X. (2016). Facile preparation of lightweight high-strength biodegradable polymer/multi-walled carbon nanotubes nanocomposite foams for electromagnetic interference shielding, *Carbon*, **105**, pp. 305–313.

74. Nam, I. W., Lee, H. K., Sim, J. B., Choi, S. M. (2012). Electromagnetic characteristics of cement matrix materials with carbon nanotubes, *ACI Mater. J.*, **109**, p. 363.

75. Nam, I. W., Kim, H. K., Lee, H. K. (2012). Influence of silica fume additions on electromagnetic interference shielding effectiveness of multi-walled carbon nanotube/cement composites, *Constr. Build. Mater.*, **30**, pp. 480–487.

76. Yehia, S., Tuan, C. Y., Ferdon, D., Chen, B. (2000). Conductive concrete overlay for bridge deck deicing: mixture proportioning, optimization, and properties, *Mater. J.*, **97**, pp. 172–181.

77. Chen, P., Chung, D. (1995). Improving the electrical conductivity of composites comprised of short conducting fibers in a nonconducting

matrix: the addition of a nonconducting particulate filler, *J. Electron. Mater.*, **24**, pp. 47–51.

78. Xie, P., Gu, P., Beaudoin, J. J. (1996). Electrical percolation phenomena in cement composites containing conductive fibres, *J. Mater. Sci.*, **31**, pp. 4093–4097.

79. Sun, M., Mu, X., Wang, X., Hou, Z., Li Z. (2008). Experimental studies on the indoor electrical floor heating system with carbon black mortar slabs, *Energy Build.*, **40**, pp. 1094–1100.

80. Yao, W., Zuo, J., Wu, K. (2013). Microstructure and thermoelectric properties of carbon nanotube-carbon fiber/cement composites, *J. Funct. Mater.*, **13**, p. 24.

81. Kim, G. M., Naeem, F., Kim, H. K., Lee, H. K. (2016). Heating and heat-dependent mechanical characteristics of CNT-embedded cementitious composites, *Compos. Struct.*, **136**, pp. 162–170.

82. Galao, O., Baeza, F. J., Zornoza, E., Garcés, P. (2014). Self-heating function of carbon nanofiber cement pastes, *Mater. Constr.*, **64**, pp. 1–11.

83. Peyvandi, A., Soroushian, P., Balachandra, A. M., Sobolev, K. (2013). Enhancement of the durability characteristics of concrete nanocomposite pipes with modified graphite nanoplatelets, *Constr. Build. Mater.*, **47**, pp. 111–117.

84. Brown, L., Sanchez, F., Kosson, D., Arnold, J. (2013). Performance of carbon nanofiber-cement composites subjected to accelerated decalcification, *Proc. EPJ Web of Conferences*, **56**, p. 02005.

85. Du, H., Gao, H. J., Pang, S. D. (2016). Improvement in concrete resistance against water and chloride ingress by adding graphene nanoplatelet, *Cem. Concr. Res.*, **83**, pp. 114–123.

86. Zuo, J., Yao W. (2015). Cathodic protection of reinforced concrete based on smart characteristics of cement-based materials, *J. Build Mater.*, p. 1.

Chapter 6

Analysis and Modeling of Electromechanical Properties of Cement-Based Nanocomposites

Siqi Ding,[a,b] Liqing Zhang,[c] Xun Yu,[d] Yiqing Ni,[a,b] and Baoguo Han[e]

[a] *Department of Civil and Environmental Engineering, Hong Kong Polytechnic University, Hung Hom, Kowloon, Hong Kong*
[b] *Hong Kong Branch of National Rail Transit Electrification and Automation Engineering Technology Research Center, Z105, Block Z, Hong Kong Polytechnic University, Hung Hom, Kowloon, Hong Kong*
[c] *School of Civil Engineering and Architecture, East China Jiaotong University, Nanchang 330013, China*
[d] *Department of Mechanical Engineering, New York Institute of Technology, New York 11568, USA*
[e] *School of Civil Engineering, Dalian University of Technology, Dalian 116024, China*
hithanbaoguo@163.com; hanbaoguo@dlut.edu.cn

6.1 Introduction

Cement-based nanocomposites are modified by using nanoscale fillers (e.g., carbon nanotubes, carbon nanofibers, nanocarbon black, graphene, and spiky spherical nickel powders with sharp nanotips)

Nanotechnology in Cement-Based Construction
Edited by Antonella D'Alessandro, Annibale Luigi Materazzi, and Filippo Ubertini

ISBN 978-981-4800-76-1 (Hardcover), 978-0-429-32849-7 (eBook)
www.jennystanford.com

in order to have multifunctional/smart properties, including enhanced mechanical and durable properties, electrically conductive properties, thermal properties, and electromagnetic and electromechanical properties. The analysis and modeling of electromechanical properties are crucial issues in research and application for multifunctional/smart cement–based nanocomposites. Electromechanical properties, namely the self-sensing properties of cement-based nanocomposites, are coupled relationships between electrical and mechanical properties, that is, a conductive network inside the nanocomposites and a force field on the nanocomposites. Particularly, cement-based nanocomposites demonstrate self-sensing abilities, whereby their electrical response is modulated by their mechanical state. In terms of analytical techniques, the mechanical properties (e.g., external force, stress, deformation, crack, strain, and damage) can be measured by using the existing measurement methods, while the sensing signal measurement is, technically, a more difficult task. Cement-based nanocomposites are multiphase and heterogeneous composites with complex electrical properties. The parameters to describe electrical properties, including electrical resistance, electrical capacitance, and dielectric characteristics, in varying degrees are numerous indeed [1–6]. Therefore, the characterization of electrical properties of cement-based nanocomposites is obviously different from other composites such as metal-based, polymer-based, or ceramic-based composites/nanocomposites. The same issues are related to the analysis and control of the electromechanical properties, mechanism, and modeling of cement-based nanocomposites. To date, most of the literature on cement-based nanocomposites has focused on fabrication and experimental characterization, and theoretical studies. The investigations about the physical principles underlying their complex behaviors are limited. The electromechanical property results from a change in the conductive network inside the composite under an external force or deformation. The mechanisms of electrical conduction and electromechanical properties of cement-based nanocomposites are similar to those of other conductive composites such as polymer-based or ceramic-based composites. However, there are some differences in the conductive characteristics of matrix materials, distribution characteristics of fillers in the matrix, and bonding characteristics between matrix and filler. The conductive characteristics of cement-based materials depend on their compositions and structures. The distribution characteristics of fillers in a cement-based matrix depend on the composition and processing technology of the composites. The bonding characteristics between cement-based matrix and filler depend on the composition of the composites, the surface conditions of fillers, and the processing technology of the composites.

In this chapter, the basic principles of electrical conduction and electromechanical effects of cement-based nanocomposites are first described, and then the electromechanical properties of cement-based nanocomposites are analyzed. Finally, several models of electromechanical effects based on different conduction mechanisms are introduced.

6.2 Electrically Conductive and Electromechanical Mechanisms

6.2.1 Basic Principles of Electrical Conduction

The basic types of electrical conduction of cement-based nanocomposites include electronic and/or hole conduction (i.e., contacting conduction, tunneling conduction, and/or field emission conduction) and ionic conduction. Electrons and/or holes come from nanoscale fillers, while ions come from the cement-based matrix [7–10].

6.2.1.1 Contacting conduction

This type of conduction is due to the direct contact of neighboring nanoscale fillers, thus forming conductive links. It is associated with the motion of electrons and/or holes through the conductive paths formed by nanoscale fillers, which are tiny and in contact with each other. The microstructural observation of cement-based nanocomposites has provided direct evidence of the existence of contacting conduction. The contacting conduction has been widely used in explaining the conductive behavior of cement-based nanocomposites with different nanoscale fillers [11, 12].

6.2.1.2 Tunneling conduction and/or field emission conduction

Tunneling conduction takes place when electrons jump through the energy barriers between functional fillers in a cement-based matrix [7, 13, 14]. Some researchers theorize that field emission is a manifestation of the tunneling effect [15, 16]. However, because field emission is induced by a local strong electric field, other researchers consider that filed emission is different from quantum tunneling to some extent [17–19]. Both tunneling conduction and field emission conduction are associated with the transmission conduction of electrons among the disconnected but close enough fillers. The tunneling conduction contributes to the electrical conductivity of cement-based nanocomposites with different nanoscale fillers [7, 13, 14], whereas the field emission conduction is not widely investigated, because conventional nanoscale fillers cannot generate a strong electric field to induce field emission at applied low voltages. However, some nanoscale fillers (e.g., carbon nanotubes and carbon nanofibers) with unique morphology can induce a localized increase of the electric field at sharp tips, which effectively reduces the barrier's width and allows field emission conduction to occur. By now, many researchers have ascribed some conductive behaviors of cement-based nanocomposites to the tunneling conduction and field emission conduction mechanism theoretically and some observed experimental results supporting this theory [7–9, 20, 21].

6.2.1.3 Ionic conduction

The hydrated cement paste, in addition to the calcium silicate hydrate gel and other solid phases, contains a variety of voids. The water filling these voids or pores can dissolve ionic species (mainly Ca^{2+} and OH^{-}) from the solid phases, resulting in some ionic conduction through the interconnected capillary pores. Since the ionic conduction is associated with the motion of ions in pore solution, ionic conductivity varies in a particularly wide range when cement contains a substantial amount of free water. In dry conditions, the cement matrix approximates an insulating material [22, 23]. In addition, cement-based nanocomposites with a filler concentration below the percolation threshold generally involve ionic conduction [12, 24].

It should be noted that the actual electrical conduction mechanism of cement-based nanocomposites is very complex in nature. The above-motioned conduction types coexist in the composite and influence each other. The DC electrical resistance–time relationship can indicate the relationship between the electronic hole conduction and the ionic conduction, which dominates in the electrical conductivity of cement-based nanocomposites. When the ionic conduction is the dominant, the DC electrical resistance increases obviously with measurement time due to the polarization effect; in the meanwhile, the AC electrical resistance is constant [25]. When the electronic hole conduction is dominant, the DC electrical resistance basically keeps stable with measurement time. In addition, the current–voltage relationship can give an indication of whether the electrical conductivity of cement-based nanocomposites is due to the tunneling and field emission conduction or the direct contact of neighboring functional fillers. A linear current–voltage relationship indicates that the direct contact of neighboring nanoscale fillers is the dominant conduction mechanism. In contrast, tunneling and field emission would induce a nonlinear power law current–voltage relationship in the electrical conductivity of cement-based nanocomposites.

6.2.2 Electrically Conductive Mechanisms

Cement-based nanocomposites are fabricated by adding nanoscale fillers into a cement-based matrix, and their conductive characteristics are closely related to the concentration of nanoscale fillers. The change of electrical resistivity of cement-based nanocomposites along with the filler concentration, that is, the conductive characteristic curve of cement-based nanocomposites, is graphically represented in Fig. 6.1. The conductive characteristic curve describes the percolation phenomenon, which can be divided into three sections: zone A with high resistivity is called the insulation zone, zone B

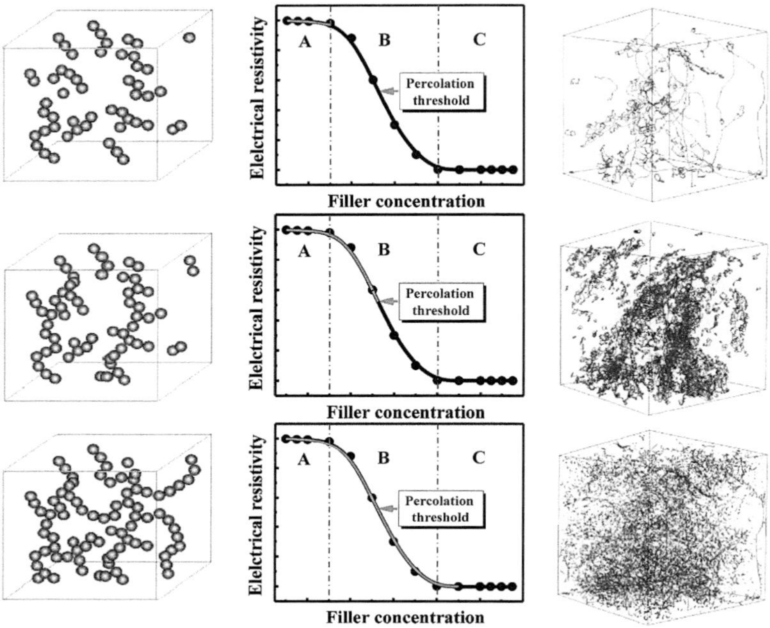

Figure 6.1 Change in electrical resistivity along with filler concentration (the top row is in zone A, the middle row is in zone B, and the bottom row is in zone C). Reprinted from Ref. [28], Copyright (2015), with permission from Elsevier.

with sharply decreasing resistivity is called the percolation zone, and zone C with stabilized low resistivity is called the conductive zone [26–28].

In zone A, the filler concentration in the concrete matrix is much lower than the percolation threshold, the spacing between fillers is large, and the filler gathering is less, so the conductive path is hard to form; the electrons are hard to move between fillers, and then the composite exhibits almost the same high resistivity as the matrix. The electrical conductivity of the matrix (i.e., the ionic conduction) dominates the electrical properties of the composite. In zone B, fillers form a conductive link and start forming a conductive path and the spacing among adjacent fillers decreases. The probability of the electronic transition greatly rises, resulting in a sharp increase in the conductivity of the composite. The contacting conduction, tunneling conduction and/or field emission conduction, and ionic conduction are all dominant factors in the electrical conductivity of the composite when the filler concentration is below the percolation threshold. However, when the filler concentration exceeds the percolation threshold, the tunneling conduction and/or field emission conduction plays a leading role in the composite conductivity, in addition to the direct contact of functional fillers. In zone C, the filler concentration is much higher than the percolation threshold, and fillers can be approximately regarded as totally contacting each other.

It should be noted that the patterns of the relationship curve between electrical resistivity and filler concentration have some differences for different functional fillers. Generally, fibrous fillers, having a high aspect ratio, can modify the electrical conductivity of cement-based nanocomposites at a much lower concentration level compared to particle fillers.

6.2.3 Electromechanical Mechanisms

The electrical resistivity of cement-based nanocomposites would change when the cement-based materials deform under loading. Several factors may contribute to the change in electrical resistivity:

- Change of intrinsic resistance of nanoscale fillers. When an external force is applied to the composites, the fillers are deformed, resulting in a change of their intrinsic resistance [9, 29].
- Change of bonding between functional filler and matrix. When the composite is subjected to an external force, the bonding between filler and matrix will alter, thus causing the contact resistance to change.
- Change of contact between nanoscale fillers. The load will lead to the direct contact or separation of nanoscale fillers, thus causing decrease or increase of the contact resistance. This factor has been considered as one of the main contributors to sensing behavior of cement-based nanocomposites.
- Change of tunneling distance between nanoscale fillers. Under an applied external force, thickness and/or microstructures of

the insulating concrete matrix between adjacent fillers may change considerably. This causes a change in the tunneling distance (i.e., a change in the tunneling barrier's width and height), thus changing the resistance of composites. This has been considered as the most important factor contributing to the electromechanical behavior of cement-based nanocomposites [7, 12, 16, 21].

- Change of capacitance. Nanoscale fillers can be regarded as capacitance plates due to the presence of ionic conduction in cement-based nanocomposites. When the composite is subjected to an external force, the capacitance plate distance and the relative dielectric constant of the cement-based matrix will change, thus causing a change in capacitance. This would lead to a change of current inside the composites and further alter the electrical resistance of the composite.

The above-mentioned factors may work together toward the electromechanical behavior of cement-based nanocomposites, but only one or some of them are leading at different zones of the conductive characteristic curve.

In zone A, the conductive path is hard to form, even though an external force is applied to the composites. As a result, the composites present high initial electrical resistivity and possess no or poor sensing response to an external force or deformation. At the beginning of zone B, the change of capacitance, the change of intrinsic resistance of fillers, and the change of bonding between filler and matrix are the dominant factors. Near the percolation threshold, the change of tunneling distance between fillers becomes the leading factor. At the end of zone B, the change of contact between fillers, the change of tunneling distance between fillers, and the change of intrinsic resistance of fillers play leading roles. It can be seen from the above analysis that several factors are in charge of the electromechanical behavior of cement-based nanocomposites together in each section of zone B. Therefore, the composites at zone B have good electromechanical properties. In zone C, the nanoscale fillers become crowded and are more likely to come into contact with each other. The change of contact between fillers and the change of intrinsic resistance of fillers (if existing) become the dominant factors. The conductive network inside the composites stabilizes and becomes hard to change under loading. As a result, the composites have low initial electrical resistivity and will present more stable electromechanical properties and low sensing sensitivity.

In general, low electrical resistivity is desirable in the analysis of electromechanical properties of cement-based nanocomposites since low electrical resistivity is helpful for enhancing the signal-to-noise ratio. However, high sensing sensitivity and low filler concentration are difficult to obtain at the same time with low electrical resistivity. Fortunately, there is a balance among high sensing sensitivity, low filler concentration, and low electrical resistivity near the percolation threshold. Therefore, the percolation threshold is an important parameter for designing and optimizing electromechanical properties of cement-based nanocomposites. Generally, a filler concentration above the percolation threshold is beneficial for sensing sensitivity under tension, while that below the percolation threshold is beneficial for sensing sensitivity under compression [8, 30–32].

6.3 Analysis of Electromechanical Properties

The analysis of electromechanical properties of cement-based nanocomposites is necessary to describe mechanical properties through electrical signals. By now, researchers have found some usable electrical signals to characterize the electromechanical behavior of cement-based nanocomposites under different loadings (e.g., compression, tension, flexure, and impact), which include electrical resistance or resistivity, electrical reactance, capacitance, relative dielectric constant, and electrical impedance tomography (EIT).

6.3.1 Electrical Resistivity

The electromechanical property of cement-based nanocomposites stems from the change of conductive network inside composites, so the volume electrical resistivity of cement-based nanocomposites is mostly fully able to characterize their sensing behavior [33, 34].

According to Ohm's law, the volume electrical resistivity of cement-based nanocomposites can be expressed as

$$\rho = R \times \frac{S}{L}, \tag{6.1}$$

where R is the volume electrical resistance of the cement-based nanocomposites, S is the sectional area of the cement-based nanocomposites, specimen and L is the space between two voltage electrodes.

After differential calculation, we can obtain

$$\frac{d\rho}{\rho} = \frac{dR}{R} - (1 + 2\mu)\frac{dL}{L}, \tag{6.2}$$

where μ is Poisson's ratio of the cement-based nanocomposites.

As the deformation of cement-based nanocomposites under an external force is very small, the changes in L can be neglected. Correspondingly, the change of volume electrical resistivity can be denoted as

$$\frac{\rho - \rho_0}{\rho_0} = \frac{R - R_0}{R_0} \text{ (i.e., } \Delta\rho/\rho_0 = \Delta R/R_0\text{)}, \tag{6.3}$$

where ρ_0 and R_0 are the initial volume electrical resistivity and the initial volume electrical resistance, respectively, of cement-based nanocomposites subjected to no external force, and $\Delta\rho$ and ΔR are the absolute changes in volume electrical resistivity and volume electrical resistance, respectively, of cement-based nanocomposites before and after the application of the external force.

It can be seen from Eq. 6.3 that the change in volume electrical resistance of cement-based nanocomposites under an external force is the same as that in volume electrical resistivity [35]. Therefore, the volume electrical resistance is also often used as a characteristic parameter to describe the electromechanical property of cement-based nanocomposites in many studies. It should be noted that the volume electrical resistance or resistivity can be classified in the stress direction (i.e., the direction parallel to the stress axis, also called volume electrical resistance or resistivity in the longitudinal direction) and in the transverse direction (i.e., the direction perpendicular to the stress axis) [36, 37].

In addition, as a constant electrical field is applied during the electrical resistance measurement, the movement and aggregation

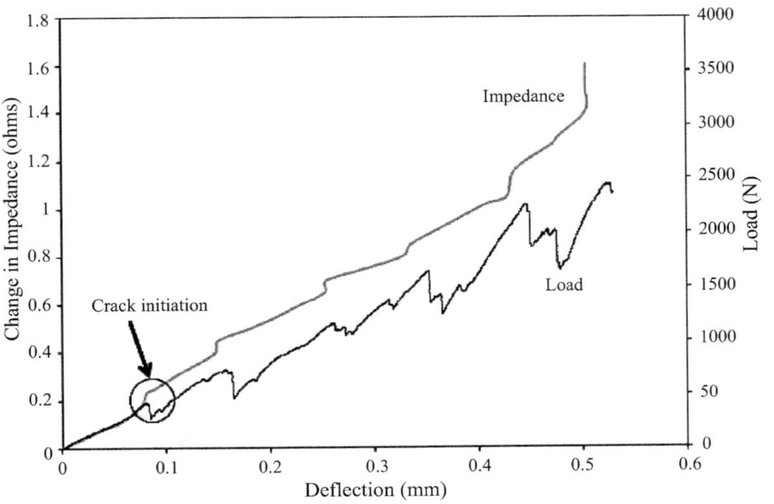

Figure 6.2 Change in impedance vs. deflection response of cement-based nanocomposites with carbon nanotubes under bending. Reprinted from Ref. [39], Copyright (2013), with permission from Elsevier.

of the ions in the cement concrete matrix will lead to an electrical polarization in the cement-based nanocomposites. The signal processing method can be used to eliminate the noise signal caused by polarization effect and extract the electromechanical property of the cement-based nanocomposites [38].

6.3.2 Impedance or Electrical Reactance

Under AC rather than DC conditions, the impedance Z consists of the electrical resistance R (real part of Z) and the electrical reactance X (imaginary part of Z) [25], that is,

$$Z = R + iX. \tag{6.4}$$

Saafi et al. (2013) observed that impedance can be used as a sensing signal to describe the sensing behavior of geopolymer cement concrete with carbon nanotubes under bending (as shown in Fig. 6.2) [39].

6.3.3 Electric Capacitance

Electric capacitance C can be expressed as

$$C = \frac{k\varepsilon_0 S}{l}, \tag{6.5}$$

where k and ε_0 are the relative dielectric constant and the vacuum dielectric constant, respectively [40].

Some researchers observed that capacitance can reflect the electromechanical property of cement-based composites [40, 41]. However, Han et al. (2012) stated that the capacitance of cement-based nanocomposites with carbon nanotubes is insensitive to an external force [33].

6.3.4 Electrical Impedance Tomography

EIT is a novel, soft-field tomography technique for estimating the cement-based nanocomposites' spatial conductivity (or resistivity) distributions. On the basis of voltage measurements taken at the body boundary when an AC signal is applied, EIT can reconstruct a 2D map of the specimen conductivity by solving the Laplace equation inversely. Since electrical conductivity at every location in the composite is sensitive to damage, the EIT conductivity maps can visualize the severities, locations, and physical attributes of damage occurring inside the composite [42, 43].

Hallaji and Pour-Ghaz (2014) investigated a thin layer of electrically conductive materials that is painted on the surface of concrete elements used as a sensing skin to detect and locate cracking and damage in the concrete substrate using EIT. They found that EIT is able to produce electrical resistivity maps that indicate the locations and severities of damage (as shown in Fig. 6.3) [43].

6.4 Modeling of Electromechanical Properties

Some constitutive models have been successively developed for verifying the proposed generation mechanism and quantitatively describing the electromechanical behavior of cement-based nano-

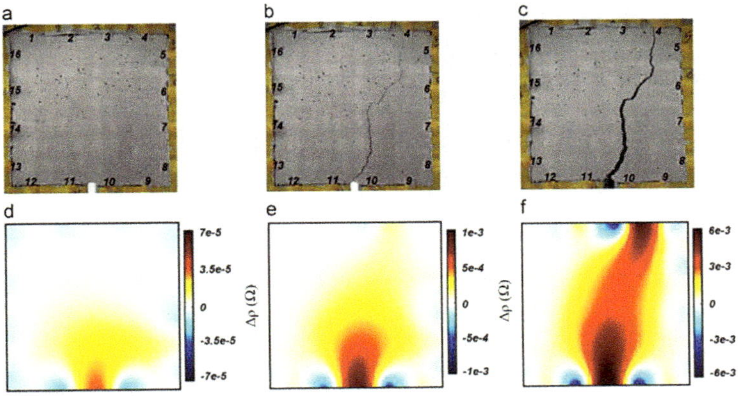

Figure 6.3 (a–c) Photographs of the sensing skin at three different load levels and (d–f) EIT images of sensing skins on concrete at three different load levels. Reprinted from Ref. [43], Copyright (2014), with permission from Elsevier.

composites in recent years. In general, their modeling principles are based on three basic types of electrical conduction mechanisms: tunneling conduction, field emission conduction, and a lumped circuit.

6.4.1 Model Based on Tunneling Conduction

Xiao et al. (2006) proposed a tunnel effect theory–based sensing model to predict strain-sensing property of carbon black cement paste when the carbon black concentration is near the percolation threshold and the conductive mechanism is dominated by the tunneling effect (as shown in Figs. 6.4 and 6.5). The proposed model is able to predict the resistance behavior of carbon black cement paste under various loading and environmental conditions [44].

6.4.2 Model Based on Field Emission Conduction

Han et al. (2009) established a constitutive model relating the change in the electrical resistivity of cement-based nanocomposites

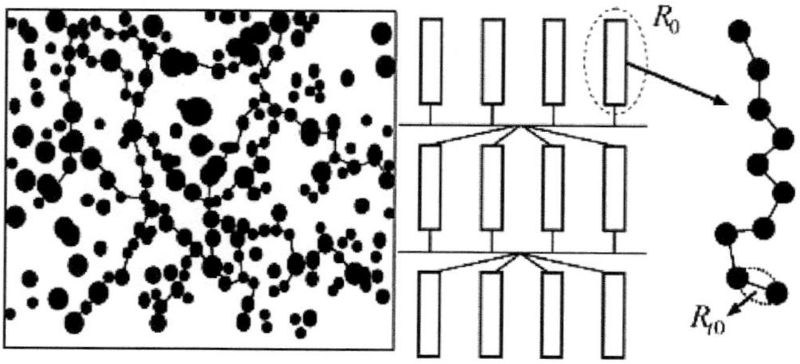

Figure 6.4 Schematic of conductive network in concrete with carbon black. Reprinted from Ref. [44], Copyright (2006), with permission from Elsevier.

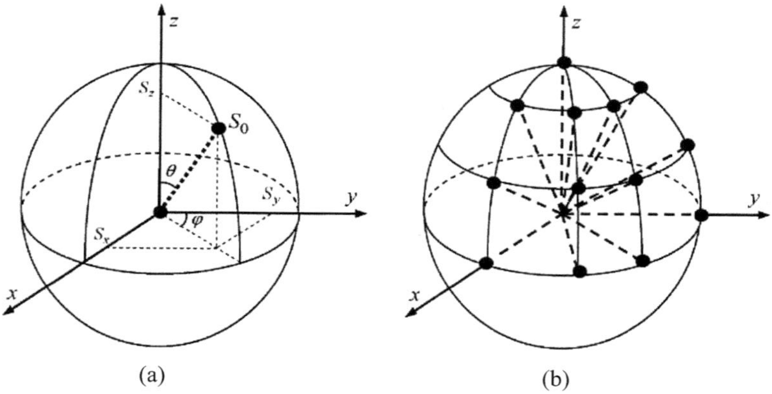

Figure 6.5 Schematic of (a) orientation and (b) distribution of R_{t0}. Reprinted from Ref. [44], Copyright (2006), with permission from Elsevier.

containing spiky spherical nickel powders with sharp nanotips to the applied compressive stress. This model incorporates the field emission effect with the interparticle separation change of nickel powders in composites within an elastic regime under uniaxial compression (as shown in Fig. 6.6). It is successfully used to predict the electromechanical behavior of cement-based nanocomposites containing different particle sizes of spiky spherical nickel powders with sharp nanotips [45].

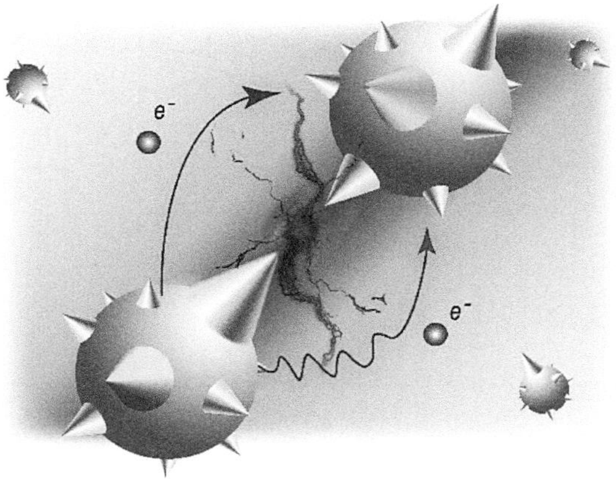

Figure 6.6 Schematic view of a conductive network in cement-based nanocomposites containing spiky spherical nickel powders with sharp nanotips.

6.4.3 Model Based on a Lumped Circuit

D'Alessandro et al. (2014) presented a linear lumped-circuit electromechanical model to predict the electrical behavior of carbon nanotube cement-based composites. The model consists of a capacitor and a resistor in parallel, and then it connects with a resistor in series (as shown in Fig. 6.7). The capacitor is related to the internal polarization of the composite, and the two resistors are associated with the contact resistance of the electrodes and the internal resistance of the composite between the two electrodes, respectively. When the composite is loaded, assuming only the internal resistance is changed as the distance between electrodes changes, the electrical current through the composites varies as a function of the applied axial strain. The proposed model is experimentally validated by measuring the step response of carbon nanotube cement-based composites with different distances between the electrodes. Model predictions are in good agreement with experimental results and model fitting, indicating that utilizing the proposed model to characterize the electromechanical properties of carbon nanotube cement-based composites has good feasibility (as shown in Fig. 6.8) [46, 47].

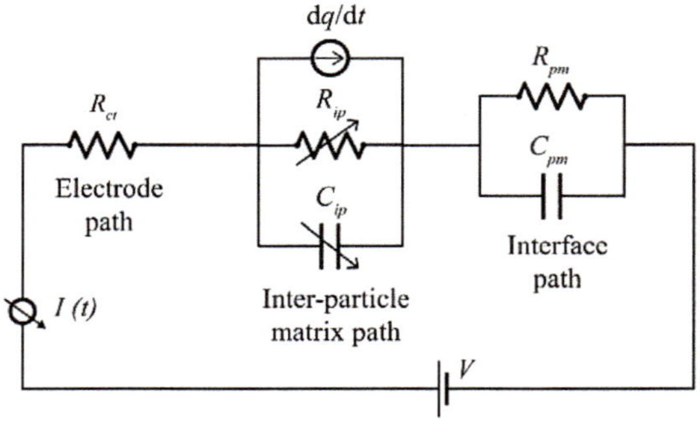

Figure 6.7 Schematic of a lumped-circuit electromechanical model of carbon nanotube cement-based composites. Reprinted from Ref. [47], Copyright (2017), with permission from Elsevier.

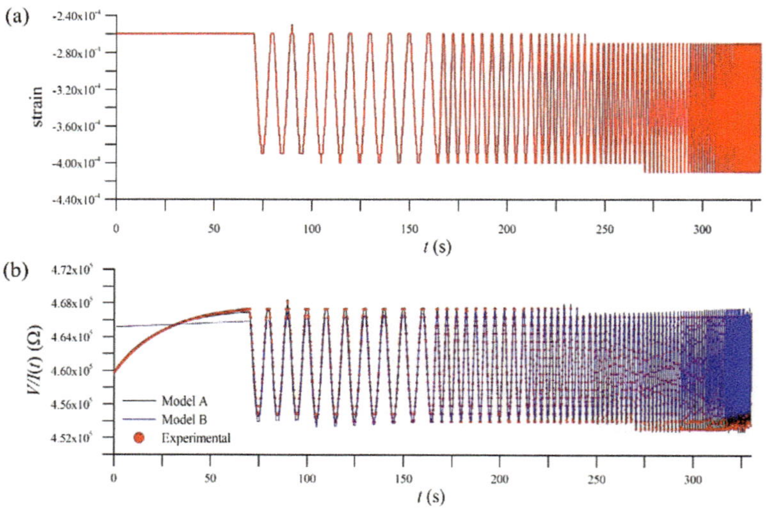

Figure 6.8 Strain series (a) and comparison between experimental results and analytical predictions of the response of carbon nanotube cement-based composites for forward sweep test from 0.1 to 3.5 Hz (b). Models A and B correspond to the newly proposed piezoresistive/piezoelectric and previously published piezoresistive modeling, respectively. Reprinted from Ref. [47], Copyright (2017), with permission from Elsevier.

6.5 Conclusion

The analysis and modeling of the electromechanical properties of cement-based nanocomposites are key problems in their research and application. The electrical conduction mechanism of cement-based nanocomposites is very complex in nature. There are mainly two types of electrical conduction, which include electronic conduction from conductive functional fillers and ionic conduction from the cement-based matrix. Furthermore, electrical conduction consists of contacting, tunneling conduction, and/or field emission. These conduction types coexist in cement-based nanocomposites and interrelate with each other. Their contribution to electrical conductivity of the composites closely depends on the concentration of functional fillers. This phenomenon can be described by percolation theory. Under an external force or deformation, the change in electrical resistivity can be attributed to several factors such as change of intrinsic resistance of functional fillers, change of bonding between functional filler and matrix, change of contact between functional fillers, change of tunneling distance between functional fillers, and change of capacitance. These factors work together to contribute to electromechanical properties of the composites, but only one or some of them are leading at certain zones of the percolation curve. The percolation threshold is an important parameter to design and optimize the sensing property of the nanocomposites. The analysis and modeling of the electromechanical properties of cement-based nanocomposites have a direct relationship with the electrical signal measurement. Six types of electrical parameters (including electrical resistance or resistivity, impedance, electrical reactance, capacitance, relative dielectric constant, and EIT) have been taken as electrical signals to characterize the electromechanical properties of cement-based nanocomposites. The direct current volume electrical resistivity or resistance is the most common electrical parameter because it is fully able to describe the electromechanical behavior of cement-based nanocomposites. The mechanism and electrical signal measurement provide the principle and evidence for establishment of constitutive models. The effective constitutive models may help to not only verify the mechanism of electromechanical properties but also describe and predict the electromechanical properties of cement-based nanocomposites.

Acknowledgments

The authors thank funding support from the National Science Foundation of China (51578110) and the Fundamental Research Funds for the Central Universities in China (DUT18GJ203).

References

1. Shi, M. L. (2001). *AC Impedance Spectroscopy Principles and Applications*, National Defense Industry Press, China.
2. Wansom, S., Kidner, N. J., Woo, L. Y., Mason, T. O. (2006). AC-impedance response of multi-walled carbon nanotube/cement composites, *Cem. Concr. Compos.*, **26**, pp. 509–519.
3. Zhang, X., Ding, X. Z., Ong, C. K., Yang, J. (1996). Dielectric and electrical properties of ordinary Portland cement and slag cement in the early hydration period, *J. Mater. Sci.*, **31**, pp. 1345–1352.
4. Tuan, C. Y. (2004). Electrical resistance heating of conductive concrete containing steel fibers and shavings, *ACI Mater. J.*, **101**(1), pp. 65–71.
5. Farrar, J. J. (1978). Electrically conductive concrete, *GEC J. Sci. Technol.*, **45**(1), pp. 45–48.
6. Whittington, H., McCarter, W., Forde, M. C. (1981). Conduction of electricity through concrete, *Mag. Concr. Res.*, **33**(114), pp. 48–60.
7. Mao, Q. Z., Zhao, B. Y., Sheng, D. R., Li, Z. Q. (1996). Resistance changement of compression sensible cement speciment under different stresses, *J. Wuhan Univ. Technol.*, **11**, pp. 41–45.
8. Han, B., Yu, X., Ou, J. (2014). *Self-Sensing Concrete in Smart Structures*, Elsevier, USA.
9. Li, G. Y., Wang, P. M., Zhao, X. H. (2007). Pressure-sensitive properties and microstructure of carbon nanotube reinforced cement composites, *Cem. Concr. Compos.*, **29**, pp. 377–382.
10. Sun, M. Q., Li, Z. Q., Liu, Q. P. (2000). The electromechanical effect of carbon fiber reinforced cement, *Carbon*, **40**(12), pp. 2263–2264.

11. Wang, Y. L., Zhao, X. H. (2005). Positive and negative pressure sensitivities of carbon fiber-reinforced cement-matrix composites and their mechanism, *Acta Mater. Compos. Sinica*, **22**(4), pp. 40–46.
12. Han, B. G., Yu, X., Ou, J. P. (2011). Multifunctional and smart carbon nanotube reinforced cement-based materials, in Gopalakrishnan K., Birgisson B., Taylor P., Attoh-Okine N. O. (eds.) *Nanotechnology in Civil Infrastructure*, Springer, Germany, pp. 1–47.
13. Li, C. T., Qian, J. S., Tang, Z. Q. (2005). Study on properties of smart concrete with steel slag, *China Concr. Cem. Prod.*, **2**, pp. 5–8.
14. Han, B. G., Ou, J. P. (2007). Embedded piezoresistive cement-based stress/strain sensor, *Sens. Actuators, A*, **138**(2), pp. 294–298.
15. Zhang, Z. M. (2007). *Nano/Microscale Heat Transfer*, McGraw-Hill, New York.
16. Wan, Y., Wen, D. J. (1999). Conducting polymer composites and their special effects, *Chin. J. Nat.*, **21**(3), pp. 149–153.
17. Chen, K., Xiong, C. X., Li, L. B., Zhou, L., Lei, Y., Dong, L. J. (2008). Conductive mechanism of antistatic poly (ethylene terephthalate)/ZnOw composites, *Polym. Compos.*, **30**, pp. 226–231.
18. Celzard, A., Mcrae, E., Furdin, G., Mareche, J. F. (1997). Conduction mechanisms in some graphite-polymer composites: the effect of a direct-current electric field, *J. Phys.: Condens. Matter*, **9**, pp. 2225–2237.
19. Peng, N., Zhang, Q., Lee, Y. C., Huang, H., Tan, O. K., Tian, J. Z., Chan, L. (2008). Humidity and temperature effects on carbon nanotube field-effect transistor-based gas sensors, *Sens. Lett.*, **6**(6), pp. 796–799.
20. Han, B. G., Yu, X., Ou, J. P. (2010). Effect of water content on the piezoresistivity of CNTs/cement composites, *J. Mater. Sci.*, **45**, pp. 3714–3719.
21. Han, B. G., Han, B. Z., Yu, X., Ou, J. P. (2011). Ultrahigh pressure-sensitive effect induced by field emission at sharp nano-tips on the surface of spiky spherical nickel powders, *Sens. Lett.*, **9**(5), pp. 1629–1635.
22. Wittmann, F. H. (2000). *Materials for Buildings and Structures*, Wiley, USA.
23. Xu, J., Zhong, W. H., Yao, W. (2010). Modeling of conductivity in carbon fiber-reinforced cement-based composite, *J. Mater. Sci.*, **45**, pp. 3538–3546.
24. Wen, S. H., Chung, D. D. L. (2006). The role of electronic and ionic conduction in the electrical conductivity of carbon fiber reinforced cement, *Carbon*, **44**, pp. 2130–2138.

25. Zheng, L. X., Song, X. H., Li, Z. Q. (2005). Investigation on the method of AC measurement of compression sensitivity of carbon fiber cement, *J. Huazhong Univ. Sci. Technol.*, **22**(2), pp. 27–29.
26. Huang, Y., Xiang, B., Ming, X. H., Fu, X. L., Ge, Y. J. (2008). Conductive mechanism research based on pressure-sensitive conductive composite material for flexible tactile sensing, *Proceeding of the 2008 IEEE International Conference on Information and Automation*, pp. 1614–1619.
27. Mohammed, H. A., Uttandaraman, S. (2009). A review of vapor grown carbon nanofiber/polymer conductive composites, *Carbon*, **47**(1), pp. 2–22.
28. Han, B., Ding, S., Yu, X. (2015). Intrinsic self-sensing concrete and structures: a review, *Measurement*, **59**, pp. 110–128.
29. Han, B. G., Han, B. Z., Yu, X. (2010). Effects of content level and particle size of nickel powder on the piezoresistivity of cement-based composites/sensors, *Smart Mater. Struct.*, **19**, p. 065012.
30. Wang, X. F., Wang, Y. L., Jin, Z. H. (2002). Electrical conductivity characterization and variation of carbon fiber reinforced cement composite, *J. Mater. Sci.*, **37**, pp. 223–227.
31. Saafi, M. (2009). Wireless and embedded carbon nanotube networks for damage detection in concrete structures, *Nanotechnology*, **20**, p. 395502.
32. Pushparaj, V. L., Nalamasu, O., Manoocher, M. (2010). Carbon nanotube-based load cells, Patent US2010/0050779 A1.
33. Han, B., Zhang, K., Yu, X., Kwon, E., Ou, J. (2012). Electrical characteristics and pressure-sensitive response measurements of carboxyl MWNT/cement composites, *Cem. Concr. Compos.*, **34**, pp. 794–800.
34. Mao, Q., Sun, M., Chen, P. (1997). Study on volume resistance and surface resistance of CFRC, *J. Wuhan Univ. Technol.*, **19**(2), pp. 65–67.
35. Kovacs, G. (1998). *Micromachined Transducers Sourcebook*, McGraw-Hill, USA.
36. Wen, S. H., Chung, D. D. L. (2001). Uniaxial compression in carbon fiber-reinforced cement, sensed by electrical resistivity measurement in longitudinal and transverse directions, *Cem. Concr. Res.*, **31**, pp. 297–301.
37. Wen, S. H., Chung, D. D. L. (2000). Uniaxial tension in carbon fiber reinforced cement, sensed by electrical resistivity measurement in longitudinal and transverse directions, *Cem. Concr. Res.*, **30**, pp. 1289–1294.

38. Meehan, D. G., Wang, S. K., Chung, D. D. L. (2010). Electrical-resistance-based sensing of impact damage in carbon fiber reinforced cement-based materials, *J. Intell. Mater. Syst. Struct.*, **21**(1), pp. 83–105.
39. Saafi, M., Andrew, K., Tang, P. L., McGhon, D., Taylor, S., Rahman, M., Yang, S. T., Zhou, X. M. (2013). Multifunctional properties of carbon nanotube/fly ash geopolymeric nanocomposites, *Constr. Build. Mater.*, **49**, pp. 46–55.
40. Zheng, L. X., Song, X. H., Li, Z. Q. (2004). Study on the compression sensibility of CFRC under quasi-triaxial compression, *Bull. Chin. Ceram. Soc.*, **4**, pp. 40–43.
41. Wen, S. H., Chung, D. D. L. (2002). Cement-based materials for stress sensing by dielectric measurement, *Cem. Concr. Res.*, **32**(9), pp. 1429–1433.
42. Gupta, S., Gonzalez, J. G., Loh, K. J. (2016). Self-sensing concrete enabled by nano-engineered cement–aggregate interfaces, *Struct. Health Monit.*, **2016**, pp. 1–15.
43. Hallaji, M., Pour-Ghaz, M. (2014). A new sensing skin for qualitative damage detection in concrete elements: rapid difference imaging with electrical resistance tomography, *NDT & E Int.*, **68**, pp. 13–21.
44. Xiao, H., Li, H., Ou, J. (2006). Modeling of piezoresistivity of carbon black filled cement-based composites under multi-axial strain, *Sens. Actuators A: Phys.*, **160**(1–2), pp. 87–93.
45. Han, B. G., Han, B. Z., Yu, X., Ou, J. P. (2009). Piezoresistive characteristic model of nickel/cement composites based on field emission effect and inter-particle separation, *Sens. Lett.*, **7**(6), pp. 1044–1050.
46. D'Alessandro, A., Ubertini, F., Materazzi, A. L., Laflamme, S., Porfiri, M. (2014). Electromechanical modelling of a new class of nanocomposite cement-based sensors for structural health monitoring, *Struct. Health Monit.*, **14**(2), pp. 1–11.
47. García-Macías, E., Downey, A., D'Alessandro, A., Castro-Triguero, R., Laflamme, S., Ubertini, F. (2017). Enhanced lumped circuit model for smart nanocomposite cement-based sensors under dynamic compressive loading conditions, *Sens. Actuators, A*, **260**, pp. 45–57.

Chapter 7

Evaluation of Mechanical Properties of Cement-Based Composites with Nanomaterials

Pedro de Almeida Carísio, Oscar Aurelio Mendoza Reales, and Romildo Dias Toledo Filho

Center of Sustainable Materials-NUMATS, Federal University of Rio de Janeiro, Av. Athos da Silveira Ramos - Cidade Universitária, Rio de Janeiro, Rio de Janeiro, 21941-590, Brazil

toledo@coc.ufrj.br

Different types of nanomaterials are able to modify the properties of cement-based composites in fresh and hardened states, with mechanical performance enhancement being one of the most sought after modifications. Due to their small particle size and high specific surface area, nanomaterials interact with the hydration reaction of cement through four known physicochemical mechanisms: (i) filler effect, (ii) nucleation, (iii) pozzolanic activity, and (iv) bridging effect; these mechanisms lead to a less porous and more resistant matrix. The aim of this chapter is to summarize the influence of a set of nanomaterials on the mechanical performance of cement-based composites, discussing their application, potential, and limitations.

Nanotechnology in Cement-Based Construction
Edited by Antonella D'Alessandro, Annibale Luigi Materazzi, and Filippo Ubertini

ISBN 978-981-4800-76-1 (Hardcover), 978-0-429-32849-7 (eBook)
www.jennystanford.com

7.1 Introduction

Although the discussion around nanotechnology emerged in 1960 [1], it was only in the past few decades that nanotechnology became of interest for the construction materials industry. Advances made by nanotechnology in the field of nanosciences made it possible to study the structure of hydrated cement on the atomic level [2, 3], clarifying its complex nanostructures and their relationship with the behavior of the bulk material [4]. In addition to understanding cement nanostructures, these advances have also helped to comprehend the mechanisms by which nanoinclusions interact with a cement-based matrix, finding that some nanoparticles are capable of improving the mechanical and chemical performance of cement-based materials [5, 6].

While investigating the effects of nanoparticles on the properties of cement-based matrices, it was found that besides improving the intrinsic properties of the matrix, nanoparticles also provide a new range of functionalities such as self-cleaning [7], self-sensing [8], self-damping [9], energy harvesting [10], and electromagnetic shielding/absorbing [11], among others. These new functionalities brought cement-based composites into the smart and multifunctional materials category [12].

Regardless of the new functionalities, the main use of cement-based composites is in structural elements; thus, the effect of nanomaterials on the mechanical performance of such composites is the focus of most of the research currently being carried out. Properties such as compressive strength [13], flexural strength [14], tensile strength [15], and elastic modulus [16] are the most common parameters by which the performance enhancement induced by nanoparticles in cement matrices is measured.

The mechanisms by which nanomaterials modify the mechanical properties of a cement matrix are consequence of both chemical and physical phenomena, which are enabled by their small particle size, chemical reactivity, and high specific surface area (SSA) [17]. These mechanisms can be listed as follows:

(1) Filler effect: During the hydration process cement grains react with water to form hydrated minerals, which do not

fill homogeneously the space available and create a porous structure within the matrix. This porous structure can be classified by size in gel pores (smaller than 10 nm), micro- or capillary pores (10 nm to 50 nm), macropores (larger than 50 nm), and air voids (entrapped air voids with a size up to 3 mm and entrained air voids ranging from 50 to 200 μm) [18]. Due to their small particle size (up to 100 nm) and their tendency to agglomerate, nanoparticles accumulate in the micro- and macropores, acting as fillers and refining the pore structure, that is, developing a denser and less permeable structure [19].

(2) Nucleation: The formation of hydrated minerals from cement hydration can occur by two mechanisms: through-solution precipitation and topochemical reaction. In the through-solution precipitation mechanism the anhydrous cement particles dissolve into their ionic constituents and precipitate as hydrates from the solution. In the topochemical mechanism the reactions occur directly at the anhydrous cement surface without the compounds going into solution [18]. Due to their small particle size and high SSA, nanoparticles can act as heterogeneous nuclei for hydrated minerals in the through-solution reaction, working as kernels that will be enveloped by hydrates and later precipitate. The main effect of nucleation is the enhancement of the hydration reaction kinetics, increasing the amount of hydrated products precipitated at very early ages, with little effect on the amount precipitated at later ages. The precipitation of more hydrates improves the performance of a cement-based matrix by densifying it and decreasing the amount and interconnectivity of its pore structure. Moreover, increases in the amount of calcium silicate hydrate (C–S–H) precipitated due to nucleation can be linked directly to improvements in the cement matrix strength [20].

(3) Pozzolanic activity: Pozzolans are amorphous materials based on SiO_2 or Al_2O_3, which in the presence of water consume calcium hydroxide (CH) to form C–S–H or calcium aluminate hydrate (CAH) compounds [21]. CH is a large crystal with a low SSA that has a limited contribution to the overall strength of a cement-based matrix. When pozzolans consume CH and produce C–S–H or CAH, the amount of hydrated products

in the matrix increases; given that the amounts of pozzolan and CH are balanced, the strength of a cement-based matrix will be improved due to a refinement or densification of its microstructure [18]. This is not an exclusive property of nanoparticles, but studies have shown that due their high SSA, the pozzolanic reaction of SiO_2- or Al_2O_3-based nanoparticles is more efficient [22].

(4) Bridge effect: When cement-based materials are submitted to tensile load, uncontrolled crack formation and propagation process take place until failure of the matrix is achieved. The use of nanofibers or nanotubes mixed into the cement matrix can help to control this cracking process by working as bridges for tensile load across submicrometric cracks [23], that is, cracks of width compatible with the nanofiber or nanotube length. The reinforcing effect achieved by these nanoparticles should not be regarded as a replacement for traditional fiber or steel rebar reinforcement but as a complement to achieve crack propagation control at a scale where other types of more traditional reinforcements are not effective [24]. The efficiency of the bridging effect of nanofibers and nanotubes depends on their aspect ratio, tensile strength, and interfacial bonding with the cement-based matrix [25].

Even though a small particle size and a high SSA accelerate the precipitation of hydration products and enhance the pozzolanic activity of nanoparticles, they also introduce a rheological challenge by increasing the water demand of the matrix [26] and consequently decreasing its workability by increasing the total amount of wettable area. Increasing the water content in the mixture to compensate for the loss of workability increases the matrix porosity, leading to a strength and durability decrease [18]. Thus, the use of high doses of plasticizer additives is necessary to obtain good performing cement-based matrices modified with nanoparticles. Additionally, since nanoparticles are prone to agglomerate in highly alkaline environments such as the one generated by the hydration of cement [27], chemical agents have to be used as dispersants; however, some of them can affect directly the hydration process of cement, increasing its setting time and decreasing its workability [28].

In light of the aforementioned mechanisms, this chapter presents a review of the effects of different nanomaterials on the mechanical properties of cement-based matrices. The chapter is divided by nanomaterial type and is concluded with a general discussion on the application potential and limitations of nanomaterials to improve the mechanical performance of cement-based composites.

7.2 Nanosilica

Nanosilica (NS) particles are spherical nanoparticles with a high surface area, composed mainly of amorphous SiO_2, and with high pozzolanic activity. Due to its high reactivity and chemical compatibility, NS has been found to increase the mechanical strength and durability of cement-based matrices by accelerating the hydration reaction of cement, enhancing the amount of C–S–H precipitated, and decreasing the porosity of the matrix [29–31]. NS is the most commonly used nanomaterial in cement-based matrices [32], being available in an industrial scale as a powder or colloidal suspension.

NS particles improve the compressive strength [22, 33–36], flexural strength [36–38], and tensile strength [36, 38] of cement pastes, mortars, and concretes. It has been reported that at 7 days the enhancement on compressive strength is higher than at 28 days [33, 35] due to its high reactivity, consequence of a high SSA, and a high proportion of atoms with free and unsaturated bonds on the particle surface [22].

Despite knowing that NS improves the mechanical properties of cement composites, there is not a consensus regarding the optimum amount of NS that maximizes such improvement. Two very distinct approaches can be found in the literature, one that recommends to use up to 1% of NS by mass of cement [39, 40] and another that uses up to 10% of NS [37, 41, 42]. It has been found that SSA plays an important role in defining the optimum amount of NS. The higher the SSA, the less the amount of NS that should be blended into the mixture to obtain a given performance [33].

The effect of NS on the mechanical properties of the cement matrix has been associated with a decrease in porosity induced through physical and chemical phenomena. The chemical phenomenon identified was a pozzolanic reaction due to the amorphous SiO_2 composition of NS, which consumes CH and yields C–S–H [18]. The physical phenomena identified were the filler effect and nucleation. NS particles fill the capillary pores in the matrix and induce nucleation sites that accelerate the hydration reaction [35, 43] and make for a more compact matrix by restricting the CH crystals' growth [36]. Furthermore, NS has been found to modify the CH crystals' orientation index at the paste–aggregate interface more efficiently than more traditional pozzolans due to its nanometric nature [22]. A lower orientation index of CH crystals has been related to higher mechanical properties of the matrix [44].

Both physical and chemical phenomena help to develop a denser matrix with a reduced capillary porosity and lower overall permeability [32, 36, 45], preventing the entry of aggressive agents such as CO_2, Cl^-, and water [46], thus increasing the durability of the composite.

7.3 Nanotitania

Nanotitania (NT) is a titanium dioxide (TiO_2) nanoparticle that can be found as three crystalline polymorphs: anatase, rutile, and brookite. NT is one of the most used nanoparticles in human life, being found in sunscreens, biomedical applications, and photovoltaic devices [47], among others. Due to its photocatalytic activity, that is, its capacity of degrading organic compounds into radicals when excited by UV radiation, NT has been used for self-cleaning, air pollution reduction [48], bactericidal [49, 50], antifogging, and aesthetic durability [7] purposes. In cement-based matrices NT has been applied to generate self-cleaning and bactericidal road pavements and building facades, which are capable of air depollution and have improved aesthetic durability [51, 52].

Despite its extraordinary properties, few studies have been performed to characterize the effects of NT on the mechanical behavior of cement-based materials. In the fresh state NT has been found to decrease the setting time [53, 54] and workability [44, 55] of cement-based matrices, while in the hardened state increases in compressive strength have been identified [54, 56].

In mortars, increases of 45% on early-age compressive strength (1 day) have been found using 5% and 10% of NT as cement replacement by mass. However, at later ages (28 days) a decrease in strength of up to 19% was identified for the same NT amounts [44]. Additionally, a synergic effect between NT and rice husk ash was found in the compressive strength of mortars with up to 30% rice husk ash and 1% NT as cement replacement by mass. In concretes, increases in flexural and tensile strength [57], and in flexural fatigue performance [20] associated with the use of NT have also been identified [56].

The most adequate NT amount to obtain compressive strength improvements of cement-based matrices has been found to be low, as expected for all nanoparticles. While some authors have found in mortars that at least between 1.5% [56] and 3% [55] NT is necessary to improve the strength of the matrix, experimental results point out that cement replacements of more than 5% of NT in mortars are ineffective, since they do not lead to significant improvements on mechanical properties and density and worsen the workability in the fresh state [44, 53].

Literature reports indicate that the mechanisms responsible for the improvements of mechanical properties induced by NT are nucleation and the filler effect, which accelerate the hydrates precipitation and refine the porous structures [53–57]. Some authors have suggested also that a size decrease and the orientation index modification of $Ca(OH)_2$ crystals also play an important role [44].

7.4 Nanoalumina

After calcium, silica and alumina are the two main elements involved in the hydration of cement. While silica-based components have a direct influence on strength, alumina-based components act on the setting time of cement, which affects indirectly the mechanical properties of the matrix at early ages [19].

An experimental study using partial replacements of cement by 1%, 2%, 4%, and 6% of nanoalumina (NA) was performed with and without a superplasticizer as a dispersant agent for cement in pastes. The effect of NA on the specimens without a superplasticizer was that when the nanoparticle content increased from 1% up to 6%, the compressive strength values decreased but still was higher than that of plain cement paste. In the case of cement paste with NA and a superplasticizer, the compressive strength values were higher than those of cement paste without a superplasticizer due to a higher formation of hydrated products, such as C–S–H, CAH, and calcium alumina silicate hydrate (CASH), and to the dispersing effect of the superplasticizer that helped to create a denser matrix through an increase in workability [58].

The mechanical properties of mortars with NA were studied using nanoparticle additions of 3%, 5%, and 7% and a fixed water-to-cement (w/c) ratio of 0.40. It was found that the compressive strength of mortars at an early age (3 and 7 days) was slightly higher than that of the control sample, whereas after 28 days the strength of mortars with 3% and 5% was lower, and with 7% it was slightly higher than the control. The modulus of elasticity was also measured, finding that the mortar with 5% NA increased by 143% at 28 days when compared to plain mortar, followed by 7% and 3% NA with an increase of 108% and 39%, respectively [16].

Concrete specimens with 0.5%, 1.0%, 1.5%, and 2.0% of NA as partial substitution of cement with a water/binder (w/b) ratio of 0.40 were produced to verify the effect of NA on their compressive strength when cured in water and limewater. According to the results, the compressive strength of concrete increased with the increase of NA content up to 1.0% when cured in water and up to 2.0% when cured in limewater, presenting enhancements of 9% and 29% at 90 days, respectively [59]. In a different work with the same mix design, tensile and flexural strength of concretes blended with NA were tested, finding a similar behavior than when tested for compressive strength [60].

An extensive literature survey has pointed out that the optimal NA content to improve the mechanical properties of cement-based materials varies among authors but seems to be between 1% and 2% [61]. Such improvements have been related to the acceleration of the hydration reaction and enhancements in the amount of hydration products [62] and to filler effects that improve the interfacial transition zone (ITZ) [16] and decrease the overall porosity of the matrix [58]. Drops in mechanical properties found when the amount of NA surpassed 2.0% have been related to the use of NA amounts higher than those required to react with the CH liberated during cement hydration, leading to an excess of alumina leaching out and not contributing to the matrix strength [63]. Another explanation given to the drop in mechanical performance is the arising of weak zones due to a poor dispersion degree of NA in the cement matrix associated with the use of high amounts of nanoparticles [59].

7.5 Nano–Iron Oxide

Iron oxides are a family of compounds composed of iron and oxygen atoms that exist in different stoichiometric combinations. Fe_2O_3, which can be found in the α, β, γ, and ε polymorphs, and Fe_3O_4, which has magnetic properties, are the most common forms of iron oxide used as iron nanoparticles (NI) to modify the properties of cementitious composites. NI are recognized for modifying the mechanical properties of cement composites and conferring self-sensing capabilities to cement-based matrices by changing their volumetric electric resistance when an applied mechanical load is varied [64]. Nevertheless, following the same behavior of others nanoparticles, addition of NI worsens the rheology of the cement-based matrix [65] due to its high SSA [66].

NI additions as Fe_2O_3 have been tested in cement mortars and concretes, finding that they can enhance the compressive, flexural, and tensile strength of cement-based matrices [67]. For reinforcing purposes it has been established that the addition of NI should be less than 10% by mass of cement to obtain a compressive strength enhancement [37, 64]. Moreover, it has been reported that at early ages the addition of NI presented lower mechanical performance [68], while at later ages it has been found that the compressive strength of mortars with NI would increase over the years when compared to plain mortars [69].

The mechanisms behind the mechanical improvements have been reported to be nucleation and the filler effect [37, 64]. Nucleation induced by NI particles has been associated with increased formation of hydrated products at early ages [67], while the refinement of the porous structure has been associated with a restricted growth of CH and ettringite crystals, which has been found to be favorable for the development of mechanical strength [64].

NI additions as Fe_3O_4, that is, nanomagnetite, have also been used in mortars and pastes, indicating that they do not affect the hydration rate of cement nor the consistency of mortars when blended in amounts of up to 5% by mass of cement [70]; nevertheless, when exposed to a magnetic field, NI particles have been found to align with the magnetic field isolines and modify the rheological behavior of cement paste [71]. Nanomagnetite has also been found to modify the compressive and flexural strength of mortars and pastes by increasing the amount of hydration products formed [70, 72, 73].

7.6 Nanoclay

Belonging to a wide clay mineral group, nanoclay (NC) is a fine-grained layered mineral silicate that has attracted researchers' attention over the past few decades due to its high availability [74]. With a basic layered structural unit, NC can be encountered as kaolinite (Al_2O_5Si), bentonite ($Al_2H_2O_{12}Si_4$), hectorite ($H_2LiMgNaO_{12}Si_4^{-2}$), montmorillonite ($Al_2H_2O_{12}Si_4$), halloysite ($Al_2H_8O_8Si_2$), and organically modified clays, among others. Since it is composed of aluminosilicates, NC has the potential to present pozzolanic activity when blended in cement-based matrices.

It has been found that NC worsens the workability of cement-based matrices [75], not only due to its high SSA, but also because NC is highly hydrophilic, increasing the water demand of the matrix [17]. In mortars, NC has been reported to enhance compressive strength by 24% when using 3% of clay as cement replacement [75], tensile strength by 49% when using 8% NC [76], impact strength by 29%, fracture toughness by 31%, and Rockwell hardness by 31% when using 1% NC [77]. A hybrid of 2.5% NC and 20% waste glass powder as cement replacement has also been found to achieve 29% of improvement in flexural strength [78]. In cement paste, a significant enhancement of compressive strength has been found only after 56 days of curing when 0.6% of NC was used; nevertheless, a significant decrease in permeability showed great potential to increase the durability of the cement matrix [79].

The mechanisms associated with these improvements are the pozzolanic reaction and the filler effect [77, 80]. Additionally, depending on their morphology of platelets or tubes, NC acts as a bridge through the microcracks [76], increasing the flexural strength of the matrix. The use of calcined NC has also been studied. In comparison to NC, calcined NC presents higher pozzolanic activity [77] due to its amorphous structure that is achieved during a controlled burning process.

7.7 Nanocarbon Materials

Nanocarbon materials (NCM) have low density, chemical stability, and impressive mechanical, electrical, and electromagnetic characteristics; this makes them potential candidates to develop a new generation of cementitious composites with novel multifunctional applications [81].

Most NCM are formed by graphene, which is an allotrope of carbon organized as a thin layer or sheet of carbon atoms bonded together in a hexagonal honeycomb structure [82]. Graphene is distinguished from graphite basically in the number of dimensions that its structure has. Graphene is composed of a single layer of atoms, which can be considered a 2D structure, while graphite is formed by layers of graphene stacked on top of each other, forming a 3D structure.

NCM can be classified by their dimensional structural form in 0D, 1D, and 2D nanoparticles, depending on how many of their dimensions are outside of the nanometric size range [83]. Following this logic, the 0D group is composed of fullerenes and carbon black nanoparticles; carbon nanotubes (CNTs) and carbon nanofibers (CNFs) integrate the 1D group; and graphene nanoplatelets (GNPs) compose the 2D group.

7.7.1 Graphene Nanoplatelets

GNPs are single-layered or multiple-layered graphene sheet structures [84]. They are also called graphite nanoplatelets because GNPs can be obtained from graphite by intercalating its layers with alkali metals, followed by exfoliation [85, 86].

Enhancements in compressive strength and ductility of cement pastes have been found using GNPs; such enhancements have been associated with graphene sheets bridging cracks at the nanoscale, making them discontinuous and preventing crack propagation and further growth [87]. This leads to an increase in hardness [88] and flexural strength on pastes without any significant increase in compressive strength [85].

Despite the good mechanical performance of GNP-blended cement paste [87], the addition of GNPs to concretes and mortars does not present enhancements in compressive and flexural strengths, nor in the elastic modulus. This null effect is probably related to the GNP size, which is larger than C–S–H agglomerates but a few times smaller than the ITZ around aggregates and cement grains [15, 89], which makes them ineffective to transfer the stresses or improve weak zones.

Even though some positive results have been found using GNPs as reinforcements, in general their performance cannot be considered satisfactory. Nevertheless, GNPs have presented adequate results controlling diffusion phenomena such as water permeability, chloride diffusion, and chloride migration [15].

7.7.2 Carbon Nanofibers

CNFs are different from conventional carbon fibers not only in their size but also in their preparation method [90], which gives them superior performance [91]. CNFs are graphitic filaments composed of graphene layers helically folded along the fiber axis and with a variable angle between adjacent graphene layers [92].

CNFs present good performance as reinforcing elements for cement-based composites, enhancing their strength, Young's modulus, fracture toughness, and stiffness [93–97]. Beyond that, improvements in energy sorption, maximum deflection, impact resistance, and abrasion have also been identified [98]. It has been reported that blending 0.2% of CNFs by mass of cement in a paste enhances its tensile strength by 22% [94] and its flexural strength by 82% at 7 days [93]. CNFs have also been tested as reinforcing elements in concrete, showing an improvement of 42% in compressive strength when using 0.16% of nanofibers by volume of binder [95]. It has been observed that CNFs present a bridging effect [93], facilitated by their high aspect ratio. An increase of regions with high-density C–S–H gel has also been reported [99]. Additionally, the exposed edge plans of CNFs allows a better bond with the cement matrix [94, 99] permitting a higher load transfer between nanofibers and matrix [97], improving the mechanical performance of the composite.

For CNFs to present adequate reinforcing performance in cement matrices a good dispersion degree of the nanofibers in the cement-based matrix has to be guaranteed. The challenge of mixing CNFs into cement matrices lays on their strong tendency to agglomerate due to van der Waals interactions. Studies have been conducted to improve the dispersion of CNFs in a cement matrix using a polycarboxylate-based superplasticizer and surfactants; however, non-uniform dispersion and chemical incompatibilities with cement have been identified [100, 101]. Silica fume and acid treatments have also been used to facilitate CNF dispersion in the matrix and improve the interfacial interaction between the cement matrix and nanofibers [102, 103].

7.7.3 Carbon Nanotubes

CNTs consist of graphene sheets rolled up, forming a long hollow tube. They can be formed by a single sheet (single walled) or by multiple concentric sheets (multiwalled). Multiwalled carbon nanotubes (MWCNTs) are more commonly used as a reinforcement in cement-based composites than single-walled carbon nanotubes (SWCNTs) due their lower cost and higher availability [104].

Discovered by Iijima [105], CNTs are one of the most promising candidates to be used as reinforcement in cement matrices due to their unique properties [14] such as high tensile strength, remarkable flexibility, and resilience [106], besides outstanding electrical [107] and electromagnetic [108] properties. With an SSA higher than 400 m^2/g [81], a Young's modulus around 1 TPa, and tensile strength up to 60 GPa [109], CNTs appear to be a great choice of nanomaterial to enhance the mechanical properties of cementitious materials [17].

Improvements in mechanical properties of cement-based matrices have been reported when small amounts of CNTs are incorporated [81, 110, 111]. Enhancements in compressive [112–114], flexural [14, 114] and tensile strength [115], failure strain [116], Young's modulus [117], hardness [110, 118], and ductility [96] have been found. CNTs also increase the amount of high-stiffness C–S–H on pastes, reducing the nanoporosity of C–S–H gel [111, 117]. Nevertheless, a decrease in compressive strength has also been reported and associated with poor dispersion of nanotubes in the matrix [119] and to the presence of the aggregate-paste ITZ [120].

CNT additions greatly reduce the workability of cement-based matrices not only due to their high SSA but also due the difficulty to disperse them in the matrix. This has been reported in the literature as a major issue for using CNTs in cement matrices. CNTs are highly hydrophobic and due to their strong attractive van der Waals forces, tend to agglomerate, making them difficult to disperse. There are many chemical and physical dispersion techniques being tested to obtain a homogeneous dispersion [81, 121]; however, conflicting results for mechanical performance have been reported.

Due to the poor interaction between CNTs and the cement matrix, modifications on the nanotubes' surface, usually by attachment carboxylic (COOH–) functional groups in cement composites, provide a better link between CNTs and the matrix [23].

CNTs work as reinforcement in cement matrices as well as CNFs despite their superior strength, aspect ratio, and elastic modulus [81]. When functionalized and well dispersed, bundles of hydration products can form on the nanotubes' surface [93, 118]. Nevertheless, CNTs are not considered to have chemical affinity with hydration reactions, but still are able to modify the kinetics of cement hydration by nucleation [104], which, combined with a filler effect, can also improve the porous microstructure, modifying the matrix porosity [122].

The length of CNTs is also a variable that can affect the mechanical performance of cement-based composites. It has been reported that higher amounts of short CNTs are required to achieve the same level of mechanical performance than longer CNTs, because lower amounts of short ones are not capable of bridging submicrometric cracks, and higher amounts of longer ones are more difficult to disperse due to their higher aspect ratio [14, 111, 117, 120].

7.8 Other Nanoparticles

Some nanoparticles have not caught much research attention and only a small amount of reports of their use in cement-based matrices can be found. Some examples of these nanoparticles are nano–calcium carbonate, nano–zinc dioxide, nano–copper oxide, and nano–magnesium dioxide.

Calcium carbonate, with the chemical formula $CaCO_3$, is known to have a good chemical interaction with cement [123] and accelerate its hydration process [124]. As a nanoparticle, nano-$CaCO_3$ reduces the workability of pastes, mortars, and concretes proportionally to the amount used as cement replacement [125, 126]. Despite this, when 1% of nano-$CaCO_3$ is added to cement pastes, flexural and compressive strengths have been reported to improve with respect to plain cement samples by 108% and 111%, respectively [126]. It has also been reported that nano-$CaCO_3$ can enhance the compressive strength of concretes by up to 148% [125].

The effect of nano-ZnO_2 on the mechanical performance and microstructure of concretes has also been studied. It has been found that nano-ZnO_2 additions reduce flowability and workability of the concrete matrix in the fresh state [127], accelerate the formation of hydration products [128], and modify compressive [129, 130], flexural, and tensile strengths [127].

Nano-CuO has been found to have the potential to improve the mechanical strength of cement-based matrices [131]. Improvements in tensile strength [132] and compressive strength [68, 133] of mortars containing 4% and 3% of nano-CuO, respectively, as cement replacement have been reported. The interaction mechanisms between nano-CuO and cement are practically the same as the other nanomaterials in this section, acting as nuclei due to its high surface area, filling the pores, and controlling the CH crystal growth [131].

Magnesium oxide, MgO, is known as a shrinkage reduction agent for cement-based composites due to its expansive property. Nano-MgO has been blended in cement pastes and mortars, showing improvements in compressive and flexural strengths when small amounts of nanoparticles were used [134, 135]. The compressive strength of cement mortars blended with nano-MgO has been found to decrease at early ages (7 days) and increase at later ones (28 days). The enhancement in performance at later ages was related to the formation of $Mg(OH)_2$ crystals, which occurs by the hydration of nano-MgO when in contact with water. These crystals grow over time, explaining the delayed improvement of compressive strength of the matrix through the filler effect [136].

7.9 Future Perspective

Interest in the modification of cement-based materials using nanoparticles is constantly growing, with NS, NT, and CNTs being the most developed applications, not only due to improvements in strength and durability that can be obtained with them, but also due to the novel properties they can confer to the cement-based matrix.

The filler effect and nucleation are the most common mechanisms by which nanoparticles modify the properties of cement-based materials. These mechanisms, combined with a good dispersion degree of the nanoparticles within the cement-based matrix, densify both the porous structure generated during cement hydration and the ITZ between aggregates and cement paste; this densification increases the overall mechanical performance of the matrix. Additionally, high-SSA silicon- and aluminum-based nanoparticles, which present pozzolanic activity, are much more reactive than their homologous microparticles of similar composition, having a pronounced effect on the mechanical performance of the matrix at early ages. Fiber-like nanoparticles enhance the tensile and flexural strengths of the matrix through the bridging effect, controlling submicrometric crack propagation.

In general, even with the consistent results reported on the use of these nanomaterials, there is still a large body of knowledge open for research. It is difficult to find the optimum amount of nanoparticles to be used as a general rule, but it is practically a consensus that small amounts must be used. The appropriate amount is linked to the dispersion degree, which is the most challenging variable to control. Due to their small size and high SSA making them more reactive than microparticles, nanoparticles easily agglomerate, making them difficult to disperse. When adequate dispersion is not achieved, nonreacted areas grow and nanoparticle agglomerate, creating stress concentration zones that lead to a decrease in the mechanical performance of the matrix. In the case of materials with fiber-like morphology the aspect ratio seems to be the variable governing the optimal amount of nanoparticles.

Future research efforts should focus on the aforementioned variables; this would help to better understand the interaction between cement and nanoparticles and find the suitable amount of nanoparticles that optimizes the mechanical properties of cementitious materials. This effort should be done keeping in mind that not all nanoparticles are adequate solutions for improving the mechanical properties of cement-based materials, but are capable of conferring novel properties to the cement-based matrix. In other words, nanoparticles should not be expected to be polyvalent materials that improve multiple properties of the cement matrix at the same time, but must be used smartly to generate multifunctional cement-based composites.

References

1. Feynman, R. P. (1960). There's plenty of room at the bottom, *Eng. Sci.*, **23**(5), pp. 22–36.
2. Sobolev, K., Gutiérrez, M. F. (2005). How nanotechnology can change the concrete world, *Am. Ceram. Soc. Bull.*, **84**(10), pp. 14–18.
3. Bittnar, Z., Bartos, P. J. M., Nemecek, J., Smilauer, V., Zeman, J. (2009). *Nanotechnology in Construction 3*, Springer Berlin Heidelberg, Berlin, Heidelberg.

4. Bastos, G., Patiño-Barbeito, F., Patiño-Cambeiro, F., Armesto, J. (2016). Nano-inclusions applied in cement-matrix composites: a review, *Materials (Basel)*, **9**(12), pp. 1–30.
5. Shah, S. P., Hou, P., Konsta-Gdoutos, M. S. (2015). Nano-modification of cementitious material: toward a stronger and durable concrete, *J. Sustain. Cem. Mater.*, **5**(1), pp. 1–22.
6. Mendes, T., Hotza, D., Repette, W. (2015). Nanoparticles in cement based materials: a review, *Rev. Adv. Mater. Sci.*, **40**, pp. 89–96.
7. Folli, A., Pade, C., Hansen, T. B., De Marco, T., MacPhee, D. E. (2012). TiO_2 photocatalysis in cementitious systems: insights into self-cleaning and depollution chemistry, *Cem. Concr. Res.*, **42**(3), pp. 539–548.
8. D'Alessandro, A., Ubertini, F., Materazzi, A. L. (2016). Self-sensing concrete nanocomposites for smart structures, *Int. J. Civil, Environ. Struct. Constr. Archit. Eng.*, **10**(5), pp. 584–589.
9. Luo, J., Duan, Z., Xian, G., Li, Q., Zhao, T. (2015). Damping performances of carbon nanotube reinforced cement composite, *Mech. Adv. Mater. Struct.*, **22**(3), pp. 224–232.
10. Hosseini, T., Flores-Vivian, I., Sobolev, K., Kouklin, N. (2013). Concrete embedded dye-synthesized photovoltaic solar cell, *Sci. Rep.*, **3**, pp. 1–5.
11. Nam, I. W., Kim, H. K., Lee, H. K. (2012). Influence of silica fume additions on electromagnetic interference shielding effectiveness of multi-walled carbon nanotube/cement composites, *Constr. Build. Mater.*, **30**, pp. 480–487.
12. Han, B., et al. (2015). Smart concretes and structures: a review, *J. Intell. Mater. Syst. Struct.*, **26**(11), p. 1045389X15586452.
13. Jiang, S., et al. (2017). Comparison of compressive strength and electrical resistivity of cementitious composites with different nano- and micro-fillers, *Arch. Civ. Mech. Eng.*, **18**(1), pp. 60–68.
14. Konsta-Gdoutos, M. S., Metaxa, Z. S., Shah, S. P. (2010). Highly dispersed carbon nanotube reinforced cement based materials, *Cem. Concr. Res.*, **40**(7), pp. 1052–1059.
15. Du, H., Pang, S. D. (2015). Enhancement of barrier properties of cement mortar with graphene nanoplatelet, *Cem. Concr. Res.*, **76**, pp. 10–19.
16. Li, Z., Wang, H., He, S., Lu, Y., Wang, M. (2006). Investigations on the preparation and mechanical properties of the nano-alumina reinforced cement composite, *Mater. Lett.*, **60**(3), pp. 356–359.
17. Sanchez, F., Sobolev, K. (2010). Nanotechnology in concrete: a review, *Constr. Build. Mater.*, **24**(11), pp. 2060–2071.

18. Mehta, P. K., Monteiro, P. J. M. (2006). *Concrete: Microstructure, Properties, and Materials*, 3rd ed., McGraw-Hill.

19. Norhasri, M. S. M., Hamidah, M. S., Fadzil, A. M. (2017). Applications of using nano material in concrete: a review, *Constr. Build. Mater.*, **133**, pp. 91–97.

20. Li, H., Zhang, M.-H., Ou, J.-P. (2007). Flexural fatigue performance of concrete containing nano-particles for pavement, *Int. J. Fatigue*, **29**(7), pp. 1292–1301.

21. Hewlett, P. (2004). Lea's chemistry of cement and concrete, *Science (80–)*, **58**(10, p. 1066.

22. Qing, Y., Zenan, Z., Deyu, K., Rongshen, C. (2007). Influence of nano-SiO_2 addition on properties of hardened cement paste as compared with silica fume, *Constr. Build. Mater.*, **21**(3), pp. 539–545.

23. Mendoza, O., Toledo, R. (2016). Nanotube–cement composites, *Carbon Nanomater. Sourceb. Nanoparticles, Nanocapsules, Nanofibers, Nanoporous Struct. Nanocomposites*, **2**, pp. 573–596.

24. Sakulich, A. R., Li, V. C. (2011). Nanoscale characterization of engineered cementitious composites (ECC), *Cem. Concr. Res.*, **41**(2), pp. 169–175.

25. Abu Al-Rub, R. K., Ashour, A. I., Tyson, B. M. (2012). On the aspect ratio effect of multi-walled carbon nanotube reinforcements on the mechanical properties of cementitious nanocomposites, *Constr. Build. Mater.*, **35**, pp. 647–655.

26. Quercia, G., Hüsken, G., Brouwers, H. J. H. (2012). Water demand of amorphous nano silica and its impact on the workability of cement paste, *Cem. Concr. Res.*, **42**(2), pp. 344–357.

27. Mendoza, O., Sierra, G., Tobón, J. I. (2013). Influence of super plasticizer and Ca(OH)2 on the stability of functionalized multi-walled carbon nanotubes dispersions for cement composites applications, *Constr. Build. Mater.*, **47**, pp. 771–778.

28. Mendoza Reales, O. A., Arias Jaramillo, Y. P., Ochoa Botero, J. C., Delgado, C. A., Quintero, J. H., Toledo Filho, R. D. (2018). Influence of MWCNT/surfactant dispersions on the rheology of Portland cement pastes, *Cem. Concr. Res.*, **107**, pp. 101–109.

29. Senff, L., Labrincha, J. A., Ferreira, V. M., Hotza, D., Repette, W. L. (2009). Effect of nano-silica on rheology and fresh properties of cement pastes and mortars, *Constr. Build. Mater.*, **23**(7), pp. 2487–2491.

30. Tobón, J. I., Payá, J. J., Borrachero, M. V., Restrepo, O. J. (2012). Mineralogical evolution of Portland cement blended with silica

nanoparticles and its effect on mechanical strength, *Constr. Build. Mater.*, **36**, pp. 736–742.

31. Wang, L., Zheng, D., Zhang, S., Cui, H., Li, D. (2016). Effect of nano-SiO_2 on the hydration and microstructure of Portland cement, *Nanomaterials*, **6**(12), p. 241.
32. Singh, L. P., Karade, S. R., Bhattacharyya, S. K., Yousuf, M. M., Ahalawat, S. (2013). Beneficial role of nanosilica in cement based materials: a review, *Constr. Build. Mater.*, **47**, pp. 1069–1077.
33. Bolhassani, M., Samani, M. (2015). Effect of type, size, and dosage of nanosilica and microsilica on properties of cement paste and mortar, *ACI Mater. J.*, **112**(2), pp. 259–266.
34. Singh, L. P., Goel, A., Bhattachharyya, S. K., Ahalawat, S., Sharma, U., Mishra, G. (2015). Effect of morphology and dispersibility of silica nanoparticles on the mechanical behaviour of cement mortar, *Int. J. Concr. Struct. Mater.*, **9**(2), pp. 207–217.
35. Flores, Y. C., Cordeiro, G. C., Toledo Filho, R. D., Tavares, L. M. (2017). Performance of Portland cement pastes containing nano-silica and different types of silica, *Constr. Build. Mater.*, **146**, pp. 524–530.
36. Nazari, A., Riahi, S. (2011). The effects of SiO_2 nanoparticles on physical and mechanical properties of high strength compacting concrete, *Composite Part B*, **42**(3), pp. 570–578.
37. Li, H., Xiao, H. G., Yuan, J., Ou, J. (2004). Microstructure of cement mortar with nano-particles, *Composite Part B*, **35**(2), pp. 185–189.
38. Naji Givi, A., Abdul Rashid, S., Aziz, F. N. A., Salleh, M. A. M. (2010). Experimental investigation of the size effects of SiO_2 nano-particles on the mechanical properties of binary blended concrete, *Composite Part B*, **41**(8), pp. 673–677.
39. Heidari, A., Tavakoli, D. (2013). A study of the mechanical properties of ground ceramic powder concrete incorporating nano-SiO_2 particles, *Constr. Build. Mater.*, **38**, pp. 255–264.
40. Mendes, T. M., Repette, W. L., Reis, P. J. (2017). Effects of nano-silica on mechanical performance and microstructure of ultra-high performance concrete, *Cerâmica*, **63**.
41. Byung Wan Jo, B., Hyun Kim, C., Hoon Lim, J. (2007). Investigations on the development of powder concrete with nano-SiO_2 particles, *KSCE J. Civ. Eng.*, **11**(1), pp. 37–42.
42. Sadrmomtazi, A., Fasihi, A., Balalaei, F., Haghi, A. (2009). Investigation of mechanical and physical properties of mortars containing silica fume and nano-SiO_2, *Third Int. Conf. Concr. Dev.*, pp. 1153–1161.

43. Land, G., Stephan, D. (2012). The influence of nano-silica on the hydration of ordinary Portland cement, *J. Mater. Sci.*, **47**(2), pp. 1011–1017.
44. Meng, T., Yu, Y., Qian, X., Zhan, S., Qian, K. (2012). Effect of nano-TiO_2 on the mechanical properties of cement mortar, *Constr. Build. Mater.*, **29**, pp. 241–245.
45. Zhang, M.-H., Li, H. (2011). Pore structure and chloride permeability of concrete containing nano-particles for pavement, *Constr. Build. Mater.*, **25**, pp. 608–616.
46. Peña, I. D., Sanchez, M., Alonso, M. C. (2015). Effect of the electrochemical migration of colloidal nano-SiO_2 on the durability performance of hardened cement mortar, *Int. J. Electrochem. Sci.*, **10**, pp. 10261–10271.
47. Rehman, F. U., Zhao, C., Jiang, H., Wang, X. (2016). Biomedical applications of nano-titania in theranostics and photodynamic therapy, *Biomater. Sci.*, **4**(1), pp. 40–54.
48. Boonen, E., Beeldens, A. (2014). Recent photocatalytic applications for air purification in Belgium, *Coatings*, **4**(3), pp. 553–573.
49. de Almeida Carísio, P., Martins, J. A. L. G., de Melo, G. B., Dias, J. F. (2017). Study of a photodegradant polymeric composite containing TiO_2 and glass residue, *Open J. Civ. Eng.*, **07**(04), pp. 553–560.
50. Tsuang, Y.-H., Sun, J.-S., Huang, Y.-C., Lu, C.-H., Chang, W. H.-S., Wang, C.-C. (2008). Studies of photokilling of bacteria using titanium dioxide nanoparticles, *Artif. Organs*, **32**(2), pp. 167–174.
51. Hassan, M. M., Dylla, H., Mohammad, L. N., Rupnow, T. (2010). Evaluation of the durability of titanium dioxide photocatalyst coating for concrete pavement, *Constr. Build. Mater.*, **24**(8), pp. 1456–1461.
52. Loh, K., Gaylarde, C. C., Shirakawa, M. A. (2018). Photocatalytic activity of ZnO and TiO_2 'nanoparticles' for use in cement mixes, *Constr. Build. Mater.*, **167**, pp. 853–859.
53. Essawy, A. A., Abd, S. (2014). Physico-mechanical properties, potent adsorptive and photocatalytic efficacies of sulfate resisting cement blends containing micro silica and nano-TiO_2, *Constr. Build. Mater.*, **52**, pp. 1–8.
54. Chen, J., Kou, S. C., Poon, C. S. (2012). Hydration and properties of nano-TiO_2 blended cement composites, *Cem. Concr. Compos.*, **34**(5), pp. 642–649.
55. Zhang, R., Cheng, X., Hou, P., Ye, Z. (2015). Influences of nano-TiO_2 on the properties of cement-based materials: hydration and drying shrinkage, *Constr. Build. Mater.*, **81**, pp. 35–41.

56. Noorvand, H., Abang Ali, A. A., Demirboga, R., Farzadnia, N., Noorvand, H. (2013). Incorporation of nano TiO_2 in black rice husk ash mortars, *Constr. Build. Mater.*, **47**, pp. 1350–1361.

57. Jalal, M., Fathi, M., Farzad, M. (2013). Effects of fly ash and TiO_2 nanoparticles on rheological, mechanical, microstructural and thermal properties of high strength self compacting concrete, *Mech. Mater.*, **61**, pp. 11–27.

58. Heikal, M., Ismail, M. N., Ibrahim, N. S. (2015). Physico-mechanical, microstructure characteristics and fire resistance of cement pastes containing Al_2O_3 nano-particles, *Constr. Build. Mater.*, **91**, pp. 232–242.

59. Nazari, A., Riahi, S. (2011). Improvement compressive strength of concrete in different curing media by Al_2O_3 nanoparticles, *Mater. Sci. Eng. A*, **528**(1), pp. 1183–1191.

60. Nazari, A., Riahi, S. (2011). Al_2O_3 nanoparticles in concrete and different curing media, *Energy Build.*, **43**(6), pp. 1480–1488.

61. Rashad, A. M. (2013). A synopsis about the effect of nano-Al_2O_3, nano-Fe_2O_3, nano-Fe_3O_4 and nano-clay on some properties of cementitious materials: a short guide for civil engineer, *Mater. Des.*, **52**(1), pp. 143–157.

62. Nazari, A., Riahi, S. (2011). Effects of Al_2O_3 nanoparticles on properties of self compacting concrete with ground granulated blast furnace slag (GGBFS) as binder, *Sci. China Technol. Sci.*, **54**(9), pp. 2327–2338.

63. Nazari, A., Riahi, S., Riahi, S., Shamekhi, S. F., Khademno, A. (2010). Mechanical properties of cement mortar with Al_2O_3 nanoparticles, *J. Am. Sci.*, **6**(4), pp. 94–97.

64. Li, H., Xiao, H.-G., Ou, J.-P. (2004). A study on mechanical and pressure-sensitive properties of cement mortar with nanophase materials, *Cem. Concr. Res.*, **34**(3), pp. 435–438.

65. Nazari, A., Riahi, S., Riahi, S., Shamekhi, S. F., Khademmo, A. (2010). Benefits of Fe_2O_3 nanoparticles in concrete mixing matrix, *J. Am. Sci.*, **6**(4), pp. 102–106.

66. Nazari, A., Riahi, S. (2011). Assessment of the effects of Fe_2O_3 nanoparticles on water permeability, workability, and setting time of concrete, *J. Compos. Mater.*, **45**(8), pp. 923–930.

67. Khoshakhlagh, A., Nazari, A., Khalaj, G. (2012). Effects of Fe_2O_3 nanoparticles on water permeability and strength assessments of high strength self-compacting concrete, *J. Mater. Sci. Technol.*, **28**(1), pp. 73–82.

68. Madandoust, R., Mohseni, E., Mousavi, S. Y., Namnevis, M. (2015). An experimental investigation on the durability of self-compacting mortar containing nano-SiO_2, nano-Fe_2O_3 and nano-CuO, *Constr. Build. Mater.*, **86**, pp. 44–50.

69. Oltulu, M., Şahin, R. (2011). Single and combined effects of nano-SiO_2, nano-Al_2O_3 and nano-Fe_2O_3 powders on compressive strength and capillary permeability of cement mortar containing silica fume, *Mater. Sci. Eng. A*, **528**(22–23), pp. 7012–7019.

70. Sikora, P., Horszczaruk, E., Cendrowski, K., Mijowska, E. (2016). The influence of nano-Fe_3O_4 on the microstructure and mechanical properties of cementitious composites, *Nanoscale Res. Lett.*, **11**(1), p. 182.

71. Nair, S. D., Ferron, R. D. (2016). Real time control of fresh cement paste stiffening: smart cement-based materials via a magnetorheological approach, *Rheol. Acta*, **55**(7), pp. 571–579.

72. Amin, M. S., El-Gamal, S. M. A., Hashem, F. S. (2013). Effect of addition of nano-magnetite on the hydration characteristics of hardened Portland cement and high slag cement pastes, *J. Therm. Anal. Calorim.*, **112**(3), pp. 1253–1259.

73. Shekari, A. H., Razzaghi, M. S. (2011). Influence of nano particles on durability and mechanical properties of high performance concrete, *Procedia Eng.*, **14**, pp. 3036–3041.

74. Nazir, M. S., Mohamad Kassim, M. H., Mohapatra, L., Gilani, M. A., Raza, M. R., Majeed, K. (2016). Characteristic properties of nanoclays and characterization of nanoparticulates and nanocomposites, in Jawaid, M., et al. (eds.) *Nanoclay Reinforced Polymer Composites*, Springer Science+Business Media, Singapore, pp. 35–55.

75. Farzadnia, N., Abang Ali, A. A., Demirboga, R., Anwar, M. P. (2013). Effect of halloysite nanoclay on mechanical properties, thermal behavior and microstructure of cement mortars, *Cem. Concr. Res.*, **48**, pp. 97–104.

76. Morsy, M. S., Alsayed, S. H., Aqel, M. (2010). Effect of nano-clay on mechanical properties and microstructure of ordinary Portland cement mortar, *Int. J. Civ. Environ. Eng. IJCEE-IJENS*, **10**(1), pp. 23–27.

77. Hakamy, A., Shaikh, F. U. A., Low, I. M. (2015). Characteristics of nanoclay and calcined nanoclay-cement nanocomposites, *Composite Part B*, **78**, pp. 174–184.

78. Aly, M., Hashmi, M. S. J., Olabi, A. G., Messeiry, M., Hussain, A. I. (2011). Effect of nano clay particles on mechanical, thermal and physical

behaviours of waste-glass cement mortars, *Mater. Sci. Eng. A*, **528**(27), pp. 7991–7998.

79. Chang, T.-P., Shih, J.-Y., Yang, K.-M., Hsiao, T.-C. (2007). Material properties of Portland cement paste with nano-montmorillonite, *J. Mater. Sci.*, **42**(17), pp. 7478–7487.
80. Morsy, M. S., Alsayed, S. H., Aqel, M. (2011). Hybrid effect of carbon nanotube and nano-clay on physico-mechanical properties of cement mortar, *Constr. Build. Mater.*, **25**(1), pp. 145–149.
81. Han, B., Sun, S., Ding, S., Zhang, L., Yu, X., Ou, J. (2015). Review of nanocarbon-engineered multifunctional cementitious composites, *Composite Part A*, **70**, pp. 69–81.
82. Allen, M. J., Tung, V. C., Kaner, R. B. (2010). Honeycomb carbon: a review of graphene, *Chem. Rev.*, **110**(1), pp. 132–145.
83. Han, B., et al. (2017). Nano-core effect in nano-engineered cementitious composites, *Composite Part A*, **95**, pp. 100–109.
84. Jang, B. Z., Zhamu, A. (2008). Processing of nanographene platelets (NGPs) and NGP nanocomposites: a review, *J. Mater. Sci.*, **43**(15), pp. 5092–5101.
85. Sixuan, H. (2012). Multifunctional graphite nanoplatelets (GNP) reinforced cementitious composites, National University of Singapore, Singapore.
86. Viculis, L. M., Mack, J. J., Mayer, O. M., Hahn, H. T., Kaner, R. B. (2005). Intercalation and exfoliation routes to graphite nanoplatelets, *J. Mater. Chem.*, **15**, pp. 974–978.
87. Rehman, S. K. U., Ibrahim, Z., Memon, S. A., Javed, M. F., Khushnood, R. A. (2017). A sustainable graphene based cement composite, *Sustainability*, **9**(7), pp. 1–20.
88. Du, H., Pang, S. D. (2018). Dispersion and stability of graphene nanoplatelet in water and its influence on cement composites, *Constr. Build. Mater.*, **167**, pp. 403–413.
89. Du, H., Gao, H. J., Pang, S. D. (2016). Improvement in concrete resistance against water and chloride ingress by adding graphene nanoplatelet, *Cem. Concr. Res.*, **83**, pp. 114–123.
90. Feng, L., Xie, N., Zhong, J. (2014). Carbon nanofibers and their composites: a review of synthesizing, properties and applications, *Materials (Basel)*, **7**(5), pp. 3919–3945.
91. Mordkovich, V. Z. (2003). Carbon nanofibers: a new ultrahigh-strength material for chemical technology, *Theor. Found. Chem. Eng.*, **37**(5), pp. 429–438.

92. Teo, K. B. K., Singh, C., Chhowalla, M., Milne, W. I. (2003). Catalytic synthesis of carbon nanotubes and nanofibers, *Encycl. Nanosci. Nanotechnol.*, **10**, pp. 1–22.

93. Tyson, B. M., Al-rub, R. K. A., Yazdanbakhsh, A., Grasley, Z. (2011). Carbon nanotubes and carbon nanofibers for enhancing the mechanical properties of nanocomposite cementitious materials, *J. Mater. Civ. Eng.*, **23**, pp. 1–8.

94. Gay, C., Sanchez, F. (2010). Performance of carbon nanofiber–cement composites with a high-range water reducer, *Transp. Res. Rec. J. Transp. Res. Board*, **2142**(1), pp. 109–113.

95. Gao, D., Sturm, M., Mo, Y. L. (2011). Electrical resistance of carbon-nanofiber concrete, *Smart Mater. Struct.*, **20**(4), p. 049501.

96. Abu Al-Rub, R. K., Tyson, B. M., Yazdanbakhsh, A., Grasley, Z. (2012). Mechanical properties of nanocomposite cement incorporating surface-treated and untreated carbon nanotubes and carbon nanofibers, *J. Nanomech. Micromech.*, **2**(1), pp. 1–6.

97. Metaxa, Z. S., Konsta-Gdoutos, M. S., Shah, S. P. (2013). Carbon nanofiber cementitious composites: effect of debulking procedure on dispersion and reinforcing efficiency, *Cem. Concr. Compos.*, **36**(1), pp. 25–32.

98. Peyvandi, A., Sbia, L. A., Soroushian, P., Sobolev, K. (2013). Effect of the cementitious paste density on the performance efficiency of carbon nanofiber in concrete nanocomposite, *Constr. Build. Mater.*, **48**, pp. 265–269.

99. Barbhuiya, S., Chow, P. (2017). Nanoscaled mechanical properties of cement composites reinforced with carbon nanofibers, *Materials (Basel)*, **10**(6), p. 662.

100. Yazdanbakhsh, A., Grasley, Z. C., Tyson, B., Abu Al-Rub, R. K. (2009). Carbon nano filaments in cementitious materials: some issues on dispersion and interfacial bond, *Am. Concr. Institute, ACI Spec. Publ.* (267 SP), pp. 21–34.

101. Yazdanbakhsh, A., Grasley, Z., Tyson, B., Abu Al-Rub, R. (2010). Distribution of carbon nanofibers and nanotubes in cementitious composites, *Transp. Res. Rec. J. Transp. Res. Board*, **2142**, pp. 89–95.

102. Sanchez, F., Ince, C. (2009). Microstructure and macroscopic properties of hybrid carbon nanofiber/silica fume cement composites, *Compos. Sci. Technol.*, **69**(7–8), pp. 1310–1318.

103. Sanchez, F., Zhang, L., Ince, C. (2009). Multi-scale performance and durability of carbon nanofiber/cement composites, in *Nanotechnology*

in Construction 3, Springer Berlin Heidelberg, Berlin, Heidelberg, pp. 345–350.

104. Mendoza Reales, O. A., Dias Toledo Filho, R. (2017). A review on the chemical, mechanical and microstructural characterization of carbon nanotubes-cement based composites, *Constr. Build. Mater.*, **154**, pp. 697–710.
105. Iijima, S. (1991). Helical microtubules of graphitic carbon, *Nature*, **354**(6348), pp. 56–58.
106. Salvetat, J.-P., et al. (1999). Mechanical properties of carbon nanotubes, *Appl. Phys. A*, **69**(3), pp. 255–260.
107. Khare, R., Bose, S. (2005). Carbon nanotube based composites: a review, *J. Miner. Mater. Charact. Eng.*, **4**(1), pp. 31–46.
108. Nikfarjam, A., Rafiee, R., Taheri, M. (2016). Electrical and electromagnetic properties of isolated carbon nanotubes and carbon nanotube-based composites electrical conductivity, *Polyolefins J.*, 4(1), pp. 1–35.
109. Yu, M. F., Lourie, O., Dyer, M. J., Moloni, K., Kelly, T. F., Ruoff, R. S. (2000). Strength and breaking mechanism of multiwalled carbon nanotubes under tensile load, *Science (80–)*, **287**(5453), pp. 637–640.
110. Sáez De Ibarra, Y., Gaitero, J. J., Erkizia, E., Campillo, I. (2006). Atomic force microscopy and nanoindentation of cement pastes with nanotube dispersions, *Phys. Status Solidi Appl. Mater. Sci.*, **203**(6), pp. 1076–1081.
111. Konsta-Gdoutos, M. S., Metaxa, Z. S., Shah, S. P. (2010). Multi-scale mechanical and fracture characteristics and early-age strain capacity of high performance carbon nanotube/cement nanocomposites, *Cem. Concr. Compos.*, **32**(2), pp. 110–115.
112. Yakovlev, G., Kerienė, J., Gailius, A., Girnienė, I. (2006). Cement based foam concrete reinforced by carbon nanotubes, *Mater. Sci.*, **12**(2), pp. 147–151.
113. Cwirzen, A., Habermehl-Cwirzen, K., Penttala, V. (2008). Surface decoration of carbon nanotubes and mechanical properties of cement/carbon nanotube composites, *Adv. Cem. Res.*, **20**(2), pp. 65–73.
114. Hawreen, A., Bogas, J. A., Dias, A. P. S. (2018). On the mechanical and shrinkage behavior of cement mortars reinforced with carbon nanotubes, *Constr. Build. Mater.*, **168**, pp. 459–470.
115. Hunashyal, A. M., Tippa, S. V., Quadri, S. S., Banapurmath, N. R. (2011). Experimental investigation on effect of carbon nanotubes and carbon fibres on the behavior of plain cement mortar composite round bars under direct tension, *ISRN Nanotechnol.*, **2011**, pp. 1–6.

116. Li, G. Y., Wang, P. M., Zhao, X. (2005). Mechanical behavior and microstructure of cement composites incorporating surface-treated multi-walled carbon nanotubes, *Carbon N. Y.*, **43**(6), pp. 1239–1245.

117. Shah, S. P., Konsta-Gdoutos, M. S., Metaxa, Z. S., Mondal, P. (2009). Nanoscale Modification of Cementitious Materials, in *Nanotechnology in Construction 3*, Springer Berlin Heidelberg, Berlin, Heidelberg, pp. 125–130.

118. Makar, J. M., Margeson, J. C., Luh, J. (2005). Carbon nanotube/cement composites - early results and potential applications, in *3rd International Conference on Construction Materials: Performance, Innovations and Structural Implications*, pp. 1–10.

119. Musso, S., Tulliani, J. M., Ferro, G., Tagliaferro, A. (2009). Influence of carbon nanotubes structure on the mechanical behavior of cement composites, *Compos. Sci. Technol.*, **69**(11–12), pp. 1985–1990.

120. Hawreen, A., Bogas, J. A., Dias, A. P. S. (2018). On the mechanical and shrinkage behavior of cement mortars reinforced with carbon nanotubes, *Constr. Build. Mater.*, **168**, pp. 459–470.

121. Parveen, S., Rana, S., Fangueiro, R. (2013). A review on nanomaterial dispersion, microstructure, and mechanical properties of carbon nanotube and nanofiber reinforced cementitious composites, *J. Nanomater.*, **2013**, p. 710175.

122. Sobolkina, A., et al. (2012). Dispersion of carbon nanotubes and its influence on the mechanical properties of the cement matrix, *Cem. Concr. Compos.*, **34**(10), pp. 1104–1113.

123. Sato, T., Beaudoin, J. J. (2006). The effect of nano-sized $CaCO_3$ addition on the hydration of OPC containing high volumes of ground granulated blast-furnace slag, in *2nd International Symposium on Advances in Concrete through Science and Engineering 11–13 September 2006*, Quebec City, Canada, pp. 1–12.

124. Péra, J., Husson, S., Guilhot, B. (1999). Influence of finely ground limestone on cement hydration, *Cem. Concr. Compos.*, **21**(2), pp. 99–105.

125. Shaikh, F. U. A., Supit, S. W. M. (2014). Mechanical and durability properties of high volume fly ash (HVFA) concrete containing calcium carbonate ($CaCO_3$) nanoparticles, *Constr. Build. Mater.*, **70**, pp. 309–321.

126. Liu, X., Chen, L., Liu, A., Wang, X. (2011). Effect of nano-$CaCO_3$ on properties of cement paste, *Energy Procedia*, **16**(Part B), pp. 991–996.

127. Nazari, A., Riahi, S. (2011). The Effects of ZnO_2 nanoparticles on strength assessments and water permeability of concrete in different curing media, *Mater. Res.*, **14**(2), pp. 178–188.

128. Nazari, A., Riahi, S. (2011). The effects of zinc dioxide nanoparticles on flexural strength of self-compacting concrete, *Composite Part B*, **42**(2), pp. 167–175.

129. Vazinram, F., Jalal, M., Foroushani, M. Y. (2015). Effect of nano ZnO_2 and lime water curing on strength and water absorption of concrete, *Int. J. Mater. Prod. Technol.*, **50**(3/4), pp. 356–365.

130. Behfarnia, K., Keivan, A., Keivan, A. (2013). The effects of TiO_2 and ZnO nanoparticles on physical and mechanical properties of normal concrete, *Asian J. Civ. Eng.*, **14**(4), pp. 517–531.

131. Ghanei, A., Jafari, F., Khotbehsara, M. M., Mohseni, E., Tang, W., Cui, H. (2017). Effect of nano-CuO on engineering and microstructure properties of fibre-reinforced mortars incorporating metakaolin: experimental and numerical studies, *Materials (Basel)*, **10**(10).

132. Nazari, A. H., Riahi, S. (2011). Effects of CuO nanoparticles on microstructure, physical, mechanical and thermal properties of self-compacting cementitious composites, *J. Mater. Sci. Technol.*, **27**(1), pp. 81–92.

133. Khotbehsara, M. M., Mohseni, E., Yazdi, M. A., Sarker, P., Ranjbar, M. M. (2015). Effect of nano-CuO and fly ash on the properties of self-compacting mortar, *Constr. Build. Mater.*, **94**, pp. 758–766.

134. Moradpour, R., Taheri-Nassaj, E., Parhizkar, T., Ghodsian, M. (2013). The effects of nanoscale expansive agents on the mechanical properties of non-shrink cement-based composites: the influence of nano-MgO addition, *Composite Part B*, **55**, pp. 193–202.

135. Ye, Q., Yu, K., Zhang, Z. (2015). Expansion of ordinary Portland cement paste varied with nano-MgO, *Constr. Build. Mater.*, **78**, pp. 189–193.

136. Polat, R., Demirboğa, R., Karagöl, F. (2017). The effect of nano-MgO on the setting time, autogenous shrinkage, microstructure and mechanical properties of high performance cement paste and mortar, *Constr. Build. Mater.*, **156**, pp. 208–218.

Chapter 8

Micromechanics Modeling of Nanomodified Cement-Based Composites: Carbon Nanotubes

Enrique García-Macías,[a] Rafael Castro-Triguero,[b] and Andrés Sáez[a]

[a] *Department of Continuum Mechanics and Structural Analysis, School of Engineering, Universidad de Sevilla, Camino de los Descubrimientos s/n, E-41092-Seville, Spain*
[b] *Department of Mechanics, University of Córdoba, Campus de Rabanales, Cordoba, CP 14071, Spain*
egarcia28@us.es

8.1 Introduction and Synopsis

This chapter is concerned with the micromechanics modeling of nanomodified cement-based composites, in particular carbon nanotube (CNT)-reinforced composites. Many researchers have reported the exceptional mechanical properties of CNTs with an elastic modulus greater than 1 TPa [1] and ultimate tensile strength around 150 GPa [2], largely exceeding those of any previously existing material. However, the interest on CNTs is

Nanotechnology in Cement-Based Construction
Edited by Antonella D'Alessandro, Annibale Luigi Materazzi, and Filippo Ubertini

ISBN 978-981-4800-76-1 (Hardcover), 978-0-429-32849-7 (eBook)
www.jennystanford.com

not limited to mechanical aspects but also focused on their electrical properties, with conductivities between 1000 and 200,000 S/cm [3], that is, several orders of magnitude larger than most polymeric and cementitious materials. These composites exhibit strain-sensing capabilities by means of measurable variations of their electrical properties under applied mechanical deformations [4], opening a vast field of applications in structural health monitoring [5–7]. On the basis of the mean-field homogenization theory, it is possible to develop computationally efficient analytical approaches to estimate the electromechanical properties of these nanocomposites. Furthermore, such micromechanics approaches are particularly interesting as they offer a favorable framework to incorporate physically meaningful variables of the microstructure, such as filler aspect ratio, waviness, agglomeration, or electrical transport mechanisms. In this regard, Section 8.2 focuses on the mechanical homogenization of nanomodified composites. First, Section 8.2.1 presents a succinct overview of the fundamentals of mean-field homogenization theory, followed by an introduction of the main existing approaches on the modeling of inhomogeneous materials in Sections 8.2.2 to 8.2.5. Finally, Sections 8.2.6 and 8.2.7 report on the modeling of wavy fibers and nonhomogeneous filler dispersions.

The theoretical framework utilized for the mechanical homogenization serves as basis for the modeling of the electrical transport properties of CNT-reinforced composites in Section 8.3. After a concise introduction to the physical mechanisms underlying the conductivity of CNT-reinforced composites in Sections 8.3.1 and 8.3.2, Section 8.3.3 introduces a micromechanics approach for the modeling of the overall electrical conductivity of these composites. Finally, in the light of the previously developed micromechanics model, Section 8.3.4 poses the implementation of the strain-induced effects into a mixed micromechanics approach of the piezoresistive properties of CNT-reinforced composites, and Section 8.4 summarizes the chapter.

8.2 Micromechanics Modeling of the Mechanical Properties of Nanomodified Composites

Computational micro-macro-mechanics aims at determining the relationship between the microstructure and the macroscopic response of composite structures. The direct solution of the response of a macroscopic engineering structure by incorporating all the microscopic information into a multiscale framework demands an extremely fine spatial discretization and, therefore, computational efforts far beyond the capacity of current computing machines. Due to this fact, currently available approaches divide the problem into two steps. First, constitutive relations between volume-averaged field variables are computed, which, in the second step, can be used to study the macroscopic behavior of the structure. To this end, it is necessary to define a representative volume element (RVE) that represents the composite material as a whole from a statistical perspective. Hence, the internal fields can be averaged in the RVE, a more computationally affordable dominion, through the solution of a series of boundary value problems (BVPs) under certain test loadings. Such regularization processes are typically referred to as homogenization. In particular, mean-field homogenization schemes are conceived as an efficient way to predict the behavior of heterogeneous materials with a reasonable computational cost.

8.2.1 Fundamentals of Mean-Field Homogenization

Let Ω_μ denote the RVE of a linear elastic matrix doped with dispersed inhomogeneities. The RVE must contain a sufficient number of fillers in such a way that the overall properties of the composite are statistically represented, that is, the heterogeneous material is assumed to be spatially homogeneous and ergodic within the RVE.

The basic scheme of the multiscale problem is illustrated in Fig. 8.1 It is assumed that at each point $\mathbf{x}$ of the macroscopic continuum, it is possible to find a definite RVE. Two different systems

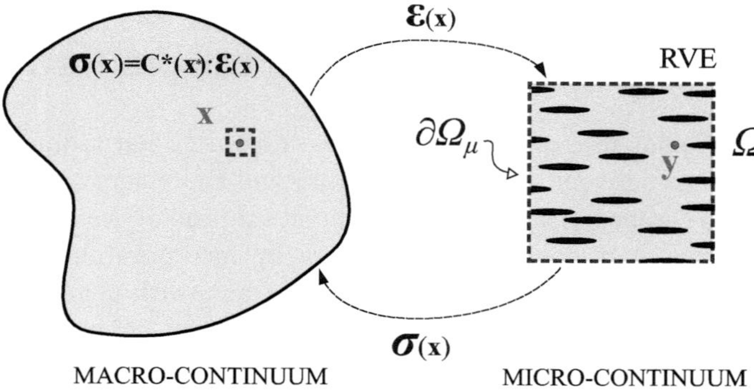

Figure 8.1 Scheme of the solving procedure for a generic multiscale problem.

of coordinates are defined: a reference system for the macroscopic scale (with position vector **x**) and another for the microscopic scale (with position vector **y**). Hence, it is considered that the macroscopic strain and stress tensors, $\varepsilon(\mathbf{x})$ and $\sigma(\mathbf{x})$, can be obtained through the volume average of the corresponding microscopic fields, $\varepsilon_\mu(\mathbf{y})$ and $\sigma_\mu(\mathbf{y})$, over the RVE:

$$\varepsilon(\mathbf{x}) = \frac{1}{V_\mu}\int_{\Omega_\mu} \varepsilon_\mu(\mathbf{y})\mathrm{d}V, \quad \sigma(\mathbf{x}) = \frac{1}{V_\mu}\int_{\Omega_\mu} \sigma_\mu(\mathbf{y})\mathrm{d}V \qquad (8.1)$$

with V_μ the total volume of the RVE. A multiscale model then consists of choosing an appropriate space of kinematically admissible displacement conditions. Afterward, for a given history of macroscopic strain/stress, the macroscopic stress/strain fields in the RVE are obtained by solving the corresponding BVP. Particularly efficient are mean-field approaches (MFAs), which are formulated in the context of linear elasticity, and thus, the solution of the BVP is known to be unique. In this case, the fields inside the domain Ω_μ can be represented though a linear dependency such as

$$\varepsilon_\mu(\mathbf{y}) = \mathbf{A}(\mathbf{y}) : \varepsilon(\mathbf{x}), \quad \sigma_\mu(\mathbf{y}) = \mathbf{B}(\mathbf{y}) : \sigma(\mathbf{x}), \qquad (8.2)$$

where $\mathbf{A}(\mathbf{y})$ and $\mathbf{B}(\mathbf{y})$ are fourth-order tensors termed "mechanical strain" and "stress concentration" tensors, respectively, and represent the solution of the respective BVP. Once the concentration tensors are known, the solution of the multiscale problem is trivial. Nevertheless, most microgeometries of real inhomogeneous materials are often too complex to obtain exact expressions for these magnitudes. MFAs approximate the microfields within each constituent by their volume averages. Hence, uniform strain and stress fields on each phase p are used as follows (the dependence on the coordinate $\mathbf{x}$ has been omitted for clarity):

$$\overline{\varepsilon}_{\mu}^{p} = \overline{\mathbf{A}}_{p} : \varepsilon, \quad \overline{\sigma}_{\mu}^{p} = \overline{\mathbf{B}}_{p} : \sigma, \tag{8.3}$$

and the homogenized strain and stress fields, $\overline{\varepsilon}_{\mu}^{p}$ and $\overline{\sigma}_{\mu}^{p}$, can be written as

$$\overline{\varepsilon}_{\mu}^{p} = \frac{1}{V_{\mu}^{p}} \int_{\Omega_{\mu}^{p}} \varepsilon(\mathbf{y}) \mathrm{d}V, \quad \overline{\sigma}_{\mu}^{p} = \frac{1}{V_{\mu}^{p}} \int_{\Omega_{\mu}^{p}} \sigma(\mathbf{y}) \mathrm{d}V. \tag{8.4}$$

Considering an N-phase composite, the macroscopic strain and stress tensors read

$$\varepsilon = \sum_{p=1}^{N} c_p \overline{\varepsilon}_{\mu}^{p} = \varepsilon^{\mathrm{o}}, \quad \sigma = \sum_{p=1}^{N} c_p \overline{\sigma}_{\mu}^{p} = \sigma^{\mathrm{o}}, \tag{8.5}$$

where $\varepsilon^{\mathrm{o}}(\mathbf{x})$ and $\sigma^{\mathrm{o}}(\mathbf{x})$ denote the imposed far-field homogeneous stress and strain tensors, respectively. V_{μ}^{p} is the volume occupied by phase p, and $c_p = V_{\mu}^{p} / \sum_{k=1}^{N} V_{\mu}^{k}$ is the volume fraction of the phase. It is important to note that MFAs assume perfect bonding between the constituents. Now, assuming elastic inhomogeneous phases and isothermal conditions, the macroscopic stress–strain relations can be written in the form

$$\varepsilon = \mathbf{C}^{*} : \sigma, \quad \sigma = \mathbf{M}^{*} : \varepsilon, \tag{8.6}$$

where $\mathbf{C}^{*}$ and $\mathbf{M}^{*}$ stand for the overall elasticity and compliance tensors of the composite. Each constituent of the multiphase material is also assumed to behave elastically so that

$$\sigma_{\mu}^{p} = \mathbf{C}_{p} : \varepsilon_{\mu}^{p}, \quad \varepsilon_{\mu}^{p} = \mathbf{M}^{p} : \sigma_{\mu}^{p}. \tag{8.7}$$

Now, considering $\sigma^{\mathrm{o}}(\mathbf{x}) = \mathbf{C}^{*}(\mathbf{x}) : \varepsilon^{\mathrm{o}}(\mathbf{x})$, the effective and compliance can be expressed as

$$\mathbf{C}^{*} = \sum_{p=1}^{N} c_p \mathbf{C}_p : \overline{\mathbf{A}}_p, \quad \mathbf{M}^{*} = \sum_{p=1}^{N} c_p \mathbf{M}^{p} : \overline{\mathbf{B}}_p. \tag{8.8}$$

For illustrative purposes, let us focus on two-phase composite materials. As shown in Fig. 8.2, the RVE is defined in this case by a hosting matrix (index m) loaded by ellipsoidal inclusions (index

f) dispersed according to an arbitrary orientation distribution. The aforementioned expressions can be rewritten for this particular case in a very simple way:

$$\overline{\varepsilon} = c_f \left\langle \overline{\varepsilon}^f \right\rangle + c_m \overline{\varepsilon}^m = \varepsilon^o, \quad \overline{\sigma} = c_f \left\langle \overline{\sigma}^f \right\rangle + c_m \overline{\sigma}^m = \sigma^o, \tag{8.9}$$

where c_f denotes the volume fraction occupied by the inclusions, and therefore, the volume fraction of the matrix is $c_m = 1 - c_f$. Assuming that all the inclusions have equal mechanical and topological characteristics but different orientations, averages are here defined over all possible orientations and denoted $\langle \cdot \rangle$. Now, the relations between the concentration tensors of the different phases can be written as

$$c_f \langle \mathbf{A}_f \rangle + (1 - c_f) \overline{\mathbf{A}}_m = \mathbf{I}, \quad c_f \langle \mathbf{B}_f \rangle + (1 - c_f) \overline{\mathbf{B}}_m = \mathbf{I}, \tag{8.10}$$

with $\mathbf{I}$ being the fourth-order identity matrix. The effective elasticity and compliance tensors of the composite in this case read

$$\mathbf{C}^* = \mathbf{C}_m + c_f \left\langle \left(\mathbf{C}_f - \mathbf{C}_m \right) : \mathbf{A}_f \right\rangle \tag{8.11}$$

$$\mathbf{M}^* = \mathbf{M}_m + c_f \left\langle \left(\mathbf{M}_f - \mathbf{M}_m \right) : \mathbf{B}_f \right\rangle. \tag{8.12}$$

In this particular case, only one concentration tensor is necessary to describe the full elastic behavior of the inhomogeneous material. Different assumptions on the tensor $\mathbf{A}$ correspond to different effective medium theories.

Finally, let us study in more detail the concept of orientational averages that are present in this type of composites. First, to describe the orientation of the fillers, a reference local coordinate system $K'' \equiv \left\{0; x_1'' x_2'' x_3''\right\}$ is fixed for each fiber. The inclusions are assumed to be ellipsoids with an aspect ratio $a_1 = a_3 < a_2$, being the major axis aligned in the local x_2'' axis. Two Euler angles, θ and γ, are required to describe the relative orientation of any ellipsoid of revolution. The filler configuration is shown in Fig. 8.3, where the polar angle θ stands for a rotation angle around the x_2 axis and transforms the global coordinate system K into an auxiliary coordinate system K', and the azimuthal angle γ denotes a rotation angle around the resultant x_1', that is

$$\theta \equiv \widehat{x_1 x_1'}, \quad \gamma \equiv \widehat{x_3' x_3''}. \tag{8.13}$$

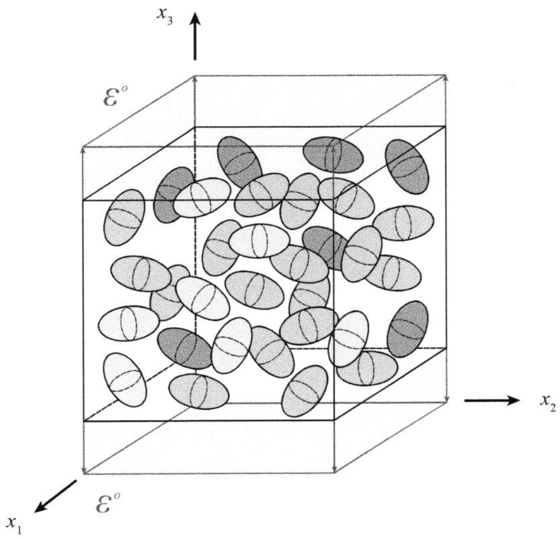

Figure 8.2 Representative volume element, including misoriented ellipsoidal inclusions.

The base vectors $\mathbf{e}_i$ and $\mathbf{e}_i''$ of the global and local coordinate systems can be related by $\mathbf{e}_i'' = g_{ij}\mathbf{e}_j$, where $\mathbf{g}$ is defined by the successive rotation around x_2 and x_1 axes as

$$\mathbf{g} = [R_1(\gamma)]\,[R_2(\theta)] = \begin{bmatrix} \cos(\gamma) & 0 & \sin(\gamma) \\ -\sin(\theta)\sin(\gamma) & \cos(\theta) & \cos(\gamma)\sin(\theta) \\ -\cos(\theta)\sin(\gamma) & -\sin(\theta) & \cos(\theta)\cos(\gamma) \end{bmatrix}. \tag{8.14}$$

The coordinate transformation of a fourth-rank tensor $\mathbf{P}$ into the local coordinate system K'' is explicitly represented in terms of $\mathbf{g}$ as $P''_{ijlk} = g_{ip}g_{jq}g_{kr}g_{ls}P_{pqrs}$.

Due to the high number of fillers contained in the RVE, the description of their orientation field is of statistical nature. The probability of a fiber lying in an infinitesimal range of angles $[\theta, \gamma] \times [\theta + \mathrm{d}\theta, \gamma + \mathrm{d}\gamma]$ is given by $p(\theta, \gamma)\sin(\theta)\mathrm{d}\theta\mathrm{d}\gamma$, with $p(\theta, \gamma)$ being the so-called orientation distribution function (ODF). Any ODF must satisfy the following normalization condition:

$$\int_0^{2\pi}\int_0^{\pi/2} p(\theta, \gamma)\sin(\theta)\mathrm{d}\theta\mathrm{d}\gamma = 1 \tag{8.15}$$

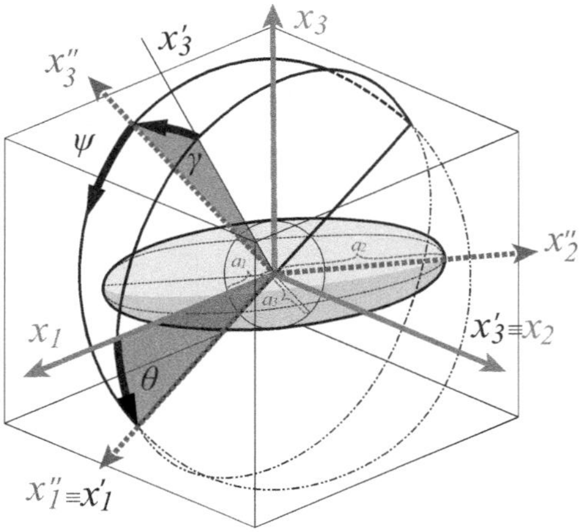

Figure 8.3 Euler angles defining the relation between the orientation of an ellipsoidal inclusion with an aspect ratio $a_1 = a_3 < a_2$ in the local coordinate system, $\{0; x_1''x_2''x_3''\}$, and the global coordinate system, $\{0; x_1x_2x_3\}$.

The integration of any ODF-weighted function $F(\theta, \gamma)$ over all possible orientations in the Euler space, also referred to as the orientational average of F, $\langle F \rangle$, is defined through

$$\langle F \rangle = \int_0^{2\pi} \int_0^{\pi/2} F(\theta, \gamma) p(\theta, \gamma) \sin(\theta) \mathrm{d}\theta \mathrm{d}\gamma. \tag{8.16}$$

8.2.2 Eshelby's Equivalent Inclusion

A large number of MFAs are based on the work of Eshelby [8], who formulated the problem of an elastic ellipsoidal inclusion embedded in an elastic matrix subjected to a uniform far-field strain. Eshelby demonstrated that if an elastic homogeneous ellipsoidal inclusion (i.e., an inclusion consisting of the same material as the matrix) in an infinite matrix is subjected to a homogeneous strain ε^*, termed as "stress-free strain" or "eigenstrain," the stress and strain states in the constrained inclusion are uniform, that is, $\boldsymbol{\sigma}^i_\mu(\mathbf{y}) = \boldsymbol{\sigma}^i_\mu$ and $\varepsilon^i_\mu(\mathbf{y}) = \varepsilon^i_\mu$. The uniform strain in the constrained inclusion, ε_c, is related to the eigenstrain, ε^*, by the expression

$$\varepsilon_c = \mathbf{S} : \varepsilon^*, \tag{8.17}$$

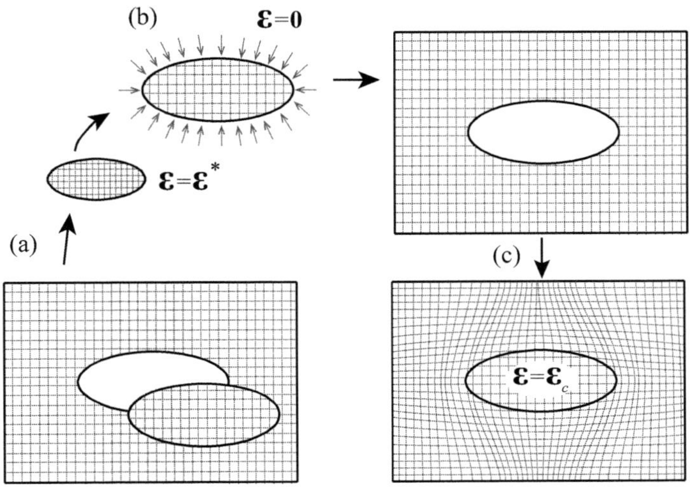

Figure 8.4 Solution of Eshelby's problem.

where tensor **S** is referred to as the Eshelby's tensor (extensive documentation on the Eshelby's tensor **S** for diverse particle shapes and matrix materials can be found in [9]). Eshelby illustrated the problem through a set of cut and weld operations shown in Fig. 8.4:

(a) The inclusion is first removed from the matrix, and it is freely deformed under an eigenstrain ε^*. This strain does not induce any stress in the inclusion neither in the solid.
(b) To return the inclusion into its original position, a stress $\mathbf{C}_m : \varepsilon^*$ has to be applied.
(c) Once in the initial configuration, the inclusion can be welded with the rest of the solid, removing the stresses until a new equilibrium configuration in the inclusion is found ε_c. Hence, the stresses in the inclusion can be readily determined as $\sigma^i_\mu = \mathbf{C}_m : [\varepsilon_c - \varepsilon^*] = \mathbf{C}_m : [\mathbf{S} - \mathbf{I}] : \varepsilon^*$.

Eshelby's result can now be applied to an inhomogeneous inclusion that is embedded in a matrix through the concept of equivalent homogeneous inclusion, as schematized in Fig. 8.5. The equivalent inclusion is made of matrix material, $\mathbf{C}_m$, and is assumed to be subjected to an eigentrain ε^* such that its stress field is the

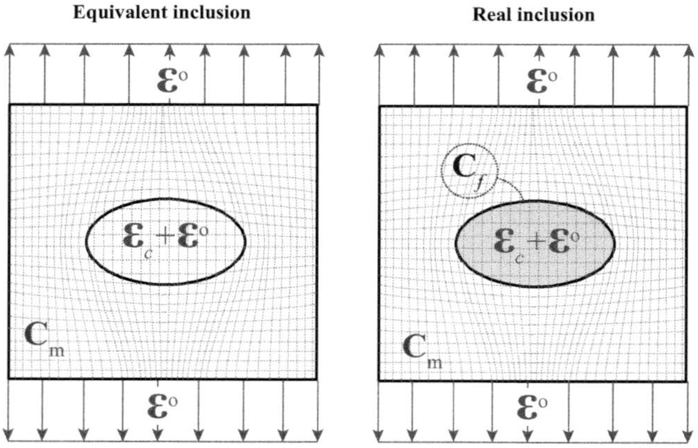

Figure 8.5 Schematic representation of Eshelby's equivalent inclusion.

same as for the actual inclusion. The stress and the strain field in the equivalent inclusion can be readily obtained as

$$\varepsilon^i_\mu = \varepsilon_c + \varepsilon^o, \quad \sigma^i_\mu = \mathbf{C}_m : [\varepsilon_c + \varepsilon^o - \varepsilon^*]. \tag{8.18}$$

The strain field in the real inclusion should be equal to that in the equivalent inclusion. Therefore, the stress field in the real inhomogeneity must be

$$\sigma^i_\mu = \mathbf{C}_f : [\varepsilon_c + \varepsilon^o]. \tag{8.19}$$

The eigenstrain ε^* is thus obtained with the condition of equal stresses in both inclusions (Eqs. 8.18 and 8.19), which yields

$$\varepsilon^* = \left[\left(\mathbf{C}_f - \mathbf{C}_m\right) : \mathbf{S} + \mathbf{C}_m\right]^{-1} : \left(\mathbf{C}_f - \mathbf{C}_m\right) : \varepsilon^o. \tag{8.20}$$

Finally, combining Eqs. 8.18 and 8.20, the strain concentration tensor $\mathbf{A}^{dil}$, also refereed to as the dilute strain concentration tensor, can be obtained as follows:

$$\varepsilon^i_\mu = \varepsilon^o + \varepsilon_c = \varepsilon^o + \mathbf{S} : \varepsilon^* = \underbrace{\left[\mathbf{I} + \mathbf{S} : \mathbf{C}_m^{-1} : \left(\mathbf{C}_f - \mathbf{C}_m\right)\right]^{-1}}_{\mathbf{A}^{dil}} : \varepsilon^o, \tag{8.21}$$

and the effective constitutive tensor of the composite can be extracted from Eq. 8.11 as

$$\mathbf{C}^* = \mathbf{C}_m + c_f \left(\mathbf{C}_f - \mathbf{C}_m\right) : \left[\mathbf{I} + \mathbf{S} : \mathbf{C}_m^{-1} : \left(\mathbf{C}_f - \mathbf{C}_m\right)\right]^{-1}. \tag{8.22}$$

8.2.3 The Mori–Tanaka Approach

It is important to note that Eshelby's approach does not account for perturbations in the stress fields produced by the neighboring inhomogeneities, and therefore, it should be strictly applied only in the dilute regime, $c_f \to 0$. To take into account the interactions among inclusions, Mori–Tanaka's (MT) method [10] was proposed as an extension of the theory of Eshelby, restricted to one single inclusion embedded in a semi-infinite elastic medium, to the case of a finite domain doped with multiple inhomogeneities. To this end, the far-field strain ε^{o} acting on the RVE is replaced by the averaged matrix strain ε_{μ}^{m}. According to Benveniste's revision [11], the strain concentration tensor derived from the MT approach $\mathbf{A}^{MT}$ reads

$$\mathbf{A}^{MT} = \mathbf{A}^{dil} : \left[(1 - c_f)\mathbf{I} + c_f \left\langle \mathbf{A}^{dil} \right\rangle\right]^{-1}, \tag{8.23}$$

and according to Eq. 8.11, the resulting constitutive tensor can be written as

$$\mathbf{C}^{*} = \mathbf{C}_m + c_f \left\langle \left(\mathbf{C}_f - \mathbf{C}_m\right) : \mathbf{A}^{MT} \right\rangle. \tag{8.24}$$

8.2.4 Self-Consistent Effective-Medium Approach

In the case of high-volume fractions of inhomogeneities, the interaction hypothesis of the MT method becomes uncertain. Typically, an alternative strategy for obtaining closer estimates for the elastic behavior of nondilute inhomogeneous materials is the SC method. The SC method approximates the interaction between phases by assuming that inclusions are embedded in an infinite volume of an effective medium with a yet unknown elastic tensor C^{**}. Hence, the effective stiffness tensor for the so-called classical SC scheme writes [12]

$$\mathbf{C}^{*} = \mathbf{C}^{**} + c_f \left\langle \left(\mathbf{C}_f - \mathbf{C}^{*}\right) : \mathbf{A}^{SC} \right\rangle, \tag{8.25}$$

where $\mathbf{A}^{SC}$ reads

$$\mathbf{A}^{SC} = \left[\mathbf{I} + \mathbf{S}^{*} \left(\mathbf{C}^{*}\right)^{-1} \left(\mathbf{C}_r - \mathbf{C}^{*}\right)\right]^{-1}. \tag{8.26}$$

Now Eshelby's tensor $\mathbf{S}^{*}$ is a function of the effective stiffness tensor $\mathbf{C}^{**}$ instead of $\mathbf{C}_m$. Since the effective stiffness tensor $\mathbf{C}^{**}$ is not known a priori, the SC method is an implicit method and must be solved by an iterative algorithm.

8.2.5 Extended Eshelby–Mori–Tanaka Approaches

A few research works in the literature report that the MT method may provide diagonally asymmetric stiffness tensors, as well as violating the theoretical Hashin–Shtrikman–Walpole bounds [13]. From Eq. 8.11, note that the effective stiffness tensor $\mathbf{C}^*$ is diagonally symmetric, that is, $C_{ijkl}^{MT} = C_{klij}^{MT}$, if and only if $\langle (\mathbf{C}_f - \mathbf{C}_m) : \mathbf{A}^{MT} \rangle$ is, for nonvanishing values of c_f. Nevertheless, several works in the literature demonstrate that this condition is not met for general biphase composites [14], what results in unacceptable stiffness tensors from an energy argument. The main reason for the asymmetry is attributed to the strain concentration tensors that are generally diagonally nonsymmetric [15]. Therefore, the use of the MT method is only justified in cases in which symmetry is guaranteed, such as

(1) Random orientation distribution of inclusions
(2) Perfect alignment of the inclusions
(3) Isotropic inclusions
(4) Spherical inclusions

A second weakness of the MT method concerns questionable predictions at high filler concentrations. Qui and Weng [13] proved that only in the case of perfect alignment, the effective stiffness tensor obtained using the MT approach preserves the symmetry and is comprised between the Hashin–Shtrikman–Walpole bounds.

In view of these limitations, some authors have proposed alternative approaches. These include those proposed by Ferrari [16] and later developed by Dunn et al. [17] for biphase composites, as well as the Schjødt-Thomsen and Pyrz (STP) approach [18].

The Dunn approach: Dunn et al. [17] proposed using the strain concentration tensor to be given by the MT concentration tensor for perfectly aligned fibers as follows:

$$\mathbf{A}^{Dunn} = \mathbf{A}^{dil} : \left[(1 - c_f)\mathbf{I} + c_f \mathbf{A}^{dil}\right]^{-1} \tag{8.27}$$

Thus, the stiffness tensor for two-phase composites with inclusions of identical shape reads

$$\mathbf{C}^* = \mathbf{C}_m + c_f \left[\left\langle \mathbf{C}_f \mathbf{A}^{Dunn} \right\rangle - (1 - c_f) \left\langle \mathbf{A}^{Dunn} \right\rangle\right]. \tag{8.28}$$

The STP approach: A second alternative is the one proposed by Schjødt-Thomsen and Pyrz [18]. This approach utilizes a direct orientational integration of the MT stiffness tensor for the case of perfectly aligned fibers:

$$\mathbf{C}^* = \langle \mathbf{C} \rangle = \int_0^{2\pi} \int_0^{\pi/2} \mathbf{C}(\theta, \gamma) p(\theta, \gamma) \sin(\theta) \mathrm{d}\theta \mathrm{d}\gamma \qquad (8.29)$$

To compare all the previously introduced approaches, Figs. 8.6a and 8.6b show, respectively, the longitudinal $E_{\parallel}$ and transverse $E_{\perp}$ Young's moduli versus the filler volume fraction c_f of polymer Epon 862/EPI cure W reinforced by uniaxially aligned (10,10) single-walled carbon nanotubes (SWCNTs). For illustrative purposes, also the theoretical bounds of Hashin–Shtrikman (HS) and Voigt/Reuss are included. It is noted that CNTs are highly anisotropic, with a longitudinal Young's modulus around 2 orders of magnitude higher than the transverse one. Due to this fact, $E_{\parallel}$ increases much more rapidly than $E_{\perp}$. Very little differences are found in $E_{\parallel}$ among the different methods, while larger differences arise in $E_{\perp}$. Eshelby's approach ignores the inclusion interaction, so it is only valid at the dilute regime. The SC method approximates the lower/upper HS bounds at low- and high-volume fractions, respectively, and displays a transition sigmoid curve in between.

Figure 8.6c shows the Young's modulus versus the filler volume fraction c_f of polymer Epon 862/EPI cure W reinforced by randomly oriented (10,10) SWCNTs. A first important conclusion is that the MT solution violates the upper HS bound at filler concentration around 67%, a fact that limits the applicability of the MT solution at higher filler contents. Nevertheless, the MT approach remains widely used in practical applications, where lower-volume fractions are often used. Finally, it is observed that the extended MT approaches (STP and Dunn) provide very similar results and are comprised between the Voigt/Reuss bounds for every filler concentration.

8.2.6 Modeling of CNT Waviness

A large number of experimental investigations have reported the wavy state of CNTs dispersed in composites [19, 20]. Due to their

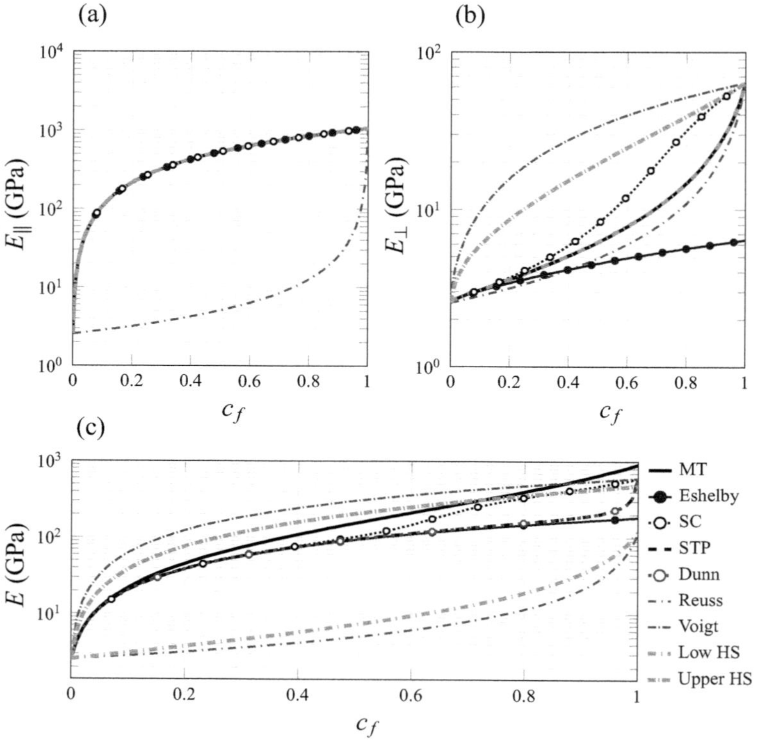

Figure 8.6 Young's modulus of Epon 862/EPI cure W reinforced by fully aligned (a, b) and randomly oriented (c) straight nanotubes. Fillers are defined as (10,10) single-walled CNTs with an infinitely long aspect ratio and transversely isotropic properties with Hill's elastic moduli [12] of $k_r = 271$ GPa, $l_r = 88$ GPa, $n_r = 1089$ GPa, $m_r = 17$ GPa, and $p_r = 442$ GPa. The polymer matrix is assumed to be isotropic with Poisson's ratio 0.2 and Young's modulus 2.55 GPa.

large aspect ratio, with typical lengths between 0.1 and 10 μm and diameters ranging from 10 to 15 nm, as well as low bending stiffness, CNTs usually present a certain degree of waviness. To characterize this curviness, different curved geometries have been proposed in the literature, such as planar sinusoidal curves [21], helixes [22, 23], or polylines with straight segments [24]. In particular, the helical approach, as sketched in Fig. 8.7, provides a suitable parametrization for micromechanics modeling. This curve is defined

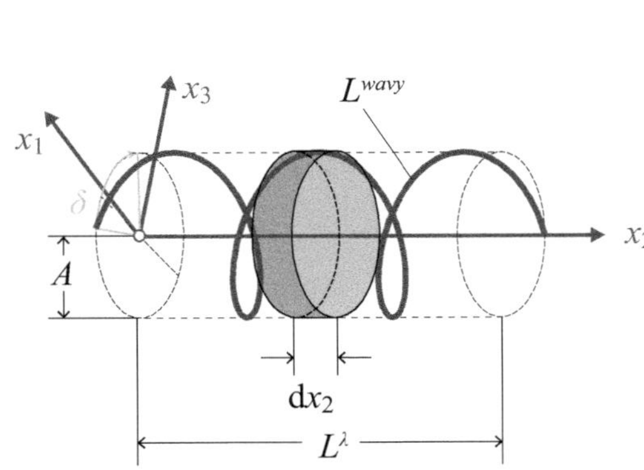

Figure 8.7 Schematics of the helical model of a curved CNT.

by its radius, A, spiral angle, θ^w, and polar angle, δ, as

$$x_1 = A\cos\delta, \quad x_2 = x_2, \quad x_3 = A\sin\delta \quad x_2 \in \left[0,\, L^\lambda\right], \qquad (8.30)$$

where the wavelength can be expressed as $L^\lambda = L^{wavy}\sin\theta^w$, $L^{wavy} = \delta A/\cos\theta^w$ being the length of the curve. The curvature is determined by the spiral angle θ^w. For instance, $\theta^w = \pi/2$ corresponds to a straight CNT, while $\theta^w = 0$ corresponds to a circular CNT.

There follows the application of three different approaches to the analysis of wavy CNTs, namely the MT method, the STP approach, and the ad hoc Eshelby's tensor approach by Yanase.

The Eshelby–Mori–Tanaka approach: The main hypothesis of this approach is that when the fiber wavelength L^λ becomes very large with respect to its diameter, the average strain in the fiber section approaches that of an infinitely long, straight fiber. It is easily seen that according to this particular geometry, the equivalent long, straight fibers are defined by a constant polar angle θ^w, and azimuthal angles that range from 0 to δ. On this basis, the application of the MT model to helical fillers reads [22]

$$\mathbf{C} = \left[\frac{c_f}{\delta}\int_0^\delta \left(\mathbf{C}_r(\theta^w, s) : \mathbf{A}^{MT}(\theta^w, s) : \mathbf{C}_m^{-1}\right) ds + c_m\mathbf{I}\right] : \left[\frac{c_f}{\delta}\int_0^\delta \left(\mathbf{A}^{MT}(\theta^w, s) : \mathbf{C}_m^{-1}\right) ds + c_m\mathbf{C}_m^{-1}\right]^{-1}. \qquad (8.31)$$

The STP approach: On the basis of the hypothesis of wavy nanotubes defined as a series of consecutive straight fibers, the direct integration of the constitutive tensor by Schjødt-Thomsen and Pyrz can be also applied in this context as

$$\mathbf{C} = \frac{1}{\delta} \int_0^{\delta} \mathbf{C}(\theta^w, s) \mathrm{d}s, \tag{8.32}$$

$\mathbf{C}(\theta^w, s)$ being the MT estimate for straight fibers with a constant polar angle θ^w and a varying azimuthal angle, s, from 0 to δ.

The ad hoc Eshelby's tensor approach: Another noteworthy approach is the one by Yanase et al. [25], who proposed an ad hoc Eshelby's tensor to account for the waviness effect. According to this model, the effective stiffness of the wavy-fiber-reinforced composite is evaluated by the following integral:

$$\overline{\mathbf{C}} = \frac{1}{\delta} \int_0^{\delta} \mathbf{C}^*(\theta^w, s) \mathrm{d}s, \tag{8.33}$$

with $\mathbf{C}^*$ being the Eshelby's equivalent inclusion solution as

$$\mathbf{C}^* = \mathbf{C}_m + c_f \left(\mathbf{C}_f - \mathbf{C}_m\right) : \mathbf{A}^{dil}. \tag{8.34}$$

To reproduce Eq. 8.33 on the basis of Eshelby's equivalent inclusion approach, an ad hoc Eshelby's tensor, $\mathbf{S}^*$, can be evaluated as follows:

$$\mathbf{S}^* = \lim_{c_f \to 0} \left[c_f \left((\mathbf{C}_m)^{-1} : \overline{\mathbf{C}} - \mathbf{I} \right)^{-1} - \left(\mathbf{C}_f - \mathbf{C}_m \right)^{-1} : \mathbf{C}_m \right] \tag{8.35}$$

To exclude the effect of far-field interaction, Yanase and coworkers proposed the limit in Eq. 8.35. It is noted that only when $A = 0$, that is, straight fibers, $\mathbf{S} = \mathbf{S}^*$. Now, the MT estimate of the overall properties of the nondilute regime reads

$$\mathbf{C}^{**} = \left(c_m \mathbf{C}_m + c_f \left\langle \mathbf{C}_r \mathbf{A}^* \right\rangle \right) : \left(c_f \mathbf{I} + c_f \left\langle \mathbf{A}^* \right\rangle \right)^{-1}, \tag{8.36}$$

where

$$\mathbf{A}^* = \left[\mathbf{I} + \mathbf{S}^* : \mathbf{C}_m^{-1} : \left(\mathbf{C}_f - \mathbf{C}_m \right) \right]^{-1}. \tag{8.37}$$

There follows a comparison of the different analyzed approaches. Figure 8.8 shows the elastic moduli of fully aligned wavy SWCNT (5,5) Epon 862/EPI cure W. A volume fraction c_f of 1% is selected. It is observed that the longitudinal modulus $E_{\parallel}$ decreases rapidly

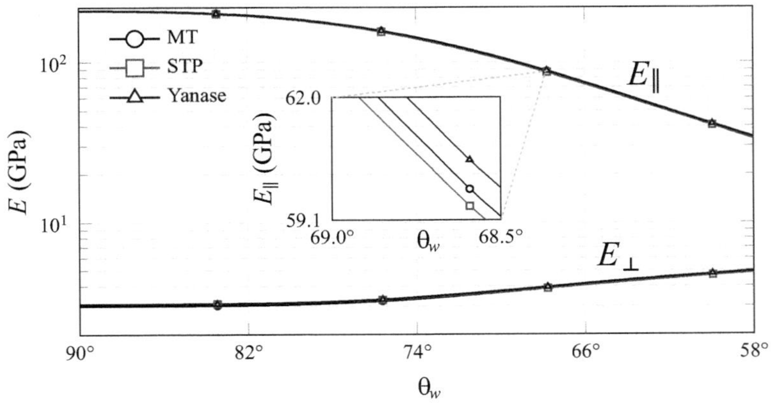

Figure 8.8 Young's modulus of Epon 862/EPI cure W reinforced by fully aligned wavy nanotubes ($c_f = 0.1$). Fillers are defined as (5,5) single-walled CNTs with an infinitely long aspect ratio and transversely isotropic properties with Hill's elastic moduli of $\mathbf{C}_f = (2k_r, l_r, n_r, 2m_r, 2p_r)$ = (1072, 184, 2143, 264, 1582) [GPa]. The polymer matrix is assumed isotropic with Poisson's ratio 0.2 and Young's modulus 2.55 GPa.

as the waviness increases, while the transverse modulus $E_{\perp}$ slightly increases with waviness. With regard to the different approaches, only slight differences are found. It can be seen that STP and Yanase approaches furnish very similar estimations. The reason for this similarity can be explained by the direct integration operations involved in the Yanase approach.

8.2.7 Modeling of CNT Agglomeration

A second important phenomenon to be taken into consideration for the simulation of CNT-nanoreinforced composites is the appearance of non-uniform spatial distributions of nanoinclusions. The difficulty in obtaining good filler dispersions stems from the tendency of CNTs to agglomerate in bundles. This effect is attributed to the electronic configuration of tube walls and their high specific surface area, which increases the van de Waals (vdW) attraction forces among nanotubes [26]. To model the effective mechanical properties of non-uniform distributions of CNTs, Shi et al. [22] proposed a two-parameter agglomeration model. That approach distinguishes two regions, one with high filler concentration (bundles) and another with low CNT concentration (surrounding composite). Therefore, the total volume of CNTs, V_r, dispersed in the RVE, V_μ, can be divided into the following two parts:

$$V_r = V_r^{bundles} + V_r^m, \tag{8.38}$$

where $V_r^{bundles}$ and V_r^m denote the volumes of CNTs dispersed in the bundles and in the matrix, respectively. To characterize the agglomeration of CNTs in bundles, two agglomeration parameters, ξ and ζ, are defined as follows:

$$\xi = \frac{V_{bundles}}{V_\mu}, \quad \zeta = \frac{V_r^{bundles}}{V_r}, \tag{8.39}$$

where $V_{bundles}$ is the volume occupied by the bundles in the RVE. The agglomeration parameter ξ represents the volume ratio of bundles with respect to the total volume of the RVE. On the other hand, ζ stands for the volume ratio of CNTs within the bundles with respect to the total volume of fillers. This pair of parameters unequivocally define the agglomeration scheme, as illustrated in Fig. 8.9. After some direct manipulations, the filler contents within the bundles and the surrounding composite, c_1 and c_2, respectively, can be expressed as

$$c_1 = c_f \frac{\zeta}{\xi}, \quad c_2 = c_f \frac{1-\zeta}{1-\xi}. \tag{8.40}$$

It can be extracted from Eq. 8.40 that $\zeta \geq \xi$ must be forced in order to impose a higher filler concentration in the clusters. The limit case $\zeta = \xi$ represents a uniform distribution of the fillers, while the heterogeneity degree grows for larger values of ζ. Hence, the homogenization process can be conducted in two steps. First, the overall constitutive tensor of the inclusions, $\mathbf{C}^{\text{in}}$, and the surrounding composite, $\mathbf{C}^{\text{out}}$, are obtained with filler concentrations c_1 and c_2, respectively. Second, the effective constitutive tensor of the composite, $\mathbf{C}^*$, is now computed by considering the surrounding composite as matrix material and bundles as inclusions.

The variations of the effective Young's and shear moduli with respect to the agglomeration parameter ζ are shown in Figs. 8.10a and 8.10b, respectively. The agglomeration parameter ξ is kept constant at a value of 0.2. It is observed that with the increase

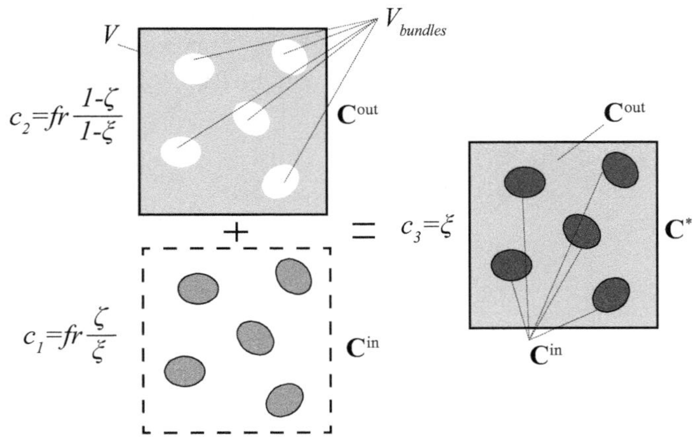

Figure 8.9 Schematic representation of the two-parameter agglomeration model.

in the relative amount of ζ, that is, a larger number of CNTs agglomerated in the clusters, the macroscopic elastic moduli experience substantial decreases. Let us remark that this weakening effect is more severe for higher filler concentrations.

8.3 Micromechanics Modeling of the Electrical Properties of CNT-Reinforced Composites

In this section, we are concerned, first, with the modeling of electrical conductivity and, second, with the piezoresistive properties of CNT-reinforced composites. On the basis of the previously introduced theoretical background, its counterpart electrical mean-field homogenization is presented. The analytical definition of the constitutive relations makes it possible to incorporate the physical mechanisms that govern the electrical transport properties of CNT-reinforced composites. Further, the piezoresistive properties of these composites, that is, measurable variations of the electrical resistance under applied loads, can be also included in the homogenization approach through a series of strain-induced effects.

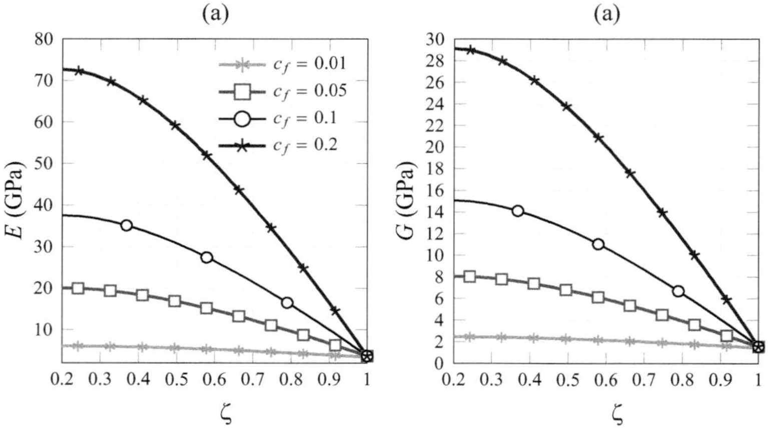

Figure 8.10 Macroscopic Young's modulus (a) and shear constant (b) of Epon 862/EPI reinforced by randomly oriented CNTs with agglomeration parameter $\xi = 0.2$ and varying agglomeration parameter ζ. Fillers are defined as (5,5) single-walled CNTs with an infinitely long aspect ratio and transversely isotropic properties with Hill's elastic moduli of $\mathbf{C}_f = (2k_r, l_r, n_r, 2m_r, 2p_r) = (1072, 184, 2143, 264, 1582)$ [GPa] and filler concentration $c_f = 0.1$. The polymer matrix is assumed isotropic with Poisson's ratio 0.2 and Young's modulus 2.55 GPa.

8.3.1 Physical Mechanisms Governing the Electrical Conductivity of CNT-Reinforced Composites

The electrical transport properties of CNT-reinforced composites are chiefly ascribed to two distinct physical mechanisms, as illustrated in Fig. 8.11, namely electron hopping and conductive networking [27, 28]. The first mechanism refers to a quantum mechanical tunneling effect by which electrons can be transferred between proximate nanotubes that are separated by an insulating matrix. On the other hand, a conductive networking mechanism relates the appearance of conductive paths formed by physically connected fillers. Most studies agree to identify the simultaneous contribution of these two mechanisms as a percolation-type behavior. Hence, when composites are loaded by filler concentrations below a percolation threshold, only the electron-hopping mechanism intervenes, since nanotubes are too separated to form conductive paths. Conversely,

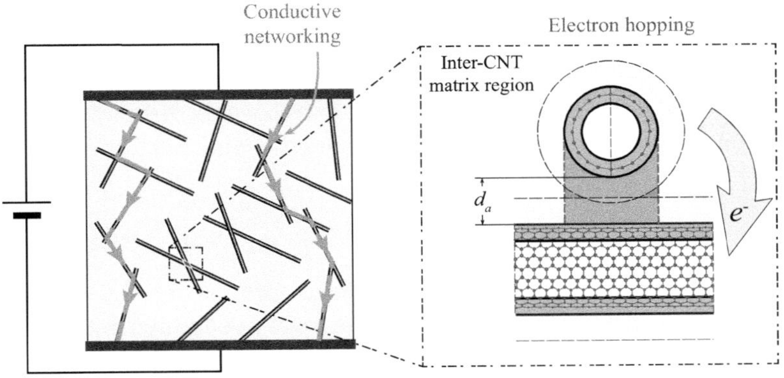

Figure 8.11 Schematic representation of the physical mechanisms governing the electrical transport of CNT-reinforced composites: electron hopping and conductive networking.

for filler concentrations above this critical concentration, nanotubes are more tightly spaced so that some conductive paths can be formed and both mechanisms act simultaneously. This percolative-type behavior is characterized by a sharp increase in the overall conductivity of composites doped with filler concentrations immediately above the percolation threshold.

8.3.1.1 Tunneling resistance: thickness and conductivity of the interface

The physical origin of the electron-hopping mechanism is a quantum tunneling effect that defines the penetration or tunneling of an electron through a potential barrier. In the case of CNT-reinforced composites, the potential barrier is defined by the insulating gap of matrix d_a between proximate nonconnected tubes. The upper limit interparticle distance can be taken as d_c = 0.5 nm [29], which is approximately the maximum possible thickness of a cementitious medium separating two adjacent MWCNTs that allows the tunneling penetration of electrons. In the case of filler volume concentrations below the percolation threshold f_c, the average separation distance between CNTs cannot be easily determined and the assumption $d_a = d_c$ is usually made on this quantity. When conductive networks are formed after percolation, several works in the literature have demonstrated that the average separation distance d_a between adjacent multiwalled carbon nanotubes (MWCNTs) follows a power-law description [30, 31]. Overall, the interparticle distance can be defined as a piecewise function of the filler volume fraction as

$$d_a = \begin{cases} d_c & 0 \le c_f < f_c \\ d_c \left(f_c / c_f \right)^{1/3} & f_c \le c_f \le 1 \end{cases} \tag{8.41}$$

Simmons [32] derived a generalized formula for the electric tunneling effect between similar electrodes separated by a thin insulating film. If we assume the thickness of the insulating film in the contact area between two crossing CNTs to be uniform and neglect the variation of the barrier height along the thickness, the formula of the resistance to electron tunneling R_{int} for a rectangular potential barrier can be employed [33]:

$$R_{int}(d_a) = \frac{d_a \hbar^2}{a e^2 \left(2mh^{1/2}\right)} \exp\left(\frac{4\pi d_a}{\hbar} (2mh)^{1/2} \right), \tag{8.42}$$

where h is the height of the tunneling potential barrier; m and e are the mass and the electric charge of an electron, respectively; a is the contact area of the MWCNTs; and $\hbar$ is the reduced Planck's constant. A reference value of 0.36 eV can be taken for the height of the tunneling potential barrier based on the experimental results of Wen and Chung [34]. The electron-hopping mechanism can be thus taken into account through a conductive interphase layer surrounding the nanotubes, whose thickness, t, and electrical conductivity, κ_{int}, are expressed correspondingly as

$$t = \frac{1}{2} d_a, \quad \kappa_{int} = \frac{d_a}{a R_{int}(d_a)}. \tag{8.43}$$

8.3.1.2 Nanoscale composite cylinder model for CNTs

As discussed before, the electron-hopping mechanism among MWCNTs distributed in the matrix can be simulated by the definition of a continuum interphase layer surrounding the nanotubes. Therefore, an equivalent solid filler (Fig. 8.12) is defined as a composite cylinder assemblage consisting of an MWCNT (length L and diameter $D = 2r_c$) and the surrounding interphase with a

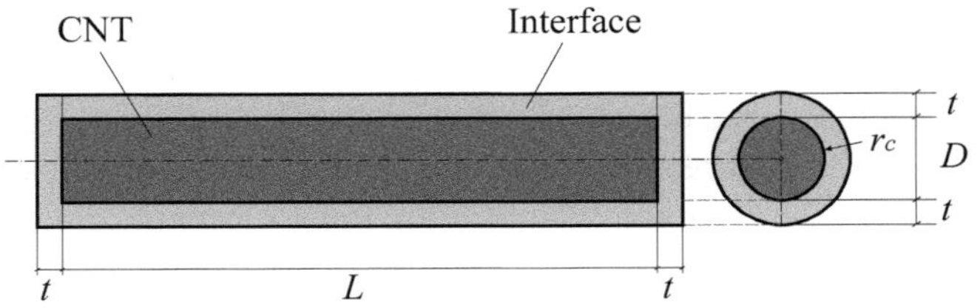

Figure 8.12 Equivalent composite cylinder assemblage.

thickness of t. MWCNTs are usually treated as solid cylinders instead of hollow tubes due to the difficulty in obtaining the actual electrical conductivity of the filler, considering its nanoscale structure. The effective longitudinal and transverse electrical conductivity of the cylinder, $\tilde{\kappa}^L$ and $\tilde{\kappa}^T$, respectively, can be obtained by applying Maxwell's equations and the law-of-mixture rule, which leads to [35]

$$\tilde{\kappa}^L = \frac{(L+2t)\kappa_{int}\left[\kappa_c^L r_c^2 + \kappa_{int}\left(2r_c t + t^2\right)\right]}{2\kappa_c^L r_c^2 t + 2\kappa_{int}\left(2r_c t + t^2\right)t + \kappa_{int} L \left(r_c + t\right)^2}, \tag{8.44}$$

$$\tilde{\kappa}^T = \frac{\kappa_{int}}{L+2t}\left[L\frac{2r_c^2\kappa_c^T + \left(\kappa_c^T + \kappa_{int}\right)\left(t^2 + 2r_c t\right)}{2r_c^2\kappa_{int} + \left(\kappa_c^T + \kappa_{int}\right)\left(t^2 + 2r_c t\right)} + 2t\right], \tag{8.45}$$

where κ_c denotes the electrical conductivity of the MWCNTs and the superscripts L and T represent the longitudinal and transverse directions, respectively. The volume fraction c_{eff} of the effective solid fillers can be obtained in terms of the volume fraction of MWCNTs as follows:

$$c_{\text{eff}} = \frac{\left(r_c + t\right)^2\left(L + 2t\right)}{r_c^2 L} c_f \tag{8.46}$$

8.3.2 Percolation Threshold Estimates

An essential point in the modeling of percolative systems is the determination of the percolation threshold f_c. This critical concentration denotes the onset of the percolation process, that is, the appearance of electrically connected fillers forming conductive networks. To implement an analytical approach apt for being included in a micromechanics approach, let us consider an RVE consisting of a matrix doped with misoriented fillers according to an ODF $p(\theta, \gamma)$. Komoro and Makishima [36] proposed a stochastic approach for the calculation of the number of filler contacts in general disordered systems doped with rod-like inclusions. According to that work, a filler A with a given orientation $p(\theta, \gamma)$ comes into contact with a second filler B with orientation $p(\theta', \gamma')$ if the center of mass of the former is located within the neighborhood region of the latter. The neighborhood region is defined when the fiber B is slid over both sides of the fiber A, as shown in Fig. 8.13. The sweepings of the axis of B define two rhombuses near both sides of A, conforming a parallelepiped. The volume of this region is $V = 2DL^2 \sin \tau$, with τ being the angle between the two nanofillers. On the basis of the defined number of contacts in a volume V along with the probability of formation of a contact, the mean distance among contacts, $\overline{b}_{KM}$, is given by

$$\overline{b}_{KM} = \frac{\pi D}{8If}, \tag{8.47}$$

where

$$I = \int_0^{2\pi} d\theta \int_0^{\pi/2} J(\theta, \gamma) p(\theta, \gamma) \sin\theta d\gamma, \tag{8.48}$$

$$J(\theta, \gamma) = \int_0^{2\pi} d\gamma' \int_0^{\pi/2} \sin\tau(\theta, \gamma, \theta', \gamma') p(\theta', \gamma') \sin\theta d\theta', \tag{8.49}$$

$$\sin\tau = \left[1 - \left\{\cos\theta \cos\theta' + \cos(\gamma - \gamma') \sin\theta \sin\theta'\right\}^2\right]^{1/2}. \tag{8.50}$$

The methodology followed by Komoro and Makishima does not consider the changes in the contact probability with successive contacts. Pan [37] raised this issue and proposed an extended approach by considering that an existing contact reduces the effective contact length of a filler, which limits the probability of forming a new contact. Pan's expression of the mean distance among inclusions $\overline{b}_{Pan}$ reads

$$\overline{b}_{Pan} = \frac{(\pi + 4f\eta)D}{8fI}, \tag{8.51}$$

where

$$\eta = \int_0^{2\pi} d\gamma \int_0^{\pi/2} J(\theta, \gamma) K(\theta, \gamma) \sin\theta d\theta, \tag{8.52}$$

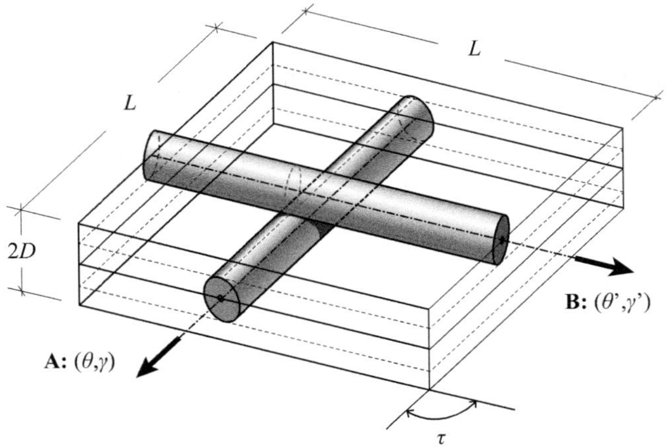

Figure 8.13 Neighborhood of the contact region of a fiber A of orientation (θ, γ) defined by the parallelepiped formed by a second fiber B of orientation (θ', γ') sliding over both sides of the former.

$$K(\theta, \gamma) = \int_0^{2\pi} \mathrm{d}\gamma' \int_0^{\pi/2} \frac{p(\theta', \gamma') \sin\theta'}{\sin\tau(\theta, \gamma, \theta', \gamma')} \mathrm{d}\theta'. \tag{8.53}$$

To be part of a conductive network in the nanocomposite, each nanofiller must have at least two contact points [38]. Alternatively, the mean distance between contacts should be at least half of the filler length to attain the percolation threshold. Kumar and Rawal [39] defined a coverage parameter, $\Gamma = b/L$, which represents the number of contacts formed on a given filler length, so that this quantity is such that $\Gamma \leq 0.5$ for percolated nanofillers. The authors also proposed a mean coverage parameter, $\overline{\Gamma} = \overline{b}/L$, as the probability of percolation of the composite. Assuming that $\overline{\Gamma}$ follows an exponential distribution, in a similar way to the distance between the contacts as reported in the literature [40], it writes

$$P(\Gamma) = (1/\overline{\Gamma})\exp\left(-\Gamma/\overline{\Gamma}\right). \tag{8.54}$$

Zheng et al. [41] reported that the statistical percolation threshold is reached when 50% of the sample percolates, that is,

$$\int_0^{0.5} P(\Gamma)\mathrm{d}\Gamma = -\exp(0.5/\overline{\Gamma}) + 1 = 0.5, \tag{8.55}$$

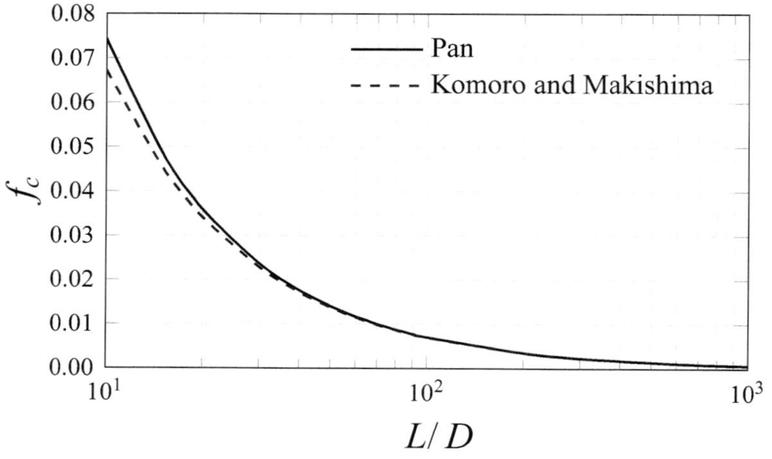

Figure 8.14 Percolation threshold f_c versus CNTs aspect ratio L/D.

where a value $\overline{\Gamma}$ of 0.72 is extracted. Finally, combining this result with Eqs. 8.48 and 8.51, the percolation threshold can be computed using the Komori–Makishima and Pan's models as

$$f_c^{KM} = \frac{\pi}{5.77sI}, \quad f_c^{Pan} = \frac{\pi}{5.77sI - 4\eta}, \tag{8.56}$$

with s being the aspect ratio of the nanofillers, that is, L/D. Figure 8.14 plots the estimation of the percolation threshold by Komoro–Makishima and Pan's models versus the nanofiller aspect ratio. In particular, a fully random distribution of fillers ($p(\theta, \gamma) = 1/2\pi$) is considered, as typically observed in unstrained CNT-reinforced composites. As expected, it is observed that the percolation threshold decreases with increasing aspect ratios; in other words, fillers with higher aspect ratios are more likely to form conductive paths. Kumar and Rawal [39] reported that there are only slight differences between the Komori–Makishima and Pan's models since the percolation threshold is often reached at significantly low nanofiller concentrations.

The percolation threshold f_c denotes the onset of the percolation process. Before this critical value, $c_f < f_c$, the only mechanism that contributes to the electrical conductivity is the electron-hopping mechanism. Nonetheless, once percolation starts, $c_f \geq f_c$, a certain percentage χ of MWCNTs get electrically connected in conductive networks, while the rest, $1-\chi$, remain unconnected. According to Deng and Zheng [42], the relative amount of percolated MWCNTs, χ, can be approximately estimated as

$$\chi = \begin{cases} 0 & 0 \leq c_f < f_c \\ \dfrac{c_f^{1/3} - f_c^{1/3}}{1 - f_c^{1/3}} & f_c \leq c_f \leq 1 \end{cases} \tag{8.57}$$

8.3.3 Micromechanics Model for the Overall Conductivity of CNT-Reinforced Composites

Let us consider an RVE of a cementitious matrix doped with randomly dispersed straight MWCNTs. The effective electric conductivity $\boldsymbol{\kappa}_{\text{eff}}$ of the RVE can be estimated by means of the electrical counterpart of the mean-field homogenization model of Eshelby–Mori–Tanaka as

$$\boldsymbol{\kappa}_{\text{eff}} = \boldsymbol{\kappa}_m + c_{\text{eff}} \left\langle (\boldsymbol{\kappa}_c - \boldsymbol{\kappa}_m) \boldsymbol{A} \right\rangle , \tag{8.58}$$

with $\boldsymbol{\kappa}_c$ and $\boldsymbol{\kappa}_m$ being the electrical conductivity tensor of the effective filler and the matrix, respectively. As previously indicated, the angle brackets $\langle\cdot\rangle$ stand for the average over all possible orientations in the Euler space. In the case of a second-order tensor F_{ij}, its orientational average, $\langle \boldsymbol{F} \rangle$, is defined in matrix notation through

$$\langle \boldsymbol{F} \rangle = \int_0^{2\pi} \int_0^{\pi/2} \boldsymbol{g}^T \boldsymbol{F} \boldsymbol{g} p(\theta, \gamma) \sin(\theta) \mathrm{d}\theta \mathrm{d}\gamma. \tag{8.59}$$

As discussed in the previous section, MWCNTs together with the surrounding interphases can be considered as equivalent solid cylinders with a transversely isotropic electrical conductivity tensor in the local coordinate system, $\boldsymbol{\kappa}_c'$, given as

$$\boldsymbol{\kappa}_c' = \begin{bmatrix} \tilde{\kappa}^L & 0 & 0 \\ 0 & \tilde{\kappa}^T & 0 \\ 0 & 0 & \tilde{\kappa}^T \end{bmatrix}, \tag{8.60}$$

with $\tilde{\kappa}^L$ and $\tilde{\kappa}^T$ being the longitudinal and transverse electrical conductivity of the effective filler, as obtained in Eqs. 8.44 and

8.45, respectively. The electric field concentration tensor $\boldsymbol{A}$ can be computed according to the MT model as

$$\boldsymbol{A} = \boldsymbol{A}^{dil}\left\{(1 - c_{\text{eff}})\,\boldsymbol{I} + c_{\text{eff}}\left\langle \boldsymbol{A}^{dil}\right\rangle\right\}^{-1}, \tag{8.61}$$

where $\boldsymbol{A}^{dil}$ is the dilute electric field concentration tensor

$$\boldsymbol{A}^{dil} = \left\{\boldsymbol{I} + \boldsymbol{S}\left(\kappa_m\right)^{-1}\left(\kappa_c - \kappa_m\right)\right\}^{-1}, \tag{8.62}$$

with $\boldsymbol{I}$ and $\boldsymbol{S}$ being the second-order identity tensor and Eshelby's tensor of the effective filler, respectively. The second-order Eshelby's tensor of a prolate spheroid ($a_2 = a_3 < a_1$) aligned in the x_1' direction is given by [43]

$$\boldsymbol{S} = \begin{bmatrix} S_{11} & 0 & 0 \\ 0 & S_{22} & 0 \\ 0 & 0 & S_{33} \end{bmatrix}, \tag{8.63}$$

where

$$S_{22} = S_{33} = \frac{A_{re}}{2\left(A_{re}^2 - 1\right)^{3/2}}\left[A_{re}\left(A_{re}^2 - 1\right)^{1/2} - \cosh^{-1} A_{re}\right] \tag{8.64a}$$

$$S_{11} = 1 - 2S_{22}, \tag{8.64b}$$

with A_{re} being the aspect ratio of the effective filler, that is, $A_{re} = (L + 2t)/(D + 2t)$.

Considering the percolative behavior of CNT-reinforced composites, both the electron-hopping and conductive networking mechanisms contribute to the electrical conductivity of the composite according to the percentage of percolated fillers χ (Eq. 8.57). Therefore, the expression of the overall electrical conductivity of CNT–cement nanocomposites from Eq. 8.58 can be extended by the sum of both mechanisms as

$$\begin{aligned} \kappa_{\text{eff}} = \kappa_m + & \underbrace{(1 - \chi)\left\langle c_{\text{eff}}\left(\kappa_{EH} - \kappa_m\right)\boldsymbol{A}_{EH}\right\rangle}_{\kappa_{N,EH}} + \\ & + \underbrace{\chi\left\langle c_{\text{eff}}\left(\kappa_{CN} - \kappa_m\right)\boldsymbol{A}_{CN}\right\rangle}_{\kappa_{N,CN}} \end{aligned}, \tag{8.65}$$

where subscripts EH and CN refer to the electron-hopping and conductive network mechanisms, respectively. In the case of

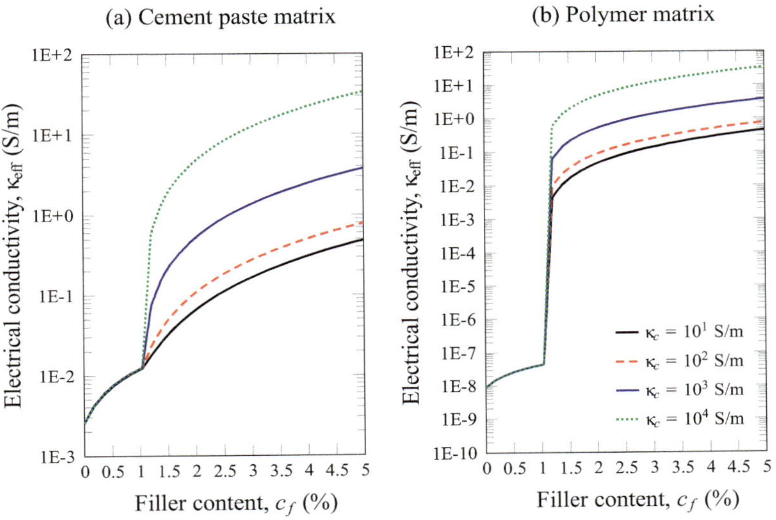

Figure 8.15 Effect of MWCNTs' electrical conductivity κ_c on the effective conductivity of MWCNT-reinforced cement paste (a) and polymer (b) nanocomposites (h = 0.36 eV; L = 1 μm; D = 15 nms; m = 9.10938291 $\times 10^{-31}$ kg; $e = -1.602176565 \times 10^{-19}$ C; $\hbar = 6.626068 \times 10^{-34} \text{m}^2\text{kg/s}$).

CNTs forming conductive networks, several fibers are electrically connected in a continuous conductive path. This effect can be approximated by considering an infinite aspect ratio of the fibers, as proposed by Seidel and Lagoudas [44]. As a result, the quantities related to the electron-hopping mechanism are defined with the real CNTs aspect ratio ($a_2 = a_3 = r_c$, $a_1 = L$), while quantities corresponding to conductive networks are estimated with an infinite aspect ratio ($a_2 = a_3 = r_c$, $a_1 \rightarrow \infty$).

Figure 8.15a demonstrates the effects of the electrical conductivity of the fillers, κ_c, on the overall conductivity of MWCNT-reinforced cement paste nanocomposites. Electrical conductivities of $\kappa_m = 2.8 \times 10^{-3}$ S/m and $\kappa_m = 9.0 \times 10^{-9}$ S/m are selected for the cement paste and polymer, respectively. It can be seen that at low filler contents below the percolation threshold, the electrical conductivity only experiences limited increases. Nevertheless, once the percolation threshold is reached, the conductive network mechanism becomes dominant and substantial increases are observed.

8.3.3.1 Waviness and agglomeration effects

As previously indicated, CNTs are usually found in a wavy state due to their large aspect ratio and low bending stiffness. The waviness approach introduced in Section 8.2.6 can be adapted here for the electrical modeling of CNT-reinforced composites. Let us recall that the helical approach is entirely defined by the diameter D_h, the spiral angle θ^w, and the polar angle δ. The length L^{wavy} of the curved CNT is defined by these parameters as

$$L^{wavy} = \frac{\delta D_h}{2\cos\theta^w}. \tag{8.66}$$

The consideration of waviness into the micromechanics modeling requires the conversion of wavy CNTs into equivalent straight fillers of length L^{str} [45]. A wavy MWCNT can be regarded as an equivalent straight fiber with the capacity to (i) conduct the same electric flux J and (ii) transport the same amount of electric charges [31]. When a wavy MWCNT is subjected to a potential difference ΔV, the electrical flux J can be approximated by [42]

$$J = \kappa_c^{wavy}\frac{\Delta V}{L^{wavy}}. \tag{8.67}$$

Hence, condition (i) dictates the effective electrical conductivity of the equivalent straight MWCNT as

$$\kappa_c^{str} = \alpha\kappa_c^{wavy}, \tag{8.68}$$

with $\alpha = L^{str}/L^{wavy} = \sin\theta^w$ the length ratio. Condition (ii) forces the electrical resistance of the wavy and the equivalent straight fillers to be the same:

$$R_{cnt}^{str} = R_{cnt}^{wavy} \tag{8.69}$$

The combination of Eqs. 8.68 and 8.69 results in a condition of equal diameters for the wavy and straight fillers, $D^{str} = D^{wavy}$. Finally, due to the reduction of the MWCNTs length from L^{wavy} to L^{str}, the volume fraction of the fillers must be updated to αc_f.

A second important feature to be accounted for is the tendency of CNTs to form agglomerates in bundles. In this case, the same two-parameter agglomeration model previously explained in Section. 8.2.7 can be retrieved. The homogenization procedure defined in Eq. 8.65 therefore has to be applied in two steps: inside the bundles and in the surrounding matrix with filler contents c_1 and c_2 (Eq. 8.40), respectively. Clusters are assumed to be ellipsoidal ($a_2 = a_3 \neq a_1$), for which the Eshelby's tensor reads

$$S_{22} = \begin{cases} \dfrac{A_r}{2\left(A_r^2 - 1\right)^{3/2}} \left[A_r \left(A_r^2 - 1^{1/2}\right) - \cosh^{-1} A_r\right]; & A_r \geq 1 \\ \dfrac{A_r}{2\left(A_r^2 - 1\right)^{3/2}} \left[\cos^{-1} A_r - A_r \left(1 - A_r^2\right)^{1/2}\right]; & A_r \geq 1 \end{cases}, \quad (8.70)$$

with $A_r = a_1/a_2$ the aspect ratio of the ellipsoid and $S_{33} = S_{22}$, $S_{11} = 1 - 2S_{22}$.

8.3.4 Micromechanics Model for the Piezoresistivity of CNT-Reinforced Composites

In light of the previous discussion on the electrical conductivity of CNT-reinforced composites, the origin of their piezoresistive properties can be ascribed to strain-induced alterations of the mechanisms that underlay the electrical conductivity. In particular, the strain-sensing capability of these composites is attributed to strain-induced (i) volume expansion and reorientation of CNTs, (ii) changes in the conductive networks, and (iii) changes in the tunneling resistance [46–48]. There follows the incorporation of these effects in a mixed micromechanics modeling of the strain-sensing properties of CNT-reinforced composites.

8.3.4.1 Volume expansion and reorientation of CNTs

Let us consider a 3D affine deformation cell containing an embedded filler before and after the application of a uniaxial strain ε, as shown in Fig. 8.16. Under the assumption of small deformations, the volume of the cell changes from $V_o = l_o^3$ to $V = l_o^3 (1 + \varepsilon)^{1-2\nu}$, with ν being the Poisson's ratio of the composite. Moreover, assuming that the deformation of the composite is mainly sustained by the matrix, the volume of the nanoinclusions remains approximately constant, and therefore, the volume expansion induces changes in the CNT volume

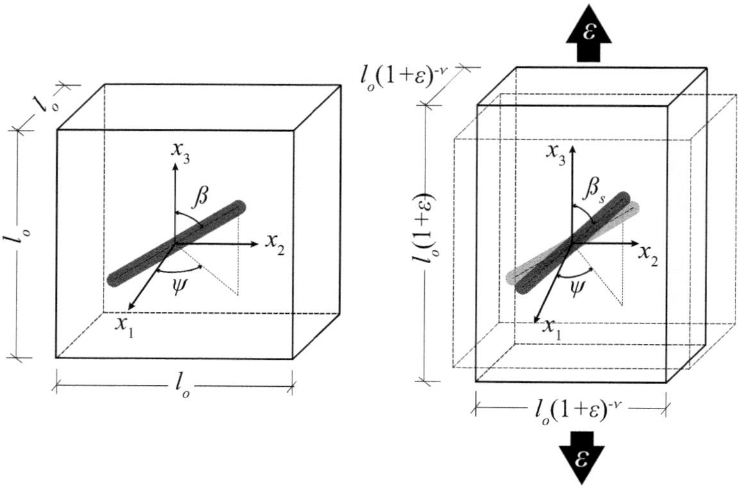

Figure 8.16 Schematic representation of the volume expansion and reorientation of a conductive filler within a deformable cell subjected to uniaxial strain ε.

fraction as follows [46]:

$$f^* = \frac{V_o f}{V} = \frac{f}{(1+\varepsilon)^{1-2\nu}} \tag{8.71}$$

It can be also noted in Fig. 8.16 that the strain ε also originates a realignment of the fiber along the strain direction x_3. This change of orientation is characterized by a variation of the polar angle from β to β_s and approximately constant azimuth angle γ [46]. The new polar angle β_s can be readily expressed in terms of its initial value $\beta = \pi/2 - \theta$ as

$$\tan\beta_s = (1+\varepsilon)^{1+\nu}\tan\beta. \tag{8.72}$$

This change of the polar angle of the fillers results in a change of the ODF from $p(\gamma,\beta)$ to $p(\gamma,\beta_s)$. After the application of the strain, the initially randomly distributed fillers $(p(\gamma,\beta) = 1/2\pi)$ tend to align in the strain direction, and therefore, the randomness of the nanofillers' distribution is reduced. The resulting ODF must preserve the number of fillers G before and after application of the strain.

To obtain the resulting ODF, the condition of a constant number of fillers before and after the application of the strain has been applied. To this end, and assuming a total number of G fillers distributed in the RVE, the number of fillers lying in the orientation range $(\beta, \beta + \mathrm{d}\beta) \times (\gamma, \gamma + \mathrm{d}\gamma)$ can be computed as

$$\mathrm{d}N_{\beta,\beta+\mathrm{d}\beta}^{\gamma,\gamma+\mathrm{d}\gamma} = \frac{1}{\pi} G p(\gamma, \beta) \sin\beta \mathrm{d}\beta \mathrm{d}\gamma. \tag{8.73}$$

Accordingly, the total number of fillers must be the same after the application of the strain within the range $(\beta_s, \beta_s + \mathrm{d}\beta_s) \times (\gamma, \gamma + \mathrm{d}\gamma)$, that is, $\mathrm{d}N_{\beta,\beta+\mathrm{d}\beta}^{\gamma,\gamma+\mathrm{d}\gamma} = \mathrm{d}N_{\beta_s,\beta_s+\mathrm{d}\beta_s}^{\gamma,\gamma+\mathrm{d}\gamma}$. Combining Eqs. 8.72 and 8.73 and considering $\beta = \pi/2 - \theta$, the resulting ODF, $p(\theta, \gamma)$, reads

$$p(\theta, \gamma) = \frac{(1+\varepsilon)^{\frac{1+\nu}{2}} / 2\pi}{\left[(1+\varepsilon)^{-(1+\nu)} \cos^2(\frac{\pi}{2} - \theta) + (1+\varepsilon)^{(1+\nu)} \sin^2(\frac{\pi}{2} - \theta)\right]^{\frac{3}{2}}}. \tag{8.74}$$

Figure 8.17 shows the variation of the ODF with the strain ε and the polar angle β. In the unloaded case, $\varepsilon = 0$, the ODF remains constant and equal to 2π, corresponding to the uniform random distribution. It is also observed that for increasing traction ($\varepsilon > 0$), the ODF gives more weight to polar angles close to 0 and π, that is, more fibers tend to realign in the direction of the strain. On the contrary, it is worth noting that for increasing compression ($\varepsilon < 0$), the ODF has higher values for polar angles around $\pi/2$, and therefore, the fibers tend to realign in the transverse direction of the strain.

8.3.4.2 Change in the conductive networks

The piezoresistivity of CNT nanocomposites is also due to the breakage of conductive networks induced by strain. As previously discussed, the reorientation of the fibers decreases the randomness of the CNT distribution as the strain increases. Hence, it can be intuitively understood that the probability of forming conductive paths must be altered, which corresponds to the change in the percolation threshold f_c. On the basis of Komoro–Makishima's model previously explained in Section 8.3.2, arbitrary for any orientation distribution of fillers, the strain-induced variation of the percolation threshold can be directly taken into account through the ODF with strain reorientation effects in Eq. 8.74. Figure 8.18 shows

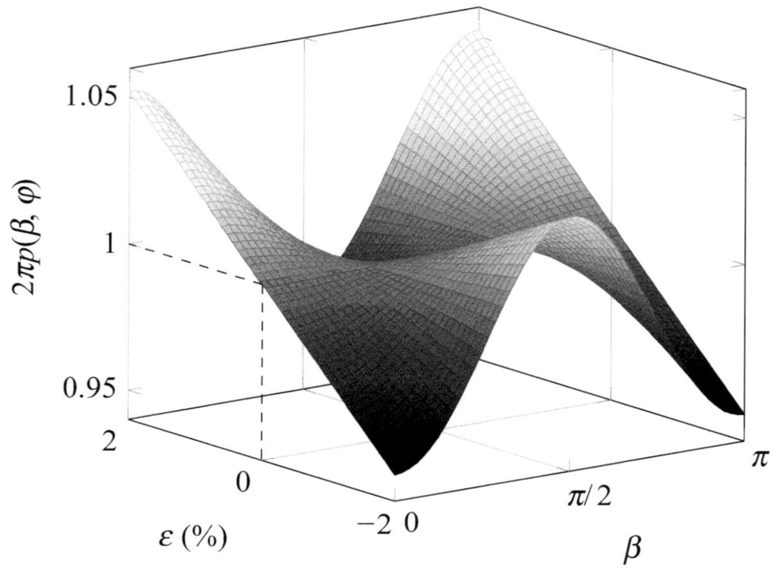

Figure 8.17 Variation of $p(\beta, \gamma)$ (ODF) with strain ε and polar angle β ($\nu = 0.3$).

the variation of the percolation threshold with respect to the strain. It can be extracted from this figure that due to the reorientation of the fillers caused by the strain, some conductive paths disappear and consequently the percolation threshold increases. In both compression and tension, the loss of randomness of the nanofillers' distribution leads to lesser probability of forming conductive paths and, thus, higher percolation thresholds.

8.3.4.3 Change in the tunneling resistance

The third mechanism originating the strain-sensing of CNT-reinforced composites refers to variation of the interparticle properties. It is assumed that the length of the nanotubes remains approximately unaltered, while a larger deformation is assumed by the internanotube matrix due to its low modulus. It has been reported in the literature that at relatively low strains ($<10^{-4}$), the interparticle distance and the height of the potential barrier change

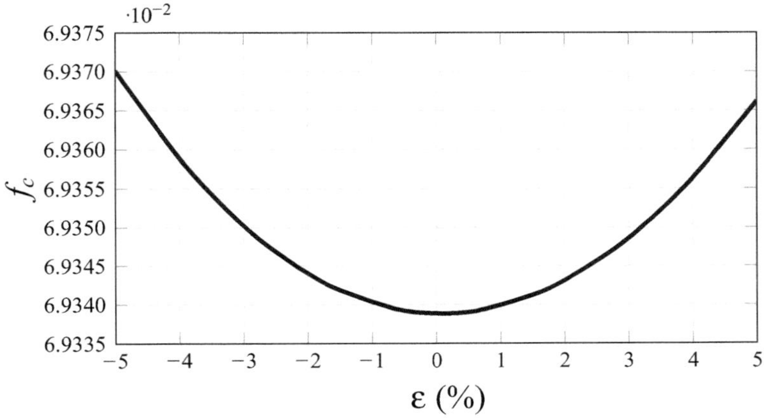

Figure 8.18 Variation of the percolation threshold f_c with respect to the strain level ε ($L = 6$ μm; $D = 10$ nm; $\nu = 0.3$).

approximately linearly with the strain [49] as follows:

$$d_a = d_{a,0}(1 + C_1\varepsilon), \quad \lambda = \lambda_0(1 + C_2\varepsilon), \tag{8.75}$$

where $d_{a,0}$ and λ_0 are the initial interparticle distance and potential height at zero strain, respectively, and C_1 and C_2 are constants. The difficulty in obtaining the constants C_1 and C_2 forces their values to be obtained by fitting some experimental data.

To illustrate the capacity of the presented micromechanics approach for the modeling of the piezoresistive properties of CNT-reinforced composites, Fig. 8.19 shows the relative resistance change $\Delta R/R_o$ under strain ε of CNT-reinforced cement paste for different values of filler conductivity. Two filler concentrations are chosen, namely $c_f = 0.695\%$ (a) and $c_f = 2\%$ (b), being two concentrations near and far from the percolation threshold, respectively. It is extracted from these figures that more conductive fillers lead to higher strain sensitivities. Another noticeable aspect is that the nonlinearities, related to the coupled effect of the volume expansion and variation of the percolation threshold, gain importance for concentrations close to the unloaded percolation threshold, as it is the case in Fig. 8.19a. On the contrary, the strain-

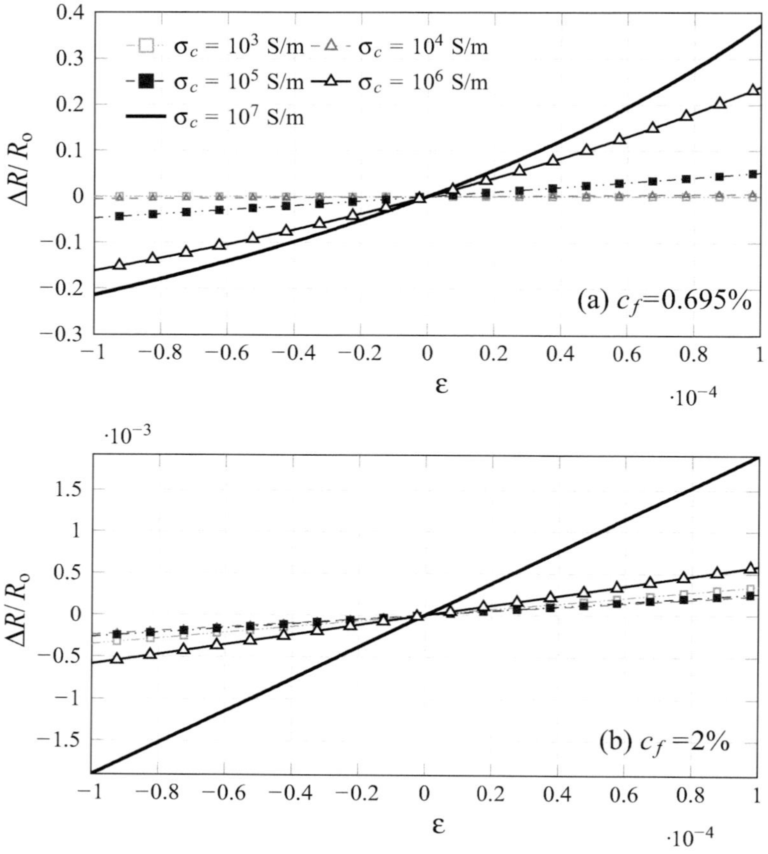

Figure 8.19 Strain-induced relative resistance change $\Delta R/R_o$ of MWCNT-reinforced cement paste for different filler conductivities κ_c and volume fraction $c_f = 0.695\%$ (a), and $c_f = 2\%$ (b) ($\nu = 0.2$; $L = 1$ µm; $D = 10$ nm; $h = 0.36$ eV; $d_c = 0.5$ nm; $C_1 = 8.9045$; $C_2 = 0.0243$).

sensing curves exhibit more linear behaviors for concentrations far from this critical concentration.

Further, an essential factor that characterizes the sensitivity of CNT-reinforced composites is the so-called gauge factor λ that relates the relative resistance change for a given strain level. In particular, two different factors, λ^- and λ^+, can be extracted from the slope of the sensitivity curves for compressive ($\varepsilon < 0$) and tensile

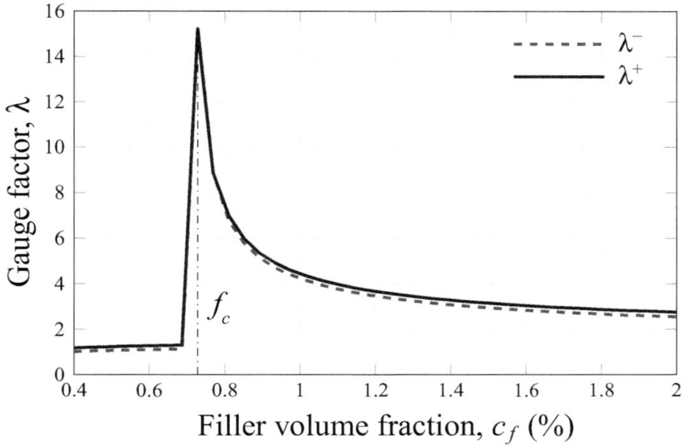

Figure 8.20 Gauge factors of MWCNT-reinforced cement paste versus filler concentration c_f under compressive λ^- and tensile λ^+ strains ($\nu = 0.2$; $L = 1$ μm; $D = 10$ nm; $h = 0.36$ eV; $d_c = 0.5$ nm; $\kappa_c = 10^5$ S/m; $C_1 = 8.9045$; $C_2 = 0.0243$).

($\varepsilon > 0$) strains, respectively, as follows:

$$\frac{\Delta R}{R_o} = \lambda^- \varepsilon, \quad \varepsilon < 0; \quad \frac{\Delta R}{R_o} = \lambda^+ \varepsilon, \quad \varepsilon > 0 \tag{8.76}$$

The variation of the gauge factor as a function of the filler concentration is plotted in Fig. 8.20. Note that the gauge factor is slightly higher in the case of tensile stresses. Moreover, it is also important to note that maximum values of both gauge factors, λ^- and λ^+, are found around the percolation threshold f_c. The sensitivity of the composites rapidly decreases as the filler content moves away from this critical threshold, especially for filler contents below the percolation threshold.

8.4 Summary

In this chapter we introduced the micromechanics modeling of nanomodified composites, in particular for CNT-reinforced composites. First, after a concise overview of the fundamentals of mean-field homogenization theory, the main approaches on the mechanical homogenization of CNT-reinforced composites were introduced. We discussed the suitability of mean-field homogenization theory for incorporating physically meaningful variables of the microstructure, such as filler topology, filler distribution, waviness, and agglomeration effects.

Second, we used the basis of mechanical homogenization for modeling of the electric transport properties of CNT-reinforced composites. By virtue of the analytical framework of mean-field homogenization theory, the electrical mechanisms governing the electrical conductivity of these composites were incorporated, including electron hopping and conductive networking. Afterward, the piezoresistive properties of the composites were analyzed by characterizing the strain-induced variations in the microstructure. Finally, a discussion of the strain-sensing capabilities of CNT-reinforced composites was made by analysis of strain sensitivity, as well as characterization of gauge factors.

References

1. Yakobson, B. I., Avouris, P. (2001). Mechanical properties of carbon nanotubes, in *Carbon Nanotubes*, Springer, pp. 287–327.
2. Demczyk, B., Wang, Y., Cumings, J., Hetman, M., Han, W., Zettl, A., Ritchie, R. (2002). Direct mechanical measurement of the tensile strength and elastic modulus of multiwalled carbon nanotubes, *Mater. Sci. Eng. A*, **334**, pp. 173–178.
3. Ebbesen, T., Lezec, H., Hiura, H., Bennett, J., Ghaemi, H., Thio, T. (1996). Electrical conductivity of individual carbon nanotubes, *Nature* **382**, pp. 54–56.
4. Galao, O., Baeza, F., Zornoza, E., Garcés, P. (2014). Strain and damage sensing properties on multifunctional cement composites with CNF admixture, *Cem. Concr. Compos.*, **46**, pp. 90–98.
5. Han, B., Han, B., Ou, J. (2009). Experimental study on use of nickel powder-filled Portland cement-based composite for fabrication of piezoresistive sensors with high sensitivity, *Sens. Actuators, A*, **149**, pp. 51–55.
6. Ubertini, F., Materazzi, A. L., D'Alessandro, A., Laflamme, S. (2014). Natural frequencies identification of a reinforced concrete beam using

carbon nanotube cement-based sensors, *Eng. Struct.*, **60**, pp. 265–275.

7. Ferreira, A. D. B., Nóvoa, P. R., Marques, A. T. (2016). Multifunctional material systems: a state-of-the-art review, *Compos. Struct.*, **151**, pp. 3–35.
8. Eshelby, J. D. (1957). The determination of the elastic field of an ellipsoidal inclusion, and related problems, *Proc. R. Soc. London, Ser. A*, **241**, pp. 376–396.
9. Mura, T. (1987). *Micromechanics of Defects in Solids*, Vol. 3, Springer Science & Business Media.
10. Mori, T., Tanaka, K. (1973). Average stress in matrix and average elastic energy of materials with misfitting inclusions, *Acta Metall.*, **21**, pp. 571–574.
11. Benveniste, Y. (1987). A new approach to the application of Mori-Tanaka's theory in composite materials, *Mech. Mater.*, **6**, pp. 147–157.
12. Hill, R. (1965). A self-consistent mechanics of composite materials, *J. Mech. Phys. Solids*, **13**, pp. 213–222.
13. Qiu, Y., Weng, G. (1990). On the application of Mori-Tanaka's theory involving transversely isotropic spheroidal inclusions, *Int. J. Eng. Sci.*, **28**, pp. 1121–1137.
14. Ferrari, M. (1991). Asymmetry and the high concentration limit of the Mori-Tanaka effective medium theory, *Mech. Mater.*, **11**, pp. 251–256.
15. Dvorak, G. J., Bahei-El-Din, A. (1997). Inelastic composite materials: transformation analysis and experiments, in *Continuum Micromechanics*, Springer Nature, pp. 1–59.
16. Ferrari, M. (1994). Composite homogenization via the equivalent poly-inclusion approach, *Compos. Eng.*, **4**, pp. 37–45.
17. Dunn, M. L., Ledbetter, H., Heyliger, P. R., Choi, C. S. (1996). Elastic constants of textured short-fiber composites, *J. Mech. Phys. Solids*, **44**, pp. 1509–1541.
18. Schjødt-Thomsen, J., Pyrz, R. (2001). The Mori-Tanaka stiffness tensor: diagonal symmetry, complex fibre orientations and non-dilute volume fractions, *Mech. Mater.*, **33**, pp. 531–544.
19. Vigolo, B. (2000). Macroscopic fibers and ribbons of oriented carbon nanotubes, *Science*, **290**, pp. 1331–1334.
20. Poelma, R. H., Fan, X., Hu, Z. Y., Van Tendeloo, G., van Zeijl, H. W., Zhang, G. Q. (2016). Effects of nanostructure and coating on the mechanics of carbon nanotube arrays, *Adv. Funct. Mater.*, **26**, pp. 1233–1242.

21. Hsiao, H., Daniel, I. (1996). Elastic properties of composites with fiber waviness, *Composites Part A*, **27**, pp. 931–941.
22. Shi, D., Feng, X., Huang, Y. Y., Hwang, K., Gao, H. (2004). The effect of nanotubewaviness and agglomeration on the elastic property of carbon nanotube-reinforced composites, *J. Eng. Mater. Technol.*, **126**, pp. 250–257.
23. García-Macías, E., D'Alessandro, A., Castro-Triguero, R., Pérez-Mira, D., Ubertini, F. (2017). Micromechanics modeling of the electrical conductivity of carbon nanotube cement-matrix composites, *Composites Part B*, **108**, pp. 451–469.
24. Dong, S., Zhou, J., Liu, H., Qi, D. (2015). Computational prediction of waviness and orientation effects in carbon nanotube reinforced metal matrix composites, *Comput. Mater. Sci.*, **101**, pp. 8–15.
25. Yanase, K., Moriyama, S., Ju, J. W. (2013). Effects of CNT waviness on the effective elastic responses of CNT–reinforced polymer composites, *Acta Mech.*, **224**, pp. 1351–1364.
26. Wernik, J. M., Meguid, S. A. (2010). Recent developments in multifunctional nanocomposites using carbon nanotubes, *Appl. Mech. Rev.*, **63**, p. 050801.
27. Li, H., Xiao, H., Ou, J. (2006). Effect of compressive strain on electrical resistivity of carbon black-filled cement-based composites, *Cem. Concr. Compos.*, **28**, pp. 824–828.
28. Chang, L., Friedrich, K., Ye, L., Toro, P. (2009). Evaluation and visualization of the percolating networks in multi-wall carbon nanotube/epoxy composites, *J. Mater. Sci.*, **44**, pp. 4003–4012.
29. Xu, J., Zhong, W., Yao, W. (2010). Modeling of conductivity in carbon fiber-reinforced cement-based composite, *J. Mater. Sci.*, **45**, pp. 3538–3546.
30. Takeda, T., Shindo, Y., Kuronuma, Y., Narita, F. (2011). Modeling and characterization of the electrical conductivity of carbon nanotube-based polymer composites, *Polymer*, **52**, pp. 3852–3856.
31. Feng, C., Jiang, L. (2013). Micromechanics modeling of the electrical conductivity of carbon nanotube (CNT)–polymer nanocomposites, *Composites Part A*, **47**, pp. 143–149.
32. Simmons, J. G. (1963). Generalized formula for the electric tunnel effect between similar electrodes separated by a thin insulating film, *J. Appl. Phys.*, **34**, pp. 1793–1803.

33. Li, C., Thostenson, E. T., Chou, T. W. (2007). Dominant role of tunneling resistance in the electrical conductivity of carbon nanotube-based composites, *Appl. Phys. Lett.*, **91**, p. 223114.
34. Wen, S., Chung, D. (2001). Effect of carbon fiber grade on the electrical behavior of carbon fiber reinforced cement, *Carbon*, **39**, pp. 369–373.
35. Yan, K., Xue, Q., Zheng, Q., Hao, L. (2007). The interface effect of the effective electrical conductivity of carbon nanotube composites, *Nanotechnology*, **18**, p. 255705.
36. Komori, T., Makishima, K. (1977). Numbers of fiber-to-fiber contacts in general fiber assemblies, *Text. Res. J.*, **47**, pp. 13–17.
37. Pan, N. (1993). A modified analysis of the microstructural characteristics of general fiber assemblies, *Text. Res. J.*, **63**, pp. 336–345.
38. Shim, W., Kwon, Y., Jeon, S., Yu, W. (2015). Optimally conductive networks in randomly dispersed CNT: graphene hybrids, *Sci. Rep.*, **5**, p. 16568.
39. Kumar, V., Rawal, A. (2016). Tuning the electrical percolation threshold of polymer nanocomposites with rod-like nanofillers, *Polymer*, **97**, pp. 295–299.
40. Toll, S. (1998). Packing mechanics of fiber reinforcements, *Polym. Eng. Sci.*, **38**, pp. 1337–1350.
41. Zheng, X., Forest, M. G., Vaia, R., Arlen, M., Zhou, R. (2007). A strategy for dimensional percolation in sheared nanorod dispersions, *Adv. Mater.*, **19**, pp. 4038–4043.
42. Deng, F., Zheng, Q. S. (2008). An analytical model of effective electrical conductivity of carbon nanotube composites, *Appl. Phys. Lett.*, **92**, p. 071902.
43. Taya, M. (2005). Electronic composites: modeling, characterization, processing, and MEMS applications, Cambridge University Press.
44. Seidel, G. D., Lagoudas, D. C. (2009). A micromechanics model for the electrical conductivity of nanotube-polymer nanocomposites, *J. Compos. Mater.*, **43**, pp. 917–941.
45. Feng, C. (2014). Micromechanics modeling of the electrical conductivity of carbon nanotude (CNT)-polymer nanocomposites, Ph.D. thesis, University of Western Ontario.
46. Feng, C., Jiang, L. (2014). Investigation of uniaxial stretching effects on the electrical conductivity of CNT-polymer nanocomposites, *J. Phys. D: Appl. Phys.*, **47**, p. 405103.

47. Hu, N., Fukunaga, H., Atobe, S., Liu, Y., Li, J. (2011). Piezoresistive strain sensors made from carbon nanotubes based polymer nanocomposites, *Sensors*, **11**, pp. 10691–10723.
48. García-Macías, E., D'Alessandro, A., Castro-Triguero, R., Pérez-Mira, D., Ubertini, F. (2017). Micromechanics modeling of the uniaxial strain-sensing property of carbon nanotube cement-matrix composites for SHM applications, *Compos. Struct.*, **163**, pp. 195–215.
49. Sobha, A., Narayanankutty, S. K. (2015). Improved strain sensing property of functionalised multiwalled carbon nanotube/polyaniline composites in TPU matrix, *Sens. Actuators, A*, **233**, pp. 98–107.

Chapter 9

Use of Carbon Cement–Based Sensors for Dynamic Monitoring of Structures

Andrea Meoni, Antonella D'Alessandro, Filippo Ubertini, and A. L. Materazzi

Department of Civil and Environmental Engineering, University of Perugia, Via G. Duranti 93, 06125 Perugia, Italy

andrea.meoni@unipg.it

9.1 Introduction

Carbon cement–based sensors represent a promising technology for the structural health monitoring (SHM) of concrete buildings. The novel technology is capable to overcome the known problems related to the use of off-the-shelf sensing technologies such as strain gauges, accelerometers, optical sensors, etc., which find limitations in their small size in comparison to the structure being monitored, durability against environmental actions, cost, and difficulty in applying and maintenance. Nanomodified cementitious materials can be used to manufacture sensors to be embedded at critical locations of reinforced concrete (RC) frames or to produce entire structural elements, enabling in both cases automated SHM.

Nanotechnology in Cement-Based Construction
Edited by Antonella D'Alessandro, Annibale Luigi Materazzi, and Filippo Ubertini

ISBN 978-981-4800-76-1 (Hardcover), 978-0-429-32849-7 (eBook)
www.jennystanford.com

The chapter is organized as follows. Section 9.2 reports a summary about the state of the art of nanomodified structures. Section 9.3 describes an example focused on the characterization of electrical and electromechanical properties of nanomodified sensors. In Section 9.4, smart cement paste sensors were used for the dynamic strain monitoring of a full-scale structural element. Section 9.5 describes results obtained performing tests on small-scale structural elements entirely made with nanomodified cementitious material, subjected to different load cases. Sections 9.6 and 9.7 conclude the work with remarks and comments.

9.2 State of the Art of Nanomodified Structures

Carbon nanotubes (CNTs), in particular multiwalled carbon nanotubes (MWCNTs), thanks to their particular aspect ratio and excellent electrical and mechanical properties, are the most promising conductive nanoinclusions for cementitious material [1–7].

One important field of research regarding the developing of this nanotechnology is represented by the study of the dispersion of nanoinclusions into cement-based materials, as concrete or cement paste [8, 9], and of the fabrication process of the composite materials [10–12]. The addition of CNTs to the base material enhances its electrical and piezoresistive properties, so a nanomodified cement-based sensor can be defined as a strain-sensing sensor where measurable changes in its electrical resistance, under static or dynamic compressive load, can be associated to the applied strain [13–21]. This attractive capability has encouraged the embedding of nanocomposite sensors inside structural elements, allowing their SHM [22, 24]. Moreover, an interesting research task with regard to the possibility of use cementitious materials doped with CNTs to detect, localize, and quantify damage [25–28] occurred at critical structural components, transforming the monitored structure into a self-sensing system enabling real-time condition assessment, reducing overall maintenance expenditures, and increasing safety.

9.3 Cement-Based Sensors for Structural Health Monitoring

The section reports the results of an investigation presented in Ref. [29], where a set of MWCNT cement-based sensors were manufactured and tested in the laboratory in order to analyze the repeatability of electrical outputs and of strain-sensing capabilities of composite nanomodified specimens using two different measurement methods.

Ten cube specimens of 5 cm sides, characterized by a cement paste matrix, were doped with 1% of MWCNTs type Arkema Graphistrength C100 [30], with respect to the weight of cement (Fig. 9.1). The dispersion of the nanoinclusions (Fig. 9.2) was achieved by performing a preliminary mechanical mixing, with a physical dispersant, followed by a sonication procedure. The obtained stable water suspension was mixed with the cement matrix and a plasticizer. The used cement was 42.5 pozzolanic, with a water/cement ratio of 0.45. Specimens were poured into oiled molds

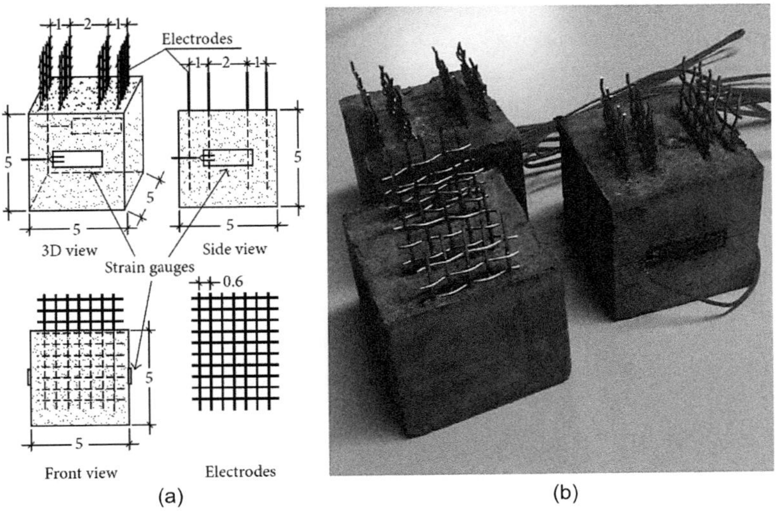

Figure 9.1 Geometry of the samples and electrodes (dimension in cm) (a) and picture of the fabricated samples with strain gauges installed onto lateral surfaces (b) [29] (open access).

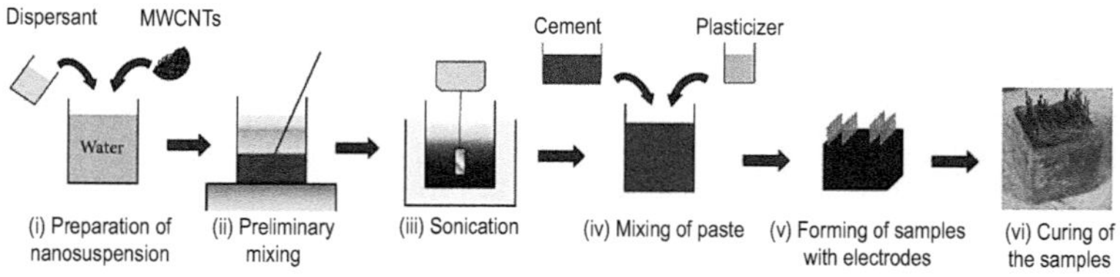

Figure 9.2 Preparation process of the cement paste samples with 1% of MWCNTs [29] (open access).

and four stainless steel meshes, used as electrodes, were embedded into each sample before solidification. The internal meshes were placed at a mutual distance of 2 cm, while the outer ones were at a distance of 4 cm. Samples were unmolded, after solidification, for a curing phase of 28 days at laboratory conditions. Two electrical strain gauges, 2 cm long, with a gauge factor of 2.1, were placed onto the center of the opposite lateral sides of each specimen.

Electrical tests in DC current were performed with both two-probe and four-probe setups in order to investigate the electrical properties of the cement-based sensors, using a NI PXIe-1073, a data acquisition system (DAQ) composed of a chassis with dedicated modules. Electric power was supplied through an NI PXI-4130 module capable of providing a four-quadrant ±20V and ±2A output in a single isolated channel, while electrical measurements were conducted using a high-speed digital multimeter, type NI PXI-4071, which acquired current in the two-probe measurement setup and voltage in the four-probe measurement setup. To reduce the signal drift due to the dielectric nature of the cementitious matrix, a polarization of 6000 s was performed. The electrical resistance was computed at the end of this phase using the first Ohm's law adapted at both measurement methods. The following equations refer to two-probe (1.0) and four-probe (2.0) methods:

$$R_{2\mathrm{P}}(t)\,|_{t=t_\mathrm{p}} = \frac{V}{I(t)\,|_{t=t_\mathrm{p}}}, \tag{9.1}$$

$$R_{4\mathrm{P}}(t)\,|_{t=t_\mathrm{p}} = \frac{V(t)\,|_{t=t_\mathrm{p}}}{I}, \tag{9.2}$$

where V and I are the applied constant voltage and current, respectively; $V(t)$ ad $I(t)$ are the measured voltage variations and current intensity over time, respectively; and t_p is the polarization time.

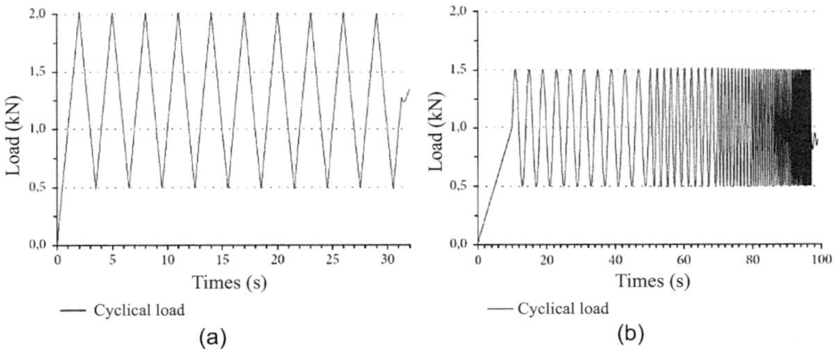

Figure 9.3 Applied cyclical (a) and dynamic (b) uniaxial loads for electromechanical tests [29] (open access).

Using the two-probe method, where the inner stainless steel meshes were used as active electrodes, the specimens were subjected to an electrical voltage of 5 V with a current measurement range of 1.0 mA. Otherwise, using the four-probe method, samples were subjected to 15 mA with a voltage measurement range of 10 V, where the current was applied to the external stainless steel meshes and the voltage drop was measured at the internal electrodes.

Electromechanical tests were performed to evaluate the strain-sensing capability of each specimen. Electrical measurements were performed with the setup used during electrical tests. An additional module PXIe-4330, with 8 channels, 24-bit resolution, 25 kHz maximum sampling rate, and anti-aliasing filters, was added to the chassis NI PXIe-1073 to acquire data from strain gauges. Two types of compression loads were applied on each sample: a cyclic load, varying from 0.5 to 2.0 kN with 1 kN/s constant speed, and a dynamic load, varying from 0.5 to 1.5 kN characterized by an increasing application frequency from 0.25 to 8.0 Hz (Fig. 9.3). Loads were applied using a servocontrolled pneumatic universal dynamic testing machine, model IPC Global UTM 14P, with a controlled temperature chamber (Fig. 9.4). A polarization phase of 10 min was performed before the application of each load history. A voltage equal to 5 V with a current measurement range of 1.0 mA was applied in the two-probe configuration, while a current of 15 mA was applied with a voltage measurement range of 10 V using the

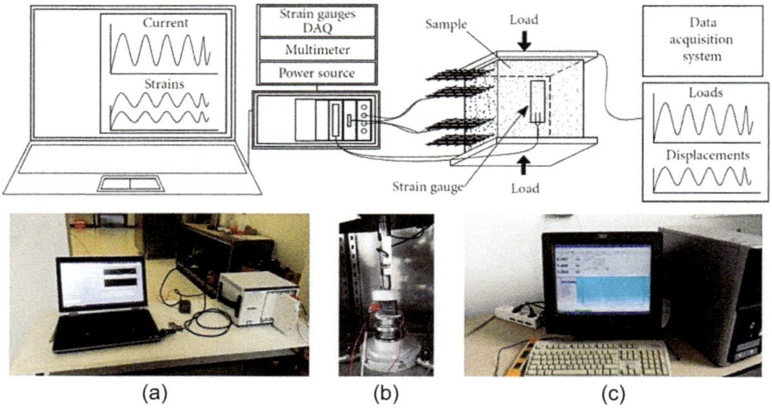

Figure 9.4 Sketch and pictures of the setup for the electromechanical tests on cement-based samples with carbon nanotubes: power source and device for acquisition of strain gauges and electrical measurements (a), a sample during test (b), and data acquisition system of the hydraulic actuator (c) [29] (open access).

four-probe method. The sampling rate was 1000 Hz. Electrical resistance was computed considering the first Ohm's law, while the gauge factor of each sensor was computed applying the following relation:

$$\frac{\Delta R}{R_0} = -\lambda\varepsilon, \tag{9.3}$$

where ΔR is the variation in electrical resistance, R_0 is the value of the resistance without applied load, λ is the gauge factor, and ε is the axial strain, positive in compression. Both electrical and electromechanical tests were carried out under a controlled temperature of 20°C.

Figure 9.5 presents results obtained from electrical tests. The average value of the electrical resistance obtained using the two-probe method is 609 Ω, while the four-probe approach has provided a lower average value of 215 Ω due to the electrical configuration, which is able to remove the electrical contact resistance from the measured value [31].

The corresponding coefficients of variation, 0.16 and 0.24, demonstrate that the repeatability of the sensors' electrical resistance is satisfying, especially considering the two-probe method for

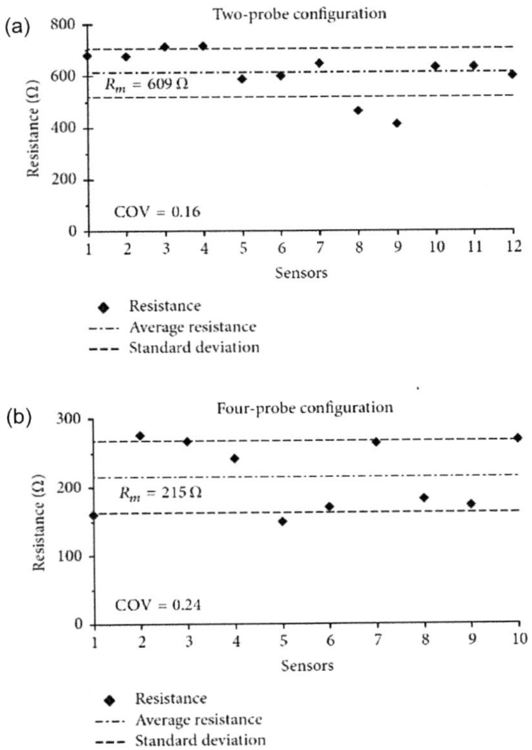

Figure 9.5 Results of the electrical measurements conducted on all of the carbon nanotube cement-based sensors (COV denotes the coefficient of variation) using two-probe (a) and four-probe (b) methods [29] (open access).

which the obtained coefficient of variation is very similar to the coefficient of variation of other properties of cementitious materials, such as their compressive strength [32]. Scattered values of the electrical resistance have been obtained owing to the non-uniform dispersion of the nanofillers during the manufacturing process.

Figures 9.6 and 9.7 show examples of time histories of the average strain, $\varepsilon(t)$, measured using strain gauges and of the normalized change in electrical resistance, ΔR, obtained from electromechanical tests performed on a sensor for both static and dynamic loads using the four-probe method. As expected,

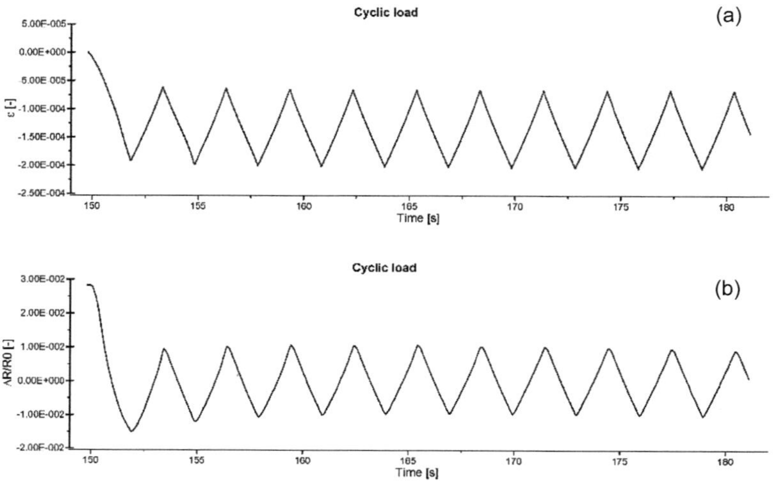

Figure 9.6 Results of electromechanical tests: measured average strain (a) and normalized variation of electrical resistance (b) on sample number 6 (typical) using the four-probe method under cyclical loads.

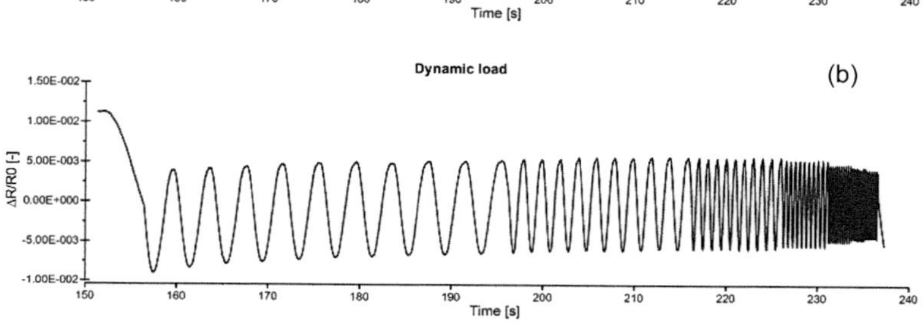

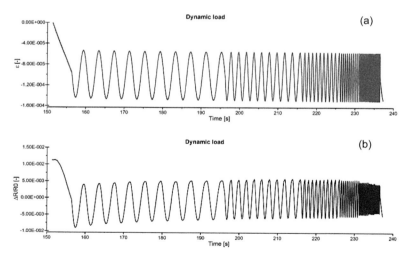

Figure 9.7 Results of electromechanical tests: measured average strain (a) and normalized variation of electrical resistance (b) on sample number 6 (typical) using the four-probe method under dynamic loads [29] (open access).

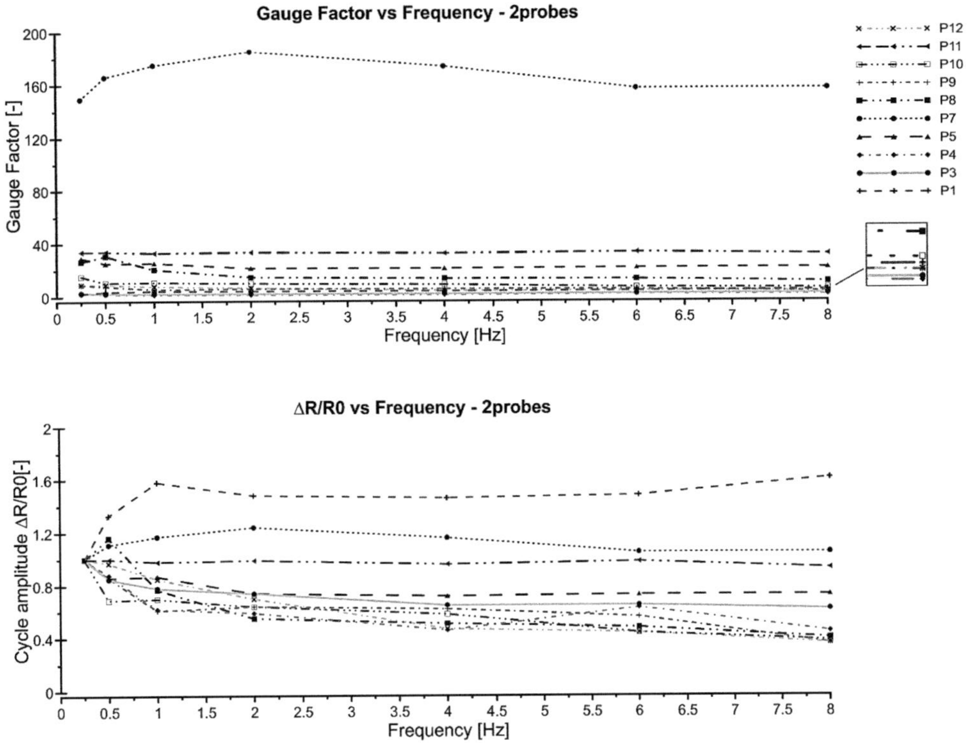

Figure 9.8 Plots of gauge factors (a) and of normalized resistance variation (b) with increasing frequency of applied load using the two-probe method.

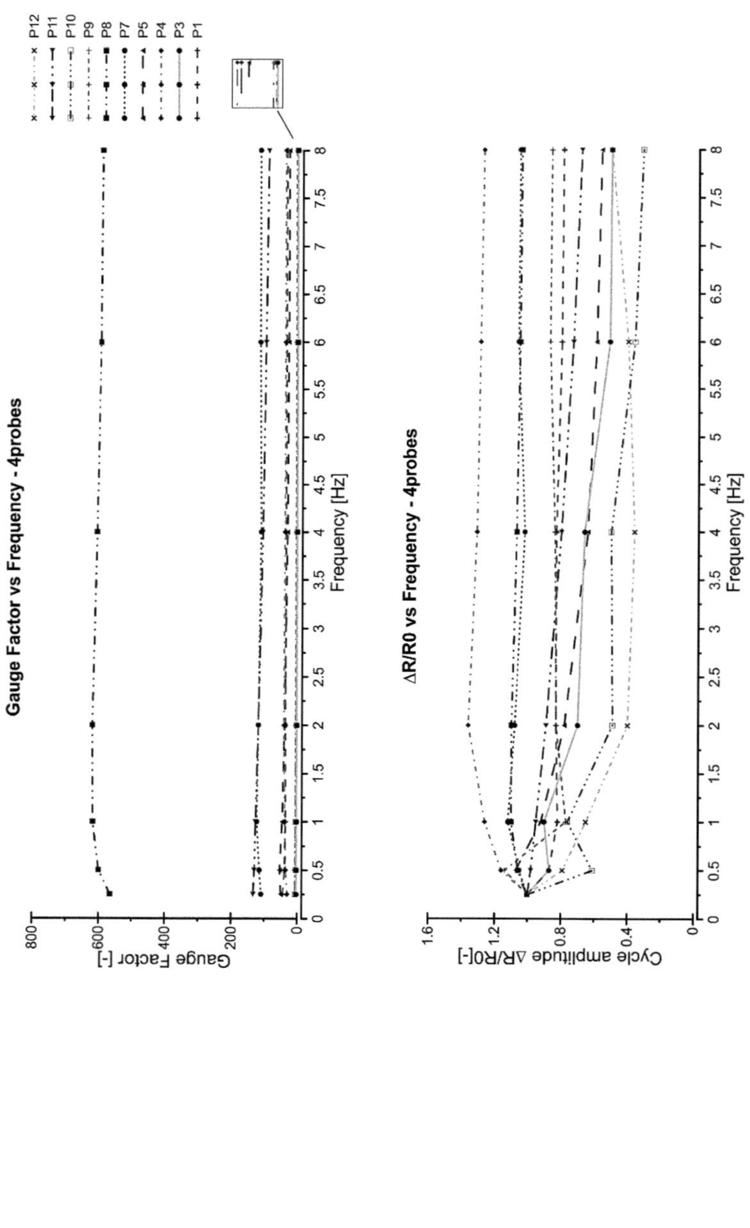

Figure 9.9 Plots of gauge factors (a) and of normalized resistance variation (b) with increasing frequency of applied load using the four-probe method.

the application of a compression load provides negative changes in electrical resistance, $\Delta R(t)$, acquired through the sensor. This behavior is due to an average reduction in the distance between nanotubes and the enhancement of the tunneling effect. As shown in the figures, this change in electrical resistance is reversible when the value of applied compression load decreases. Electrical signals are characterized by very low noise and a clear strain sensitivity can be noted in all the graphs. Figures 9.8 and 9.9 present the variations in gauge factor and normalized change in electrical resistance, varying the application load frequency, carried out for the two- and the four-probe method, respectively.

These curves are useful to assess the strain-sensing capability of the sensors at different frequencies, considering that an ideally dynamic sensor would have a horizontal frequency response function. Results show good strain-sensing capability of the sensors with more scattered values of gauge factors, which are characterized by variation coefficients of 1.72 and 1.79 computed for the two-probe and four-probe methods, respectively. This behavior is probably due to the proximity of the 1% MWCNT mix to the percolation threshold. As demonstrated in Ref. [8] the gauge factor reached the maximum value close to the percolation threshold, but it is also very sensitive to small changes in the nanofiller dispersion.

Reported results demonstrated the potential of the nanocomposite cement paste sensors for the SHM of RC buildings, thanks to their strain-sensing capability.

The scattered values of the sensors' gauge factor can be solved through the calibration of each sensor before its embedding into a structural element.

9.4 Structures with Embedded Cement-Based Sensors

To investigate the behavior of nanocomposite cement-based sensors when they are embedded in a concrete structural element, seven specimens, previously calibrated, were inserted into a full-scale RC beam during its molding.

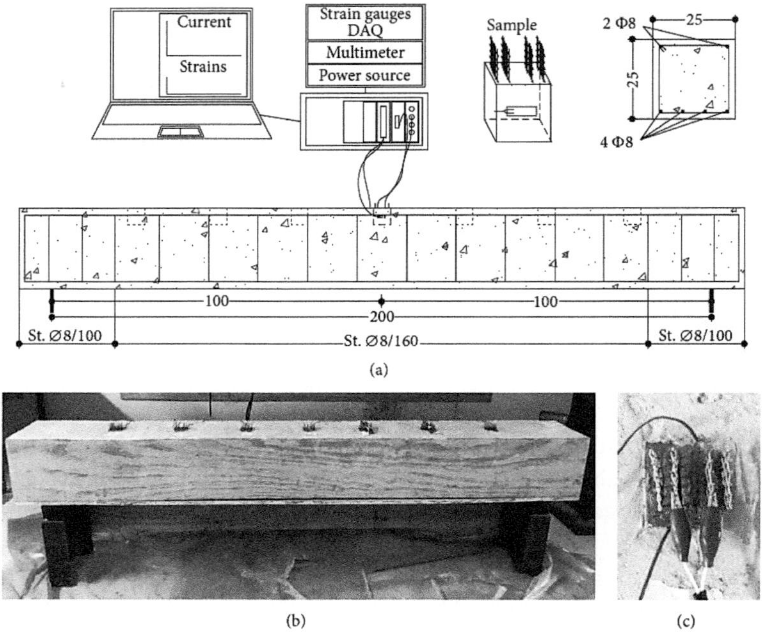

Figure 9.10 Sketch of the full-scale beam test setup (a), picture of the beam with the embedded sensors (b), and detailed view of the sensors at the midspan (c) [29] (open access).

The beam, simply supported by two steel supports installed at a distance of 200 cm, had a square cross section of 25 × 25 cm^2 and a length of 220 cm. Concrete was reinforced with 8-mm-diameter longitudinal steel rebars and stirrups.

The sensors were instrumented in order to acquire electrical output and strain gauges measurements, as shown in Fig. 9.10.

The electrical measurements were carried out using the two-probe method. Data acquisition was performed with the chassis NI PXIe-1073 characterized by the same setup used during the electromechanical tests on the sensors described in the previous section. The RC beam was involved in two different electromechanical tests, which consisted of the application of a distributed static load of 1 kN/m and in the application of random hits in time and space

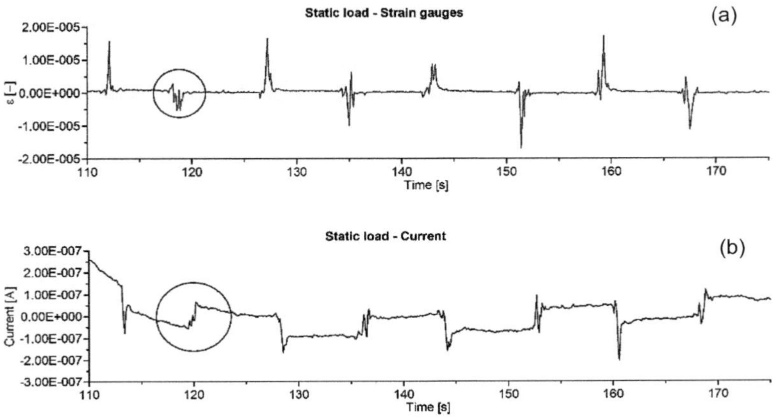

Figure 9.11 Time histories of average strain (a) and measured electrical current (b) during static loading and unloading of the full-scale RC beam [29] (open access).

using an instrumented hammer. A polarization phase of 30 min was performed before each test.

The instrumented sensor was subjected to a voltage of 2.5 V with a current measurement range of 1.0 mA. The sampling rate was 1000 Hz.

Figure 9.11 shows the results obtained from the application of the static load on the RC beam. The plots report the time histories of the average strain, measured by the strain gauges placed on the sensor embedded in the middle of the beam, and of the electrical current output provided by the sensor itself. Loading and unloading phases are clearly visible in both signals, demonstrating that the sensor detects changes in strain occurring in the beam's compression zone during the test. Figure 9.12 reports results obtained from the vibration test in which the RC beam was hit with a hammer. To eliminate the residual effect of the polarization, a high-pass filter, with a cut-off frequency of 5 Hz, was applied to the signal. Random hammer hits are clearly visible in the time history of the electrical current reported in Fig. 9.12a. Furthermore, a detailed view of a single hammer hit is shown in Fig. 9.12b in order to point out the waveform of the damped vibration of the RC beam. Afterward, an in-depth study of the strain-sensing capability

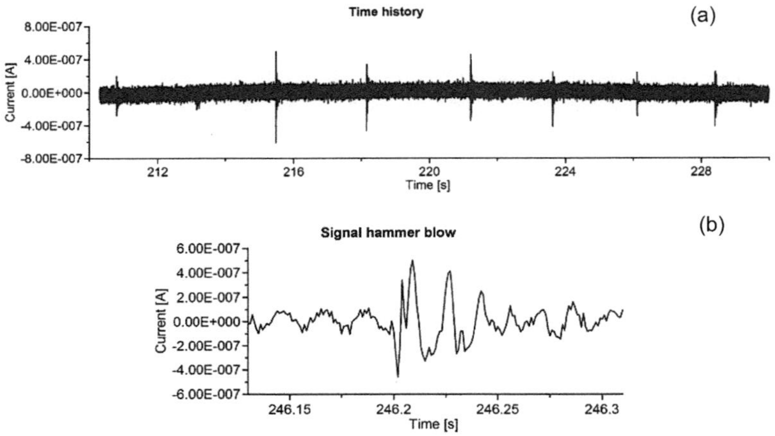

Figure 9.12 Section of the time history of the current during vibration tests (a) and detailed view of the output of a single hammer hit (b) [29] (open access).

of the sensor was performed. Figure 9.13a reports the overlapped filtered signals of the measured strain, computed considering the average value obtained from the outputs of the two strain gauges applied on the sensor and of the estimated strain, computed using Eq. 9.3 from the electrical output provided by the sensor itself.

Results reported in Fig. 9.13 show a clear correspondence between measured and computed strain, as confirmed by the signal magnifications shown in Figs. 9.13b and 9.13c. Moreover, a spectral analysis of the strain and of the electrical current outputs, respectively, provided by strain gauges and by the sensor, was performed in order to understand whether the sensors, embedded in the RC beam, can be used for output-only modal identification and vibration-based SHM of the structural element. The power spectral density (PSD) functions of the strain signal, measured by the strain gauges, and of the electrical current, obtained from the sensor, have been reported in Fig. 9.14 applying a high-pass filter with a cut-off frequency of 5 Hz. Both PSD functions are characterized by the same leading peak at a frequency of 64.45 Hz that can be associated with the first vertical vibration mode of the RC beam.

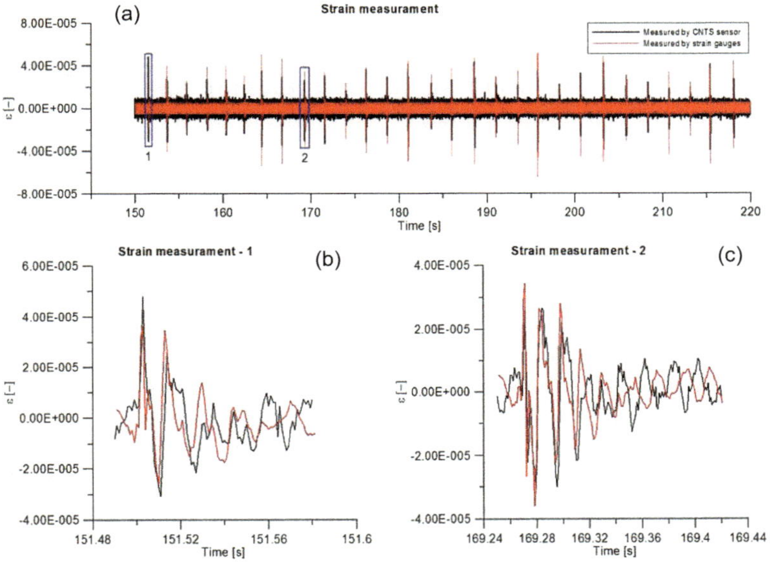

Figure 9.13 Time histories of measured average strain from strain gauges and estimated strain from electrical measurements (a). Enlarged view of measurements number 1 (b) and number 2 (c) [29] (open access).

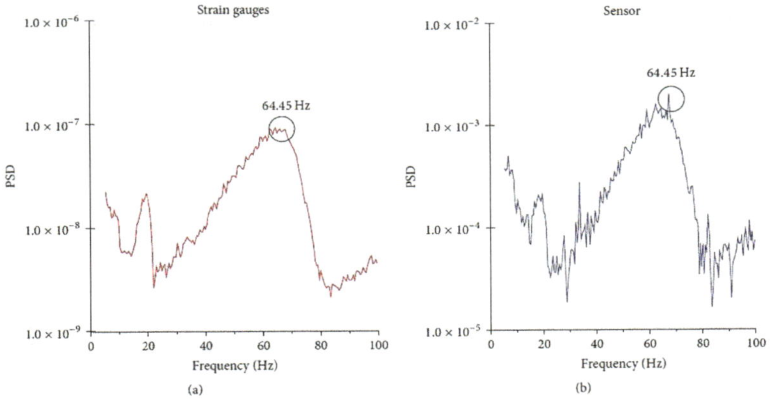

Figure 9.14 PSD of strain gauges measurements (a) and of electrical measurements (b) from the embedded sample placed at midspan [29] (open access).

Results have demonstrated that embedded MWCNTs sensors are suitable to monitor the strain conditions of a full-scale structural element subjected to a static or a dynamic load.

In addition, it has been demonstrated that embedded sensors can be used for output-only modal identification and vibration-based SHM of a structural element.

9.5 Structures Made of Nanomodified Cement-Based Materials

A very attractive solution in the field of the SHM of RC buildings concerns the possibility of manufacturing specific structural elements with nanomodified materials, which, placed at key points of the structure, are able to achieve smart diffuse monitoring.

An interesting investigation on this topic is reported in Ref. [33], where a steel-reinforced cement-based beam, made with cement paste doped with 1% of MWCNTs with respect the weight of the cement, was subjected to a four-point loading test up to failure. Furthermore, the paper presents a resistor mesh model developed in order to anticipate the electrical response of smart concrete structural elements with the purpose to allow their real-time monitoring.

Nanofillers were dispersed into deionized water with a surfactant (lignosulfonic acid sodium salt) using mechanical mixing followed by a sonication procedure. The obtained solution was mixed with type IV Portland cement. Following, the smart cement paste was cast in a prismatic mold with dimensions of $100 \times 100 \times 500$ mm^3 and 16 steel electrodes were embedded in the bottom rectangular side. The electrical tests were conducted with a DC biphasic method [34], in which a function generator was used to provide a 2 Hz square wave, ranging from -2.5 to 2.5 V, characterized by a duty cycle of 50%, to the electrodes of the specimen. This approach allows the simultaneous measurements of the voltage drop of every section of the reinforced MWCNT cement paste beam between two electrodes, eliminating also the drift effect that afflicts electrical outputs of dielectric materials, through cycles of charging and discharging. Resistance, R, was computed through the following equation:

$$R = \frac{V_{\text{sense}}}{i}, \tag{9.4}$$

where V_{sense} is the measured voltage drop and i is the flowing current between two electrodes computed considering the relation

$$i = \frac{V_{\text{drop}}}{R_{\text{in-line}}}, \tag{9.5}$$

where V_{drop} is the voltage drop across a resistor characterized by an electrical resistance of $R_{\text{in-line}}$.

The aim of the work was to propose and validate a procedure, based on a resistor mesh model, to detect, localize, and quantify cracks in self-sensing concrete structures, considering permanent local changes in electrical resistance induced by cracks. In fact,

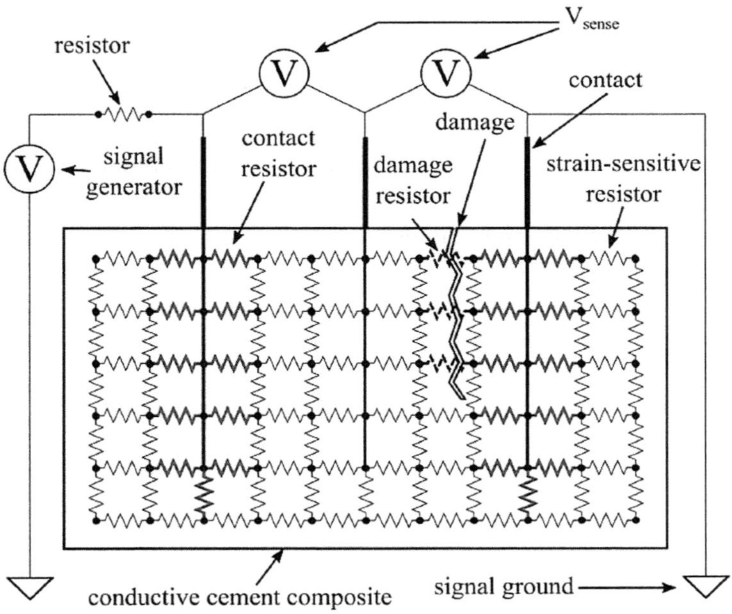

Figure 9.15 A resistor mesh model with a 10 × 5 resistor mesh for crack detection, localization, and quantification in cement composite structures with key components highlighted. Reprinted from Ref. [33], Copyright (2017), with permission from Elsevier.

when a crack opens, a greater voltage drop occurs in the element. The developed 2D resistor mesh model, fully explained in Ref. [33], is composed by strain-sensitive, contact, and damage resistors, as shown in Fig. 9.15, where an example of a resistor mesh model is illustrated. Changing in strain and damage can be detected by a resistor model through the variation of the resistive value of individual resistors. To this purpose, a finite element analysis (FEA) model of the reinforced MWCNTs cement paste beam was built to validate the behavior of the strain-sensitive resistors present in the resistor mesh model, which was solved using SPICE [35], an open source analog electronic circuit simulator. The results of the FEA model were compared with those obtained from six strain gauges placed on both the lateral surfaces of the beam.

A four-point loading test was performed applying 24 loading steps with an increment of 0.1 kN. An audible crack was heard during the application of 1.8 and 1.9 kN loads. Figure 9.16 shows the experimental setup for the reinforced MWCNT cement paste beam. A damage, localized on the intrados of the left-most monitored section and developed during the four-point loading test, has been reported in Fig. 9.16a. Minor multiple cracks, developed on the surface of the beam, were photographed to allow crack monitoring during the conduction of the test.

Figure 9.17 presents a comparison between the results obtained from experimental tests and from the analytical model for the healthy state of the beam. The experimental trend denotes a large voltage drop occurred in sections 6, 7, and 8 due to the presence of cracks and a slight cavity localized on the top of the beam.

Electrical outputs, obtained for reference sections of the beam, performing the four-point loading test are illustrated in Fig. 9.18a. In the center section 8, it is possible to note that the voltage drop increases with the increase of the applied load until reaching 1.4 kN when the value of the electric output starts to decrease. This behavior is due to the formation of several microcracks, starting at the loading step 1.4 kN, which implies a loss in the strain-sensing capability of the material. Due to the symmetric loading condition applied on the beam, each pair of sections, localized externally respect the center section, demonstrated a constant change in voltage drop, as expected. In the plot, outputs of sections 4 and 12

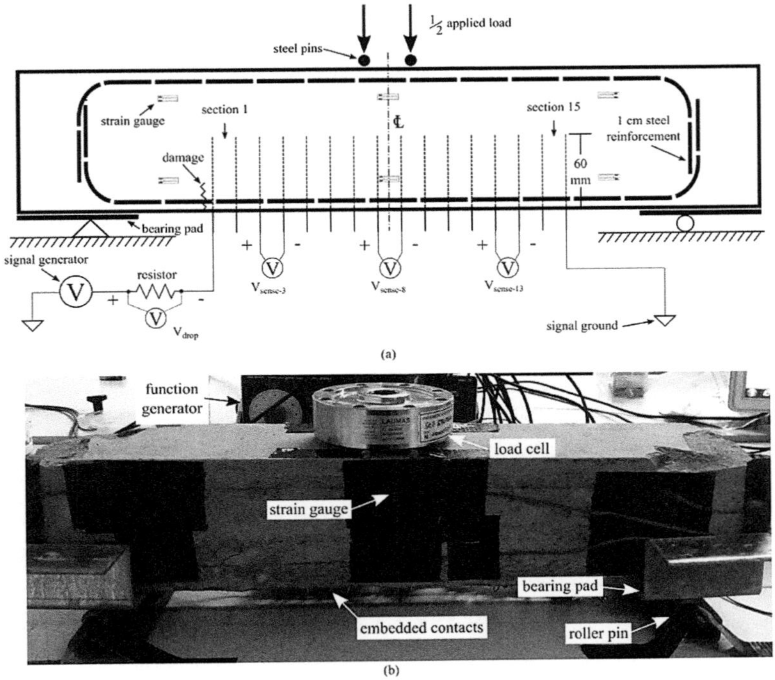

Figure 9.16 Experimental setup for the self-sensing reinforced cement paste beam: (a) annotated diagram labeling key components of the test setup and (b) experimental setup as tested (with the load-bearing clamp removed for clarity). Reprinted from Ref. [33], Copyright (2017), with permission from Elsevier.

and sections 6 and 10 confirm such behavior. The trend of the change in voltage drop outputted by the section 1 confirms the development of a crack, occurring during the loading test, at the left-adjacent section (Figs. 9.18b and 9.18d). On the other side, in section 15, the decrease of the voltage drop is a consequence of the increment that occurred in section 1. Figure 9.18c shows the strong agreement between the experimental and analytical results found in sections 1 and 15. Both results show an increment in the change of the voltage drop due to the developed of the crack at the left-most side of section 1. Damage resistors were added to the resistor mesh model in order to reproduce the observed experimental behavior. To take

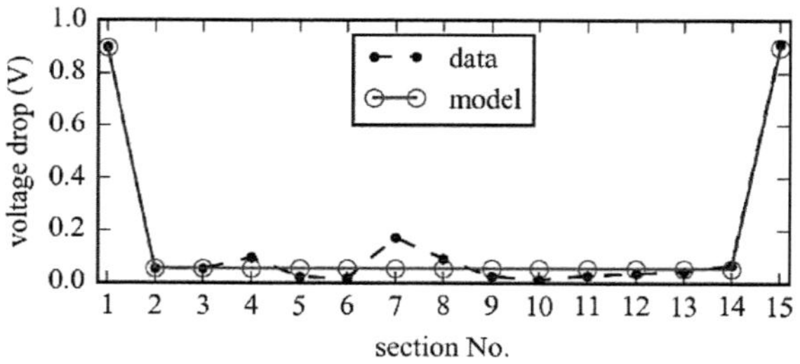

Figure 9.17 Experimental analytical model results for the healthy state of specimen III. Reprinted from Ref. [33], Copyright (2017), with permission from Elsevier.

into account the presence of a pre-existing crack of 4 mm on the outside of the left-most contact, a first resistor was added to the model. Then, another resistor was inserted to consider the 4 mm crack that occurred applying a load of 1.8 kN. A last resistor was introduced to consider the damage associated at 1.9 kN. The total simulated crack has a length of 10 mm.

Results obtained from experimental electrical measurements have demonstrated that concrete structural elements with CNTs can provide a measurable change in electrical outputs that can be related to the damage states that occur at different sections of the element. Moreover, analytical results obtained from the developed resistor mesh model, proposed and validated in Ref. [33], have demonstrated a reliable matching with the experimental results.

Another relevant example of a structural element made with nanomodified cementitious material is presented in Ref. [36]. In this work, a cement paste plate 500 × 500 × 50 mm^3, doped with 1% of MWCNTs with respect to the weight of the cement, was subjected to progressive damage developed performing repetitive impact tests. The aim of the paper was to analyze a 2D resistor mesh model developed from the resistors' system, proposed and validated in Ref. [33], adding a procedure for the automated placement of resistors using a sequential Monte Carlo method. The analytical model was validated comparing its results with those obtained from the experimental tests.

During the manufacturing of the specimen, Arkema C100 MWCNTs [30] were first mechanically mixed with water and a surfactant

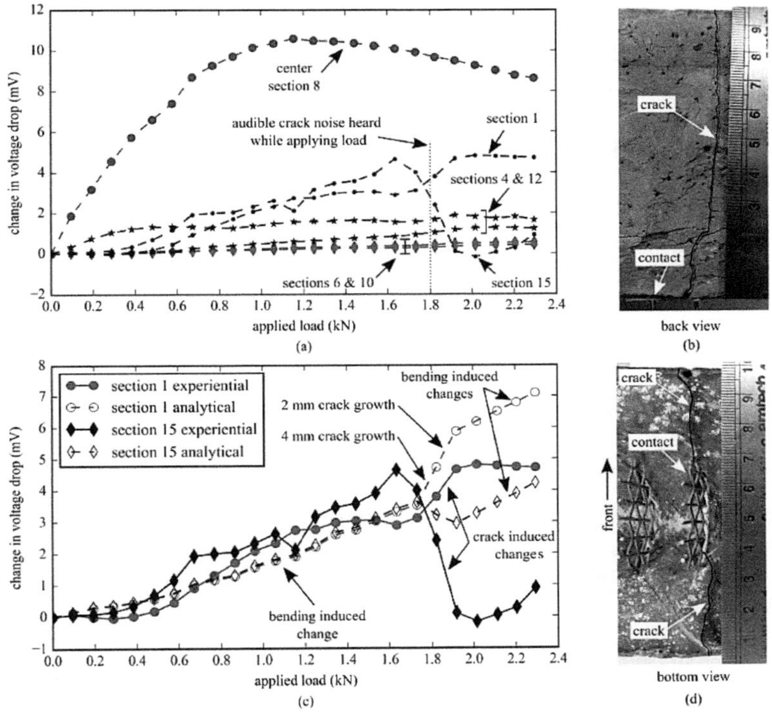

Figure 9.18 Experimental data and analytical model results for specimen III, showing: (a) measured change in voltage drop as a function of the applied load, (b) crack detected outside section 1, (c) analytical model versus experimental results for voltage drop measured at section 1, and (d) bottom view of the surface crack on the outside of section 1. Reprinted from Ref. [33], Copyright (2017), with permission from Elsevier.

and then dispersed using a sonication procedure. The obtained nanomodified suspension was mixed with cement powder to obtain a smart cement paste. The doped material was cast into a mold, adding steel nets with wires of 1.2 mm diameter, as reinforcement, arranged in a 60×60 mm^2 grid. During the curing of the plate, fifty-six 1.2 mm copper wires were embedded into the smart concrete material in a grid of 80×80 mm^2, as shown in Fig. 9.19.

Electrical measurements were carried out using a DC biphasic method [34] where a constant 20 V_{pp} square wave, with a frequency

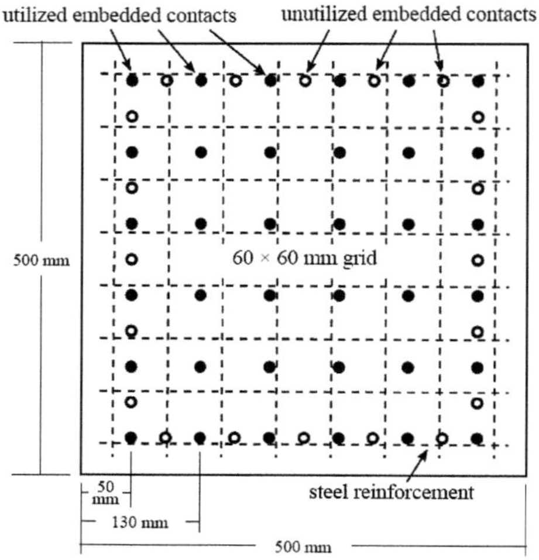

Figure 9.19 Key components and dimensions of the 500 × 500 × 50 mm^3 smart cement paste specimen used in this study. From Ref. [36].

of 1 Hz and a duty cycle of 50%, was applied by a function generator, model Rigol DG1022a, to the active copper electrodes embedded in the plate. To avoid the influence of the contact resistance between copper contacts and MWCNT cement paste, voltage measurements were directly acquired at each embedded electrode using dedicated modules, model PXIe-4302 and PXIe-6361, arranged in a chassis, type PXIe-1071. Simultaneous acquisition of the electrodes was allowed through a wire connecting them. The voltage values were taken at 80% of the constant part of the acquired voltage signal and were used as input to the resistor mesh model. Testing validation was performed acquiring the biphasic signal with an oscilloscope, model Rigol DS-1054.

Different configurations of the resistor mesh model were investigated in order to establish a suitable setup capable to mimic the conductive paths of the MWCNT cement paste plate. Figure 9.20 shows the applied model where 110 resistors were implemented to connect the electrodes of the plate.

Diagonal resistors are independent and do not form a conductive path. The procedure used to place damaged resistors in the resistor

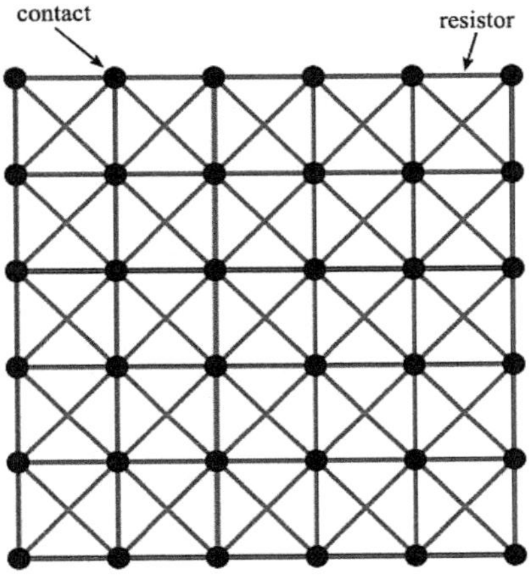

Figure 9.20 The resistor mesh model used for electrical modeling of the plate. From Ref. [36]. © IOP Publishing. Reproduced with permission. All rights reserved.

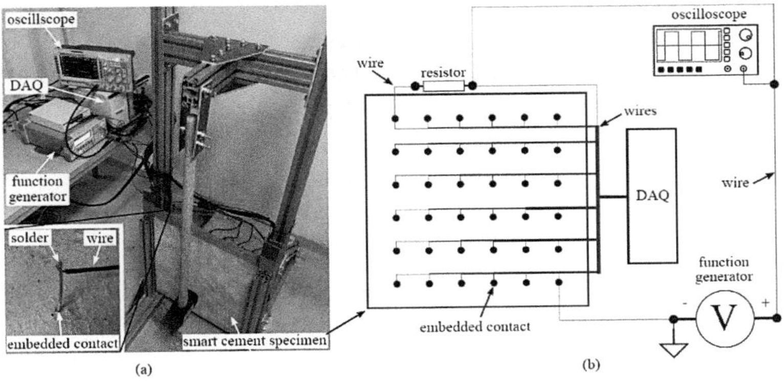

Figure 9.21 Experimental test setup showing (a) key components used in the impact testing of the smart cement paste plate with an embedded copper contact shown in the inset and (b) electrical test schematic showing the connection of the data acquisition system to the test specimen with key system components annotated. From Ref. [36]. © IOP Publishing. Reproduced with permission. All rights reserved.

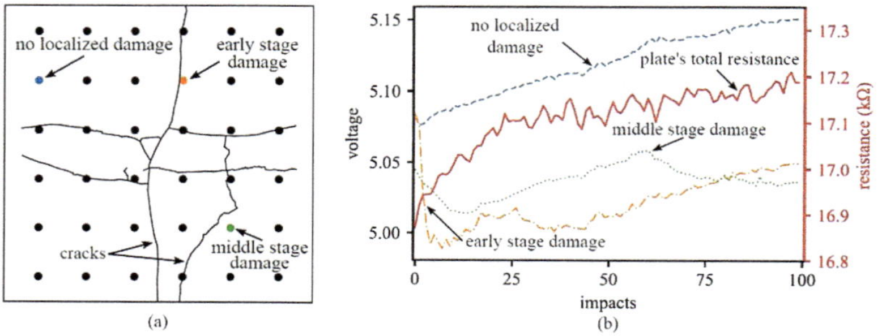

Figure 9.22 A purely data-driven damage detection analysis (no resistor mesh model) for the test plate showing (a) damage and the location of inspection points and (b) the change in the plate's total resistance and the change in voltage data taken at the three inspection points of interest, selected to show damage forming during different portions of the test. From Ref. [36]. © IOP Publishing. Reproduced with permission. All rights reserved.

mesh model is based on a bi-optimization objective function that is capable, first, to recognize points where the highest error between experimental and analytical measurements occurs, and then in a second step, to minimize the value of the mean error at the considered points. A sequential Monte Carlo algorithm is used to estimate a resistor set that best reproduces the spatial resistance distribution in the conductive nanomodified cement paste plate. More details of the used resistor mesh model and of the procedure can be found in Ref. [36].

Experimental tests were performed on an extruded aluminum frame equipped with a 3 kg hammer, as reported in Fig. 9.21, where the experimental setup is full shown. The impact of the hammer on the plate produced a 1.5 J of energy and it was carried out rotating back 20° the hammer from the initial position. The test was conducted hitting 100 times the center of the plate.

The damage state of the plate is reported in Fig. 9.22a, while Fig. 9.22b presents electrical outputs obtained from the embedded copper contacts. Plate global health was monitored assessing the total resistance value of the specimen obtained through electrical measurements with a two-probe method. The trend of the solid red line indicates that the majority of the damage occurred during the

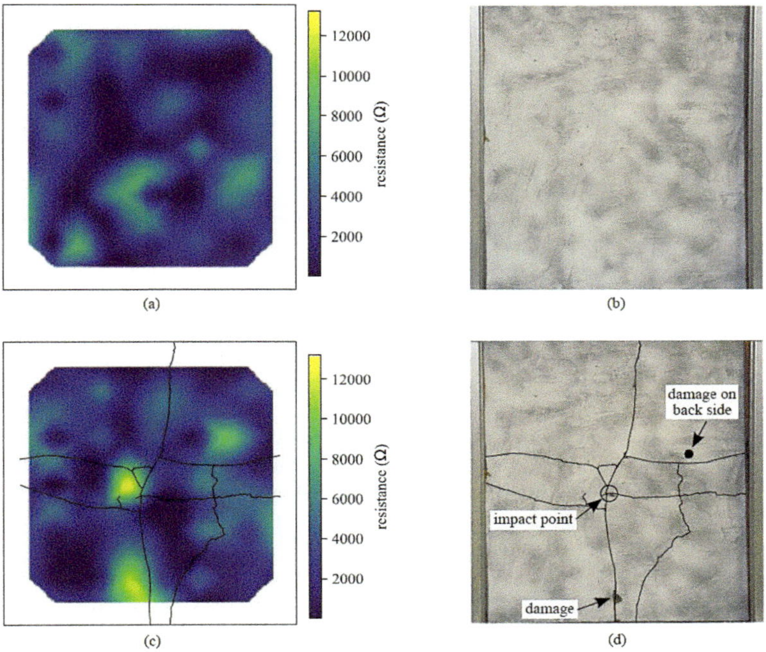

Figure 9.23 Final damage detection results showing (a) spatial distribution of high-value resistors for the healthy state, (b) healthy test specimen, (c) spatial distribution of high-value resistors for the damaged state with crack distribution overlaid to show relationship between damage and points of high resistance, and (d) damaged (100 impacts) test specimen. From Ref. [36].

first 25 hammer impacts, producing an increase in the value of the electrical resistance. Then, such a feature continued to increase with a lower rate. The damage localization was investigated monitoring the voltage at three plate electrodes placed in different positions respect to the developed crack path clearly visible in Fig. 9.22a. The dashed blue line, reported in Fig. 9.22b, corresponds to a copper contact located in an undamaged area of the plate. It can be noted that the measured voltage increases during the test probably to balance voltage drops that occur at other electrodes located in damaged areas. The orange dot-dashed line refers to the measured

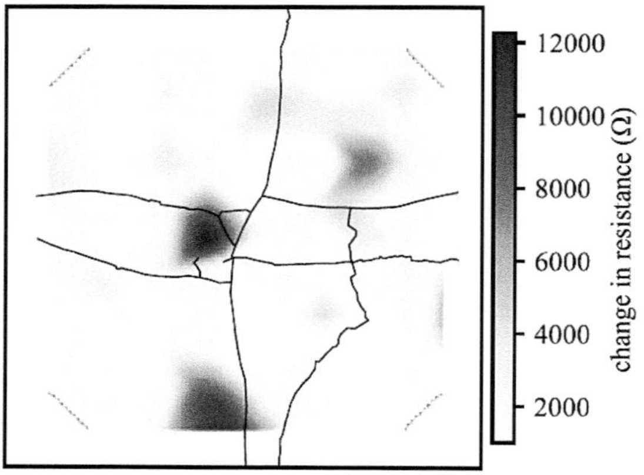

Figure 9.24 Final results for the plate showing the relative resistance change in resistance for the damaged condition when compared to the healthy condition. From Ref. [36]. © IOP Publishing. Reproduced with permission. All rights reserved.

voltage at an electrode placed where a crack visually occurs after a few hits of the hammer, as confirmed by the marked voltage drop presented at the beginning of the output. The dotted green line refers to a copper contact placed near a crack formed around hammer impact 60. Its voltage trend varies during the test in relation to the developing of the neighboring cracks.

Analytical results obtained from the resistor mesh model are presented in Fig. 9.23, compared to the healthy and damaged states of the plate.

Moreover, Fig. 9.24 reports the change in electrical resistance computed at the end of the test. The resistor mesh model, built with the sequential Monte Carlo algorithm, detects three prominent damage areas on the plate: the hammer impact point, located in the center of the specimen; a damaged area located on the bottom-center (front and back) of the plate, where a loss of material occurred after a few hammer hits; and an area above the center-right, where a piece of cement paste, on the back of the plate, was lost around hammer hit 80.

Results from experimental tests have demonstrated how a structural element made with nanomodified cement paste is able to provide electrical output useful to monitor its health state. The assessment of the plate's global health can be performed from the evaluation of the total electrical resistance, while outputs of voltage can detect and localize crack paths that occur on the structural element. Furthermore, analytical results obtained from the resistor mesh model demonstrated the possibility to enable the real-time detection, localization, and quantification of a crack path.

9.6 Comments

Several applications of carbon cement–based sensors have been presented in this chapter.

Ten cube MWCNT cement paste sensors were subjected to electrical and electromechanical tests where electrical measurements were conducted using both two-probe and four-probe methods. Results from electrical tests have demonstrated that the dispersion of MWCNTs in the cement-based material results in a key point for the repeatability and feasibility of the electrical measurement. However, even if some scattered values occur, a good repeatability of the electrical outputs has been found, especially performing measurements with the two-probe setup. Results from electromechanical tests demonstrated the strain-sensing capability of the cube sensors for both static and dynamic load cases. Scattered values of the gauge factor (GF) were due to the proximity of the filler content to the percolation threshold. This occurrence suggests the individual calibration of each sensor before its embedding in a structural element should be monitored. Results obtained from electromechanical tests have confirmed that smart sensors are capable of detecting static loads and dynamic inputs. Embedded sensors can be also used for output-only modal identification and vibration-based SHM. Experimental tests also demonstrated that cement-based nanomodified materials can be utilized as constituent materials for structural smart elements. Performed tests have demonstrated that the nanomodified material can provide a measurable change in electrical voltage that can be related to the developing of cracks in the material. Also, analytical results obtained from the model showed satisfactory agreement with those obtained from experimental tests.

Impact tests performed on a MWCNT cement paste plate produced relevant results useful to improve the interpretation of electrical outputs obtained from a smart structural element and to assess its health state. The evaluation of the electrical resistance allowed assessment of the global health state of the plate, while detecting, localizing, and quantifying the damage occurred in the structural element. An automated procedure, based on a sequential Monte Carlo algorithm, allowed the automated updating of the damage resistors in the model. The analytical results have confirmed the potential of this approach.

9.7 Conclusion

The discussed examples have demonstrated some promising applications of nanocomposite cement-based materials for civil engineering application, especially for smart SHM of concrete buildings. Nanomodified sensors can be embedded at critical positions of a concrete frame, or entire structural elements can be manufactured with such a material. Moreover, the results demonstrated that this technology can be fruitfully used to monitor changes in load and crack paths due to the application of static and dynamic loads.

References

1. Makar, J., Beaudoin, J. (2003). Carbon nanotubes and their application in the construction industry, in Bartos, P. (ed.) *Proceedings of the 1st International Symposium on Nanotechnology in Construction (NICOM 2003)*, Springer Verlag, Berlin Heidelberg, pp. 331–341.
2. Mondal, P., Shah, S. P., Marks, L. D. (2008). Nanoscale characterization of cementitious materials, *ACI Mater. J.*, **105**, pp. 174–179.
3. Yu, X., Kwon, E. (2009). Acarbon nanotube/cement composite with piezoresistive properties, *Smart Mater. Struct.*, **18**(5), p. 055010.
4. Materazzi, A. L., Ubertini, F., D'Alessandro, A. (2013). Carbon nanotube cement-based transducers for dynamic sensing of strain, *Cem. Concr. Compos.*, **37**(1), pp. 2–11.

5. Han, B., Zhang, K., Yu, X., Kwon, E., Ou, J. (2012). Electrical characteristics and pressure-sensitive response measurements of carboxyl MWNT/cement composites, *Cem. Concr. Compos.*, **34**(6), pp. 794–800.
6. Gdoutos, E. E., Konsta-Gdoutos, M. S., Danoglidis, P. A., Shah, S. P. (2016). Advanced cement based nanocomposites reinforced with MWCNTs and CNFs, *Front. Struct. Civ. Eng.*, **10**(2), pp. 142–149.
7. D'Alessandro, A., Ubertini, F., Materazzi, A. L., Laflamme, S., Porfiri, M. (2014). Electromechanical modelling of a new class of nanocomposite cement-based sensors for structural health monitoring, *Struct. Health Monit.*, **14**(2), pp. 137–147.
8. D'Alessandro, A., Rallini, M., Ubertini, F., Materazzi, A. L., Kenny, J. M. (2016). Investigations on scalable fabrication procedures for self-sensing carbon nanotube cement-matrix composites for SHM applications, *Cem. Concr. Compos.*, **65**, pp. 200–213.
9. Hilding, J., Grulke, E. A., Zhang, Z. G., Lockwood, F. (2003). Dispersion of carbon nanotubes in liquids, *J. Dispersion Sci. Technol.*, **24**(1), pp. 1–41.
10. Han, B., Ding, S., Yu, X. (2015). Intrinsic self-sensing concrete and structures: a review, *Measurements*, **59**, pp. 110–128.
11. Nadiv, R., Vasilyev, G., Shtein, M., Peled, A., Zussman, E., Regev, O. (2016). The multiple roles of a dispersant in nanocomposite systems, *Compos. Sci. Technol.*, **133**, pp. 192–199.
12. Ubertini, F., Laflamme, S., Ceylan, H., Materazzi, A. L., Cerni, G., Saleem, H., D'Alessandro, A., Corradini, A. (2014). Novel nanocomposite technologies for dynamic monitoring of structures: a comparison between cement-based embeddable and soft elastomeric surface sensors, *Smart Mater. Struct.*, **23**, p. 12.
13. Han, B., Ou, J. (2007). Embedded piezoresistive cement-based stress/strain sensor, *Sens. Actuators, A*, **138**(2), pp. 294–298.
14. Hou, T.-C., Lynch, J. P. (2005). Conductivity-based strain monitoring and damage characterization of fiber reinforced cementitious structural components, in *Proceedings of the Smart Structures and Materials 2005: Sensors and Smart Structures Technologies for Civil, Mechanical, and Aerospace Systems*, pp. 419–429.
15. Cao, J., Chung, D. D. L. (2004). Electric polarization and depolarization in cement-based materials, studied by apparent electrical resistance measurement, *Cem. Concr. Res.*, **34**, pp. 481–485.
16. Wen, S., Chung, D. D. L. (2000). Damage monitoring of cement paste by electrical resistance measurement, *Cem. Concr. Res.*, **30**, pp. 1979–1982.

17. Loh, K. J., Gonzalez, J. (2015). Cementitious composites engineered with embedded carbon nanotube thin films for enhanced sensing performance, *J. Phys. Conf. Ser.*, **628**, p. 012042.
18. Xie, P., Gu, P., Beaudoin, J. J. (1996). Electrical percolation phenomena in cement composites containing conductive fibers, *J. Mater. Sci.*, **31**, pp. 4093–4097.
19. Wen, S., Chung, D. D. L. (2007). Double percolation in the electrical conduction in carbon fiber reinforced cement-based materials, *Carbon*, **45**, pp. 263–267.
20. Azhari, F., Banthia, N. (2012). Cement-based sensors with carbon-fibers and carbon nanotubes for piezoresistive sensing, *Cem. Concr. Compos.*, **34**(7), pp. 866–873.
21. Han, B., Zhang, K., Yu, X., Kwon, E., Ou, J. (2012). Electrical characteristics and pressure-sensitive response measurements of carboxyl MWNT/cement composites, *Cem. Concr. Compos.*, **34**(6), pp. 794–800.
22. Han, B., Yu, X., Kwon, E. (2009). A self-sensing carbon nanotube/cement composite for traffic monitoring, *Nanotechnology*, **20**, p. 445501.
23. Saafi, M. (2009). Wireless and embedded carbon nanotube networks for damage detection in concrete structures, *Nanotechnology*, **20**, p. 395502.
24. Ubertini, F., Materazzi, A. L., D'Alessandro, A., Laflamme, S. (2014). Natural frequencies identification of a reinforced concrete beam using carbon nanotube cement-based sensors, *Eng. Struct.*, **60**, pp. 265–275.
25. Chen, B., Liu, J. (2008). Damage in carbon fiber-reinforced concrete, monitored by both electrical resistance measurement and acoustic emission analysis, *Constr. Build. Mater.*, **22**(11), pp. 2196–2201.
26. Vertuccio, L., Vittoria, V., Guadagno, L., De Santis, F. (2015). Strain and damage monitoring in carbon-nanotube-based composite under cyclic strain, *Composites Part A*, **71**, pp. 9–16.
27. Wang, S., Chung, D. D. L. (2006). Self-sensing of flexural strain and damage in carbon fiber polymer-matrix composite by electrical resistance measurement, *Carbon*, **44**(13), pp. 2739–2751.
28. Downey, A., Garcia-Macias, E., D'Alessandro, A., Laamme, S., Castro-Triguero, R., Ubertini, F. (2017). Continuous and embedded solutions for SHM of concrete structures using changing electrical potential in self-sensing cement-based composites, in Felix Wu, H., Gyekenyesi, A. L., Shull, P. J., Yu, T.-Y. (eds.) *Nondestructive Characterization and Monitoring of Advanced Materials, Aerospace, and Civil Infrastructure 2017*, vol. 10169, Proc. SPIE.

29. D'Alessandro, A., Ubertini, F., García-Macías, E., et al. (2017). Static and dynamic strain monitoring of reinforced concrete components through embedded carbon nanotube cement-based sensors, *Shock Vib.*, **2017**, Article ID 3648403 (11 pp).

30. McAndrew, T. P., Laurent, P., Havel, M., Roger, C. (2008). Arkema Graphistrength® multi-walled carbon nanotubes, Technical Proceedings of the 2008 NSTI Nanotechnology Conference and Trade Show, NSTI-Nanotech, *Nanotechnology*, **1**, pp. 47–50.

31. Wen, S., Chung, D. D. L. (2007). Piezoresistivity-based strain sensing in carbon fiber-reinforced cement, *ACI Mater. J.*, **104**(2), pp. 171–179.

32. NTC2008 Technical code for construction, DM 14 January 2008.

33. Downey, A., D'Alessandro, A., Baquera, M., Garcìa-Macìas, E., Rolfes, D., Ubertini, F., Laamme, S., Castro-Triguero, R. (2017). Damage detection, localization and quantification in conductive smart concrete structures using a resistor mesh model, *Eng. Struct.*, **148**, pp. 924–935.

34. Downey, A., D'Alessandro, A., Ubertini, F., Laamme, S., Geiger, R. (2017). Biphasic DC measurement approach for enhanced measurement stability and multi-channel sampling of self-sensing multi-functional structural materials doped with carbonbased additives, *Smart Mater. Struct.*, **26**(6), p. 065008.

35. Paolo Nenzi, H. V. (2014). Ngspice users manual - version 26, Paolo Nenzi, Holger Vogt, USA.

36. Downey, A., D'Alessandro, A., Ubertini, F., Laflamme, S. (2018). Automated crack detection in conductive smart-concrete structures using a resistor mesh model, *Meas. Sci. Technol.*, **29**(3).

Part II

Innovative Applications of Advanced Cement-Based Nanocomposites

Chapter 10

Cement-Based Piezoresistive Sensors for Structural Monitoring

Ilhwan You,[a] Seung-Jung Lee,[b] and Doo-Yeol Yoo[c]

[a]*School of Civil, Environmental and Architectural Engineering, Korea University, 145 Anam-ro, Seongbuk-gu, Seoul 02841, Republic of Korea*
[b]*Advanced Railroad Civil Engineering Division, Korea Railroad Research Institute, 176 Cheoldobangmulgwan-ro, Uiwang-si, Gyeonggi-do 16105, Republic of Korea*
[c]*Department of Architectural Engineering, Hanyang University, 222 Wangsimni-ro, Seongdong-gu, Seoul 04763, Republic of Korea*
ih-you@korea.ac.kr; seungjunglee@krri.re.kr; dyyoo@hanyang.ac.kr

10.1 Introduction

Concrete has been the most widely used construction material for civil infrastructure and buildings for the past several decades. Because of its excellent mechanical strength, low cost, and durability, it has been effectively applied to major structural elements, including columns, beams, slabs, etc., subjected to external loading. However, it has relatively poor tensile strength and toughness compared to those under compression, such that steel reinforcements, that is, reinforcing bars or prestressing strands, are generally incorporated, particularly where tensile stress occurs. This is because steel

Nanotechnology in Cement-Based Construction
Edited by Antonella D'Alessandro, Annibale Luigi Materazzi, and Filippo Ubertini

ISBN 978-981-4800-76-1 (Hardcover), 978-0-429-32849-7 (eBook)
www.jennystanford.com

reinforcements show very ductile tensile behavior due to their yielding plateaus, have a similar coefficient of thermal expansion, and provide good bond performance with concrete. Thus, they have been broadly adopted for concrete structures. Such reinforced-concrete (RC) structures, however, deteriorate when steel reinforcements are corroded, mainly due to penetration of detrimental materials from outside concrete through cracks. In addition, concrete creep results in secondary stresses generated in RC structures, causing damage and structural integrity problems. Accordingly, structural health monitoring (SHM) techniques have attracted attention from engineers and researchers in order to monitor and investigate the structural integrity of existing large-scale structures [1].

In most cases, SHM techniques are based on several types of attached and embedded sensors commercially available, including electric strain gauges (ESGs), fiber Bragg grating (FBG) sensors, and lead zirconate titanate (PZT)-based piezoelectric sensors [2], as shown in Fig. 10.1. Because of their advantages compared to conventional gauges—that is, high resistance to electromagnetic interference, long-term stability, light weight, and small size—numerous studies have been conducted by researchers since 1989 to measure mechanical strains and temperatures in structures. In the early 1990s, a study on PZT-based piezoelectric sensors was performed to detect damage in structures using impedance and intensive studies on the development of PZT-based piezoelectric sensors were conducted for practical applications to civil structures and buildings. Likewise, several types of sensors have been developed and intensively studied, but their practical applications have been very limited thus far. This is because, since the artificial sensors mentioned above are locally embedded and attached, only a local examination of the strain, temperature, and damage to RC

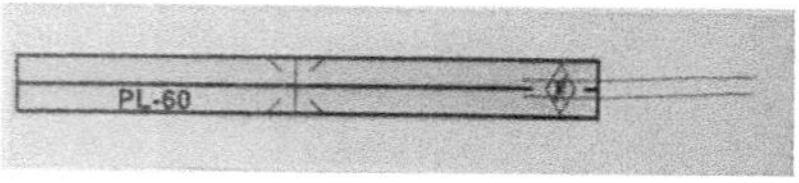

(a)

(b)

Figure 10.1 Sensors commercially available: (a) strain gauge and (b) FBG sensor.

structures is feasible. The health monitoring of entire structures, therefore, is limited, and the artificial sensors must be applied after precise prediction of the weakest parts in structures. In addition, inaccurate estimation of the weakest parts of structures may lead the sensors to be useless for SHM [3].

To overcome such drawbacks of the traditional sensors for SHM, numerous researchers [3–7] have developed a new type of self-sensing construction material, called cement-based sensors (CBS). To achieve electrical properties in a nonconductive cement composite, they have incorporated carbon-based nanomaterials, for example, carbon nanotubes (CNTs), graphene (G), carbon nanofibers (CNFs), carbon fibers (CFs), steel fibers (SFs), carbon black (CB), and nickel powder. These novel self-sensing sensors offer several advantages over conventional sensors, such as low cost, high durability, and large sensing volume, and they don't degrade the mechanical capacity of RC structures. In particular, most previous studies have focused on achieving piezoresistive properties in cement composites to monitor stress and strain variations based on changes in electrical resistivity. This special piezoresistive property could be obtained on the basis of the concept that the electrical resistivity of cement composites reinforced with conductive carbon materials varies with external force, since conductive pathways formed by connecting the carbon materials change due to deformation of the composites under force, as shown in Fig. 10.2. The electrical current can be transferred in cement composites as the conductive pathways are formed by direct connection with carbon materials. Furthermore, even though carbon materials are not in contact with each other, electrons move by the tunneling effect (Fig. 10.2c) when the distance between carbon materials becomes closer under an

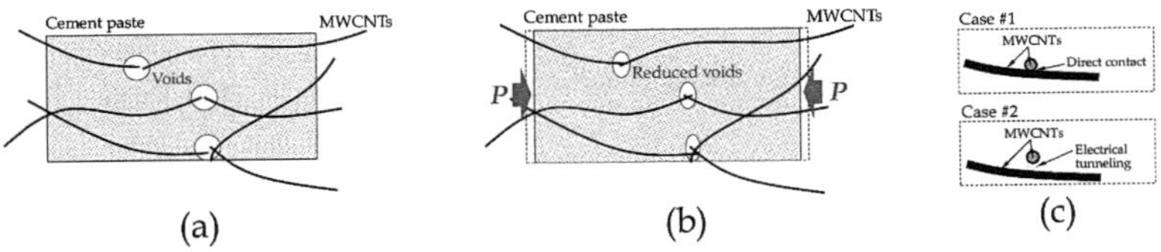

Figure 10.2 Schematic description of CNT links in composites: (a) without loading, (b) with loading, and (c) electrical tunneling effect [7].

external force. Xu et al. [8] proposed a cut-off distance between CNTs, one of the most widely used carbon nanomaterials, to achieve a tunneling effect over 0.5 nm.

Thus, to help the reader to broadly understand recent research trends on CBS for structural monitoring, we comprehensively examine the current state of knowledge. Our attention is focused on (1) conductive fillers, such as carbon-based materials; (2) their dispersion, which can depend upon the quality of CBS; (3) the CBS state of the art; and finally (4) the practical applications of CBS with the carbon-based materials for SHM of civil infrastructure and buildings.

10.2 Various Types of Cement-Based Sensors

10.2.1 Piezoresistivity

Piezoresistivity can be defined as the change in resistance of a semiconductor due to applied mechanical stress. In detail, the subjection of semiconductors to external forces can lead to deformation of their length or cross section; hence, the geometrical change in the electrical pathway results in semiconductor resistance. The change in resistance does not represent the characteristics of a specific object because resistance, having a unit Ω, is defined as the tendency of resisting the flow of current. To evaluate the electrical properties of a semiconductor, the resistivity (ρ) should be adopted. The resistivity, also called specific resistance, is a characteristic value of the electrical property of the semiconductor under consideration of its geometry; its unit is $\Omega \cdot \text{cm}$. The resistivity can be expressed as shown in Eq. 10.1:

$$\rho = \frac{RA}{L} \tag{10.1}$$

where R, A, and L are the resistance, cross-sectional area, and length of the semiconductor, respectively. The evaluation of piezoresistive properties for semiconductors depends upon the change in resistivity, $\Delta\rho$ ($\Delta\rho/\rho_0$) can be called the fractional change in resistivity

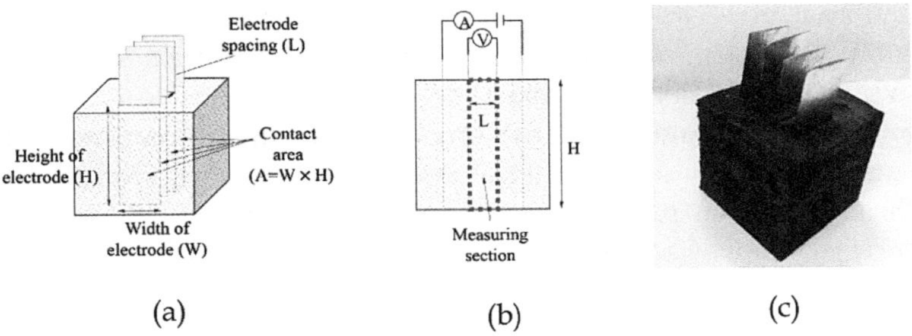

Figure 10.3 Example of cement-based sensors: (a, b) schematic diagrams and (c) true image.

(FCR) and is defined as

$$\frac{\Delta\rho}{\rho_0} = \frac{(\rho_i - \rho_0)}{\rho_0} = \text{FCR}, \tag{10.2}$$

where ρ_i is the changed resistivity of the semiconductor in the deformed state and ρ_0 is the initial resistivity before application of the external load. The FCR can be transformed in a gauge factor (GF). The GF is one of the most important characteristics for evaluating semiconductor sensitivity. It can be calculated through comparison between change in resistivity and strain (ε), as in Eq. 10.3:

$$\text{GF} = \frac{(\Delta\rho/\rho_0)}{\varepsilon} = \frac{\text{FCR}}{\varepsilon} \tag{10.3}$$

A representative CBS model based on the piezoresistive principle is introduced in Fig. 10.3. According to the measuring methods, two or four electrodes must be inserted into the CBS. In the case of a four-probe system, the inner and outer electrodes are linked with potential and voltage probes, respectively. The resistance of CBS will be measured in the section between the inner probes.

10.2.2 Cement-Based Composites

Cement composites such as cement pastes, mortars, and concretes are typically heterogeneous and are produced by mixing cement and water, with or without aggregates (sand and gravel). According to the presence and type of aggregate, cement composites are divided into cement pastes and mortars (which have only sand)

and concretes (which have both sand and gravel). Since cement composites are not electrically conductive materials on their own, a conductive filler must be incorporated inside them to induce piezoresistivity. For mortars and concretes, the presence of aggregates can disturb the electrical pathway, thereby decreasing the self-sensing performance [5]. Therefore, cement pastes are widely employed as matrices in these research fields.

10.2.3 Carbon-Based Materials (Conductive Fillers)

To change cement composites into conductive materials, conductive fillers should be incorporated at sufficient concentration. The sufficient amount is called the "percolation threshold." If the fillers are incorporated below the percolation threshold, the electrical conductivity will be governed by ionic conduction of cementitious materials and tunneling conduction, which takes place in electrical pathways with discontinuities no larger than 10 nm in length [8–10]. However, at a similar or higher concentration compared to the percolation threshold, conductive fillers are totally connected each other, meaning that continuous conductive fillers can form electrical pathways. The percolation thresholds for carbon-based materials are given in Table 10.1.

The carbon materials are frequently used as conductive fillers and can be classified according to shape, fiber, and particle. Table 10.1 indicates the physical properties of carbon fillers sorted by their shape, scale, and dimensions. Multiwalled carbon nanotubes

Table 10.1 Physical properties of carbon-based materials and their percolation threshold

	Type	Diameter (nm) [Ref.]	Length (mm) [Ref.]	Aspect ratio (length/ diameter)	Percolation threshold (%) [Ref.]
CNT		15 [3]	0.01	667	1 [5]
CNF	Fiber	200 [3]	0.01–0.03	50–150	0.5 [10]
CF		12,000–18,000 [11]	5 [11]	278–416 [11]	0.5–1 [11]
CB	Particle	120 [12]	–	–	7.22–11.39 [12]
G	Laminar	2600 [13]	–	–	2.4 [13]

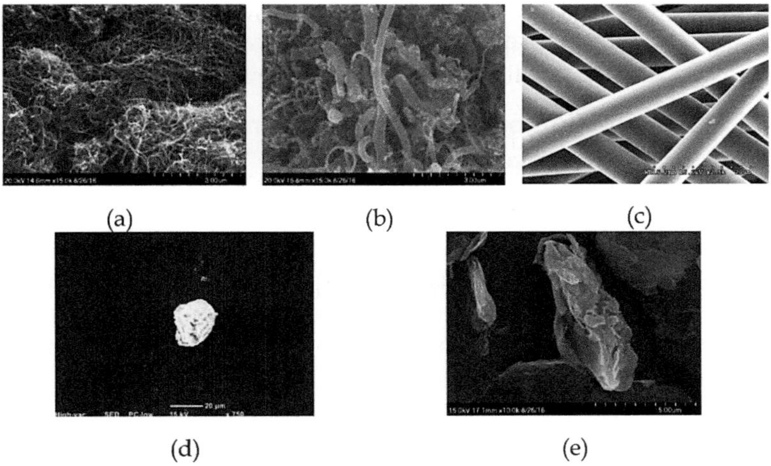

Figure 10.4 Morphologies of carbon-based materials: (a) MWCNT [3], (b) CNF [3], (C) CF [10], (d) CB [15], and (e) G [3]. (a, b, e) Reprinted from Ref. [3], Copyright (2019), with permission from Elsevier. (d) Reprinted from Ref. [15], Copyright (2017), with permission from Elsevier.

(MWCNTs), CNFs, and CFs are fiber types having an aspect ratio, CB and GP are powder types, and G and nanographite platelet (NGP) are laminar types. Figure 10.4 indicates images obtained from scanning electron microscopy on CNTs, CNFs, CFs, CB, and G.

According to previous research [9, 17–20], the self-sensing properties of CBS can be affected by the morphology of functional fillers, such as shape and size. Fiber-type fillers having high aspect ratios (i.e., ratio of length to diameter) can secure the sensing ability of concrete at lower concentrations than particle fillers. The effective concentration is not higher than 1% for fibrous fillers, whereas it is at least 5% for particle fillers. However, the dispersion of fiber-type fillers in a cement matrix is difficult since their aspect ratios can be entangled with one another. Hence, poor dispersion of carbon materials can induce mechanical defects and unbalance the self-sensing quality. Moreover, nanoscale carbon-based fillers such as CNTs, graphite nanofibers, G, and graphene oxide (GO) lead easily to self-agglomeration due to the van der Waals force. Thus, the dispersion of nanoscale carbon-based fillers is necessary in the field of self-sensing. Details concerning the dispersion of nanoscale carbon-based materials are discussed in Section 10.2.4.

10.2.4 Dispersion of Carbon-Based Nanomaterials in Cement-Based Composites

As mentioned in the previous section, the dispersion of carbon-based fillers is the most important issue in the field of CBS, as well as research areas where carbon nanomaterials are used. A great deal of research toward achieving uniform dispersion of carbon materials has been performed, and techniques can be classified as mechanical and chemical methods.

Mechanical methods consist of separating each particle physically. Ultrasound is one such simple and effective method that uses specific equipment (the sonicator; Fig. 10.5). The ultrasonic method requires a suspension in which carbon nanomaterials are dispersed to be produced. Furthermore, to achieve an effective dispersion, surfactants are usually incorporated during the process of producing the suspension. Some studies have investigated the effects of various surfactants upon the dispersion of CNTs [5, 21–23]. In the ultrasonic method, the surfactants lignosulfonate and polycarboxylate offer good dispersion of CNTs in the suspension (Fig. 10.6). However, excessive dosages of surfactants are not recommended, since they can lead to delayed cement-setting times [22], generating entrapped pores [23]. Konsta-Gdoutos et al. [24] successfully obtained a

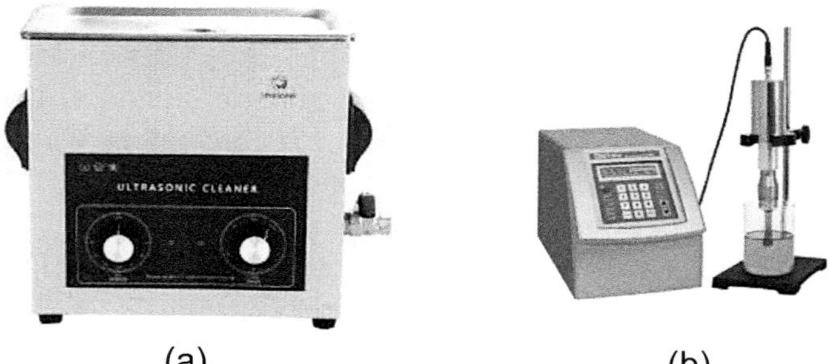

(a) (b)

Figure 10.5 Various models of sonicator [25]: (a) bath type and (b) probe type.

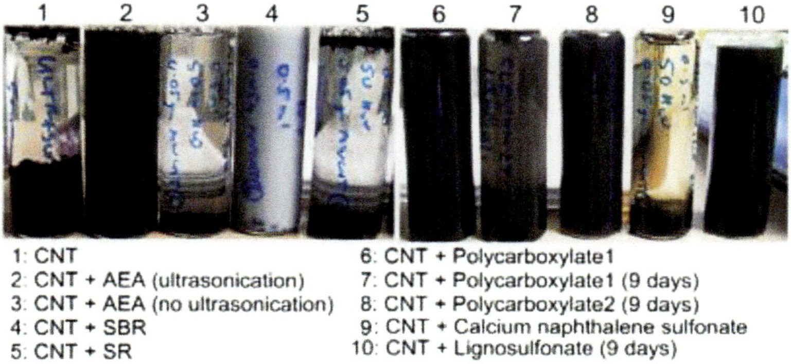

Figure 10.6 Suspension of CNTs treated by various surfactants. Reprinted from Ref. [20], Copyright (2012), with permission from Elsevier.

homogeneous suspension with surfactant/CNT weight ratios of 4.0 and 6.25 under the ultrasonic method.

Although ultrasound is useful, it suffers crucial drawbacks for making cement composites, making it unsuitable for mass production [5, 26]. To make suspensions, sonicators are necessary, and the suspension capacity is limited by the sonicator performance. According to previous studies [24, 27], incorporation of silica fumes (SFs) of similar size to CNTs can cause separate agglomerations of CNTs when only dry-mixing processes are used. Figure 10.7 shows a conceptual diagram of the dispersion processes of CNTs according to the SF contents reported Kim et al. [24].

Chemical methods can be described as surface treatment of carbon nanomaterials. There is another reason for using chemical methods on carbon nanomaterials, besides dispersion. The carbon is typically hydrophobic due to nonpolar electrical properties; hence, carbon-based materials can induce weak bonds with the cement matrix [11, 28–32]. In this method, acids are frequently used to form functional groups; one example is carboxylic acid, which can induce covalent bonding with cement hydrates. The CNTs incorporated into the cement matrix are more tightly surrounded by cement hydrates such as C–S–H and $Ca(OH)_2$ with increasing ages. Functional groups, formed on the basal plane by surface treatment carboxylic acid, form covalent bonds with C–S–H and $Ca(OH)_2$, as shown in Fig. 10.8. It can

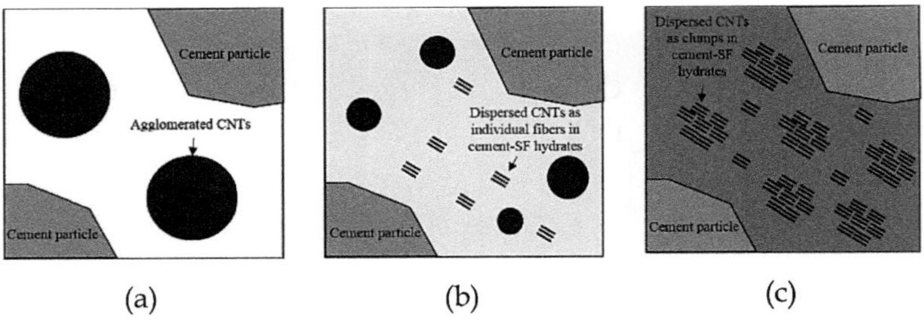

Figure 10.7 Schematics of dispersion of CNTs in a cement matrix: (a) without silica fumes, (b) with a low volume of silica fumes, and (c) with a high volume of silica fumes. Reprinted from Ref. [23], Copyright (2014), with permission from Elsevier.

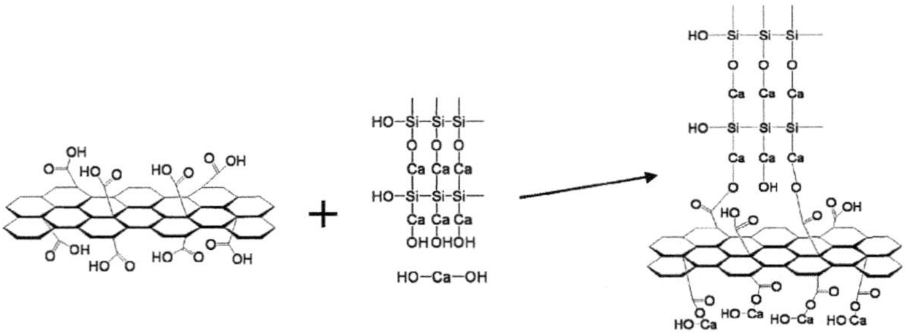

Figure 10.8 Reaction scheme between carboxylic-acid groups and the hydration productions of cement. Reprinted from Ref. [33], Copyright (2015), with permission from Elsevier.

be followed by improving the hydrophilic and bonding properties with a cement matrix.

10.2.5 Preparation of Cement-Based Sensors and Test Configurations

To measure the change of resistance in CBSs, the electrodes should take priority over the preparation of all other constituents. Electrode materials have to satisfy two basic features: low electrical resistance and stable electricity [9]. Representative materials for use as electrodes include copper and silver, which have electrical resistivity of 1.7×10^{-8} and $1.59 \times 10^{-8} \Omega \cdot \text{m}$, respectively [34]. The widths

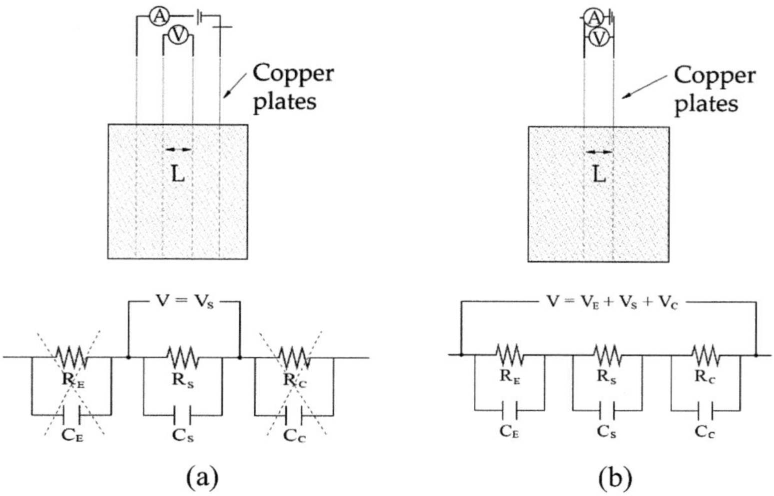

Figure 10.9 Schematic diagrams of measurement methods and principle: (a) four probes and (b) two probes [34].

and intervals of electrodes are decided according to the measuring section and dimensions of the CBS. It should be noted that improper electrode sizes can impair mechanical performance.

Two or four probes are used to measure the electrical properties of semiconductors. Figure 10.9 indicates the typical configurations of two- and four-probe systems and shows their circuit schematics, demonstrating their respective measuring principles.

The voltages in both two- and four-probe systems can be divided as coming from the electrodes (V_E), CBS (V_S), and contact resistance between the electrode and CBS (V_C), with V_E being negligible. In four-probe systems, V_C can also be eliminated, leaving only V_S [34]. Consequently, resistance measured by a two-probe system becomes higher than that for a four-probe system according to Ohm's law ($V = IR$), where V and I are voltage and current, respectively.

Chiarello et al. [35] (Fig. 10.10) discovered that with an increase in the ratio between the electrode spacing (L) and the electrode–contact area (A) ratio, the resistance of CBS increases when measured by the two-probe system. On the other hand, in the case of the four-probe system, there was no significant increase of

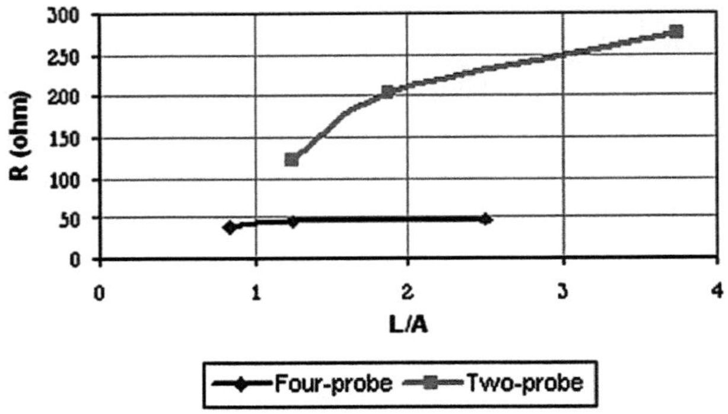

Figure 10.10 Resistance according to length-to-contact-area ratio for different probe methods. Reprinted from Ref. [35], Copyright (2005), with permission from Elsevier.

resistance. Thus, the four-probe system is preferred for electrical resistivity measurements of CBS [34].

The CBS resistance is measured differently using direct current (DC) or alternating current (AC); in the latter, AC with different frequencies can affect the resistance value. Although DC is the simplest method to measure resistance, electrical polarization, which is the phenomenon by which resistance increases consistently due to electrochemical reactions between negative ions of cement composites and the positively charged electrode, should be considered. Polarization is severe for DC because voltage is consistently supplied. To avoid polarization, the resistance of CBS should converge until complete polarization. However, because polarization depends upon specimen size, larger CBS need more time to reach complete polarization [34, 36]. Another method to reduce polarization is to employ AC signals. Polarization is also generated by AC, but it is very weak with high frequency and falls within acceptable ranges [9, 18, 34, 37, 38].

10.2.6 Self-Sensing Properties by Various Carbon-Based Materials

Various factors influence the electrical resistivity of CBS. In general, it is well known that the electrical resistivity of CBS can affect their piezoresistive properties. Such sensors, having a relatively low electrical resistivity, will show good sensitivity since lower resistivity allows an enhanced signal-to-noise ratio [9]. However, opposite results have been shown. Yoo et al. [3] investigated the influences of various carbon-based materials (MWCNTs, CFs, G, CNFs) upon electrical resistivity and the piezoresistive properties of CBS. This research showed that CBS incorporating 0.5% CFs by volume showed the lowest resistivity of about 100 $\Omega \cdot$ cm, as illustrated in Fig. 10.11a. At the same concentration as other carbon-based materials, the resistivity of CBS with MWCNTs was about 1000 $\Omega \cdot$ cm, whereas those of G and CNF composites were higher than 10,000 $\Omega \cdot$ cm, only slightly lower than the resistivity of plain paste. The trend of resistivity of carbon-based materials can be explained, as described in the previous section. Firstly, since CFs, MWCNTs, and CNFs were fiber-type nanomaterials, they could form electrical pathways more easily than G, which was laminar-type. Therefore, CNFs offered the closest particles among fiber-type materials when only the aspect ratio was considered. Under cyclic compressive loads, CBS that incorporated MWCNTs showed the best response when subjected to loads for which the FCR-versus-compressive-stress curve had a correlation coefficient of 0.85, as shown in Fig. 10.11d.

Fu et al. [57] found out that ozone treatment of CFs increased the surface oxygen concentration and changed the oxygen bond from C–O to C=O. Consequently, the incorporation of ozone-treated CFs into CBS improved the bond strength between cement paste and CFs. The sensing ability of CBS with ozone-treated CFs was improved in terms of increased GF and better repeatability, as shown in Fig. 10.12.

Yu et al. [6] investigated the change in resistance of CBS with 0.1% carboxyl-functionalized MWCNTs corresponding to cyclic compression, as shown in Fig. 10.12a. This result was compared with CBS with ordinary MWCNTs. They found that CBS with 0.1% carboxyl-functionalized MWCNTs showed a higher signal-to-noise ratio of piezoresistive response.

Numerous studies have focused on developing CBS incorporating various carbon-based materials. Table 10.2 compares the properties of such CBS. However, since the GFs depend on the loading rate [5]

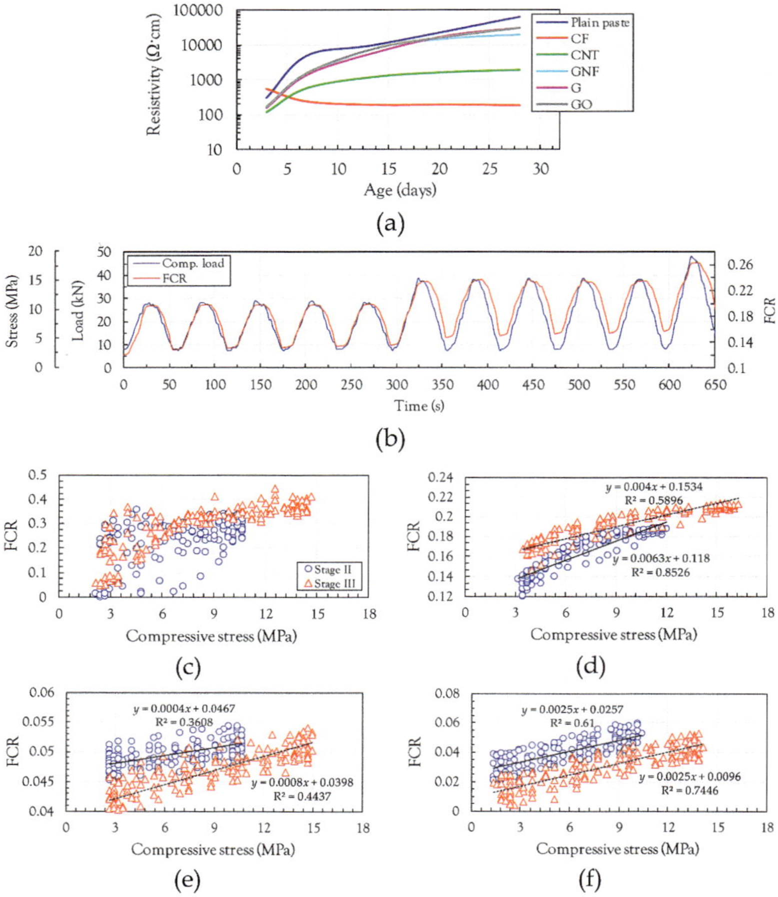

Figure 10.11 Self-sensing capacities of cement-based sensors with different carbon-based materials [3]: (a) change in resistivity with increasing age, (b) FCR by load, (c–f) correlation between FCR and loads: (c) 0.5% CF, (d) 0.5% CNT, (e) 0.5% CNF, and (f) 0.5% G.

and compositions of CBS, it is hard to compare materials directly. Moreover, there is a little deviation in these results, because there remain unsolved problems, such as the dispersion of carbon-based materials.

Although CNTs exhibit outstanding piezoresistive self-sensing properties, they are one of the most expensive carbon-based

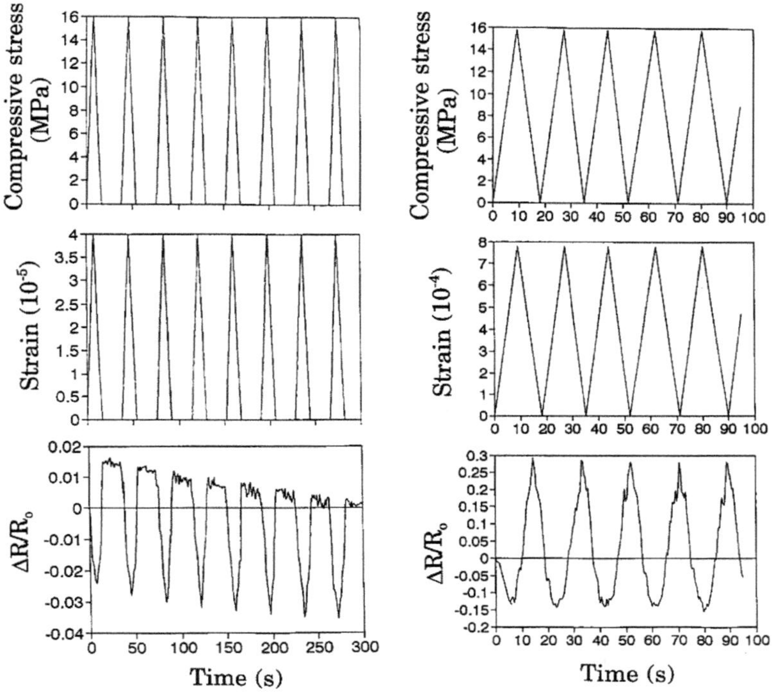

Figure 10.12 Self-sensing capacities of cement-based sensors incorporating: (a) as-received carbon fiber and (b) ozone-treated carbon fiber. Reprinted from Ref. [57], Copyright (1998), with permission from Elsevier.

materials. Therefore, they are of limited use for large-scale and field applications.

Some researchers have investigated the feasibility of replacing CNTs with CBs [44] or CFs [11, 42, 46]. Luo et al. [44] conducted experiments for stress and strain sensing of CBS incorporating both CNTs and CBs and compared them to CBS with only CNTs under cyclic loading. CBS with both CNTs and CBs showed a more sensitive response to stress and strain than did those incorporating only CNTs. CNTs wrapped in cement hydrates were found to be able to increase the probability of contact with CB particles within cement hydrates with movement under the loading condition. Azhari et al. [41] found that CBS with a hybrid system of 15% CFs and 1%

Table 10.2 Gauge factors of cement-based sensors with different carbon-based materials at concentrations

Materials	Concentration	GF	Loading rates [kN/s]	Ref.
CNT	1.0	130	Around 0.39	[5]
	1.0	166.6	0.333–0.999	[10]
CF	0.5	279	Unknown	[39]
	0.38	138	Unknown	[40]
	15	445	0.01–0.12	[41]
CB	8.79	55.28	Unknown	[12]
NGP	5	156	0.1–0.4 mm/min	[42]
GO	0.35	43.87	0.083	[43]
CNT+CF	0.5 + 0.1	160	0.333–0.666	[10]

MWCNTs were superior to CBS incorporating only 15% CFs in terms of FCR against stress and strain. According to a recent study [10], the 0.5% concentration of MWCNTs could be replaced by a 0.1% concentration of CFs. The authors analyzed the cost of incorporating carbon-based materials (CFs, MWCNTs), as well as the self-sensing properties of CBS under compressive loading and strain. One of the main CBS in this research was a hybrid material with both 0.1% CFs and 0.5% MWCNTs, which was compared to CBS with only 1% MWCNTs (Fig. 10.13).

The GFs of the two CBS were 160 and 166.6, respectively, under the same loading. The cost of carbon-based materials incorporated

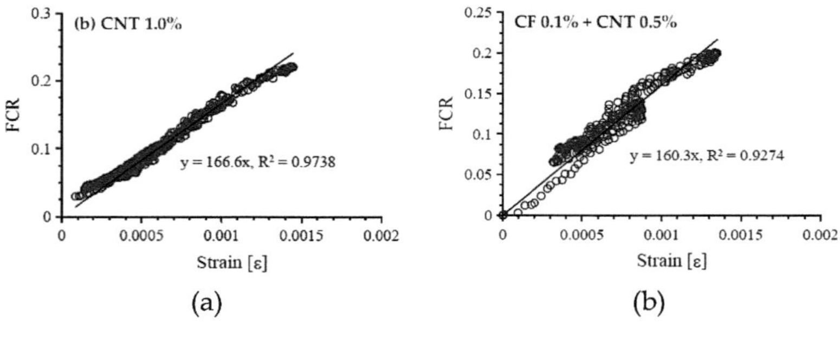

Figure 10.13 Self-sensing capacities of cement-based sensors incorporating (a) only 0.5% MWCNTs and (b) both 0.1% CFs and 0.5% MWCNTs [10].

into the CBS with 0.1% CFs was about half that of CBS with 1% MWCNTs. Thus, the economic efficiency of CBS incorporating both 0.1% CFs and 0.5% MWCNTs was verified.

10.3 Practical Applications of Cement-Based Sensors

The sensing properties of CBS can be generally calculated as the fractional change of the electrical resistivity of the composites. In order to use CBS for SHM, numerous researchers have compared fractional changes in electrical resistivity to changes in stress and strain under applied compressive loads to investigate CBS as strain sensors [5–7]. Thus, the use of CBS as crack sensors, self-heating elements, dynamic sensors, and damage sensors is explained in this chapter.

Lim et al. [46] investigated the feasibility of using CBS with CNTs as crack sensors for concrete beams. The CBS was embedded in the reinforced mortar beam under three-point bending conditions, as shown in Fig. 10.14a. The crack's width–conductivity relationship

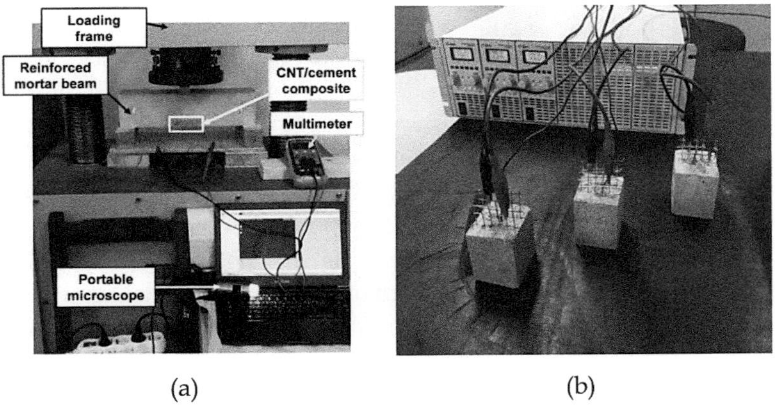

Figure 10.14 Experimental setups for SHM using cement-based sensors: (a) crack monitoring under bending [46] and (b) self-heating element [47]. (a) Reprinted from Ref. [46], Copyright (2017), with permission from Elsevier.

was studied and a simplified theoretical conductivity model was developed to compare experimental results. According to their results, the reliability of CBS for detecting cracks depended upon both crack width and moisture content.

Kim et al. [48] and Lee et al. [47] investigated the heat characteristics of CBS. Kim et al. [48] developed a CBS for use as a self-heating element for incorporation into a structural heating system. Although cement composites with 2.0 wt% CNTs can reach a temperature of 70°C within 30 min, the composite incorporating under 0.6 wt% CNTs was appropriate as a self-heating element on the basis of its heating characteristics. Lee et al. [47] analyzed the heating characteristics of cementitious composites with single-walled carbon nanotubes (SWCNTs) and MWCNTs, as shown in Fig. 10.14b. They observed that SWCNTs were more effective than MWCNTs for modifying the heat characteristics and that an applied voltage of 100 V or more yielded the desired results.

Recently, some researchers have measured dynamic properties such as impulsive loading [48–50].

Materazzi et al. [50] developed a dynamic strain sensor using cement paste with MWCNTs. On the basis of their experimental results, cement paste with 2 wt% of MWCNTs could detect the dynamic responses to a load applied with a frequency of 0.1 Hz. The output of the CBS, namely its electrical resistance, was almost identical to the input over the entire investigated frequency range of 0.1–2.0 Hz, as shown in Fig. 10.15a. A close correlation between the measured electrical resistance, axial strain of the sensor and the frequency–response function was obtained as a quasi-linear relationship, as shown in Fig. 10.15b. On the basis of their results, they developed a CBS with CNTs mounted on the RC beam, as shown in Fig. 10.16a [51]. Ubertini et al. [51] observed that the CBS could detect high frequencies with a very low level of noise and an excellent signal-to-noise ratio. Moreover, the output of the proposed sensor compared well to those of off-the-shelf strain gauges and accelerometers.

Baeza et al. [52] investigated the performance of CBSs as damage sensors, as shown in Fig. 10.16b. They tested cement composites with CNFs and CFs under different casting conditions (embedded or attached), and electrical resistance was compared to the bending

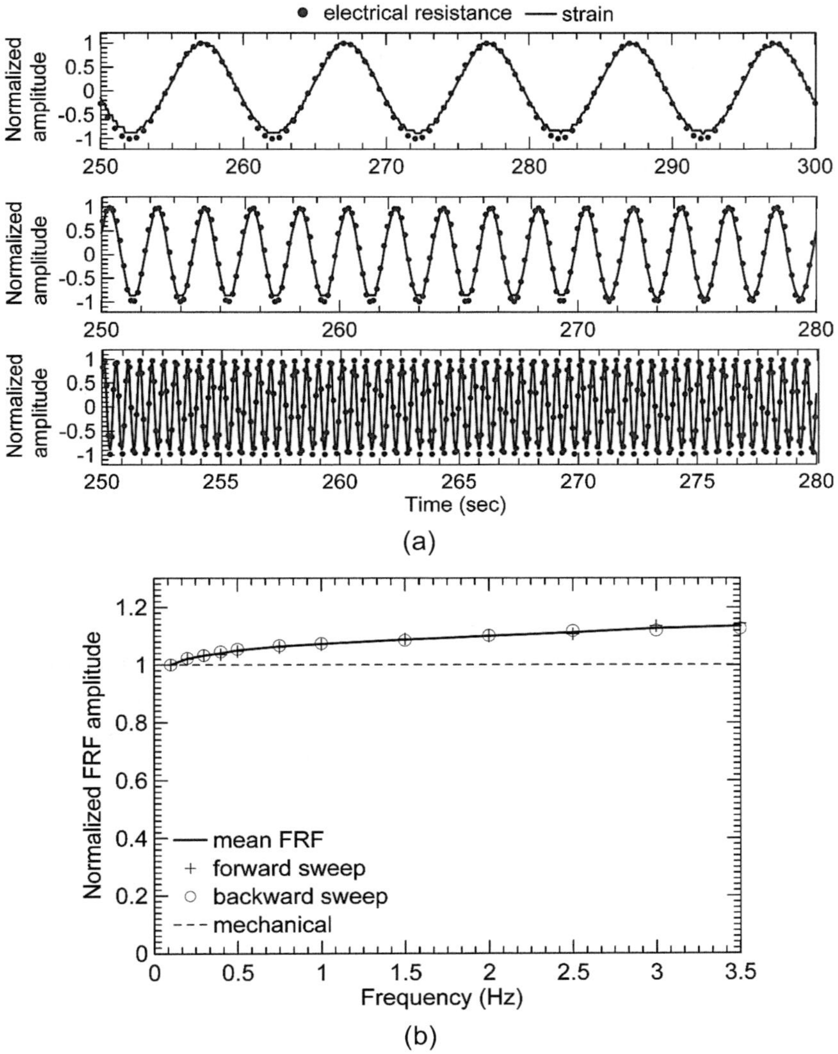

Figure 10.15 Dynamic response of cement-based sensors: (a) comparison of normalized measured strain and electrical resistance with loading frequencies 0.1, 0.5, and 2.0 Hz and (b) measured frequency–response curve. Reprinted from Ref. [50], Copyright (2013), with permission from Elsevier.

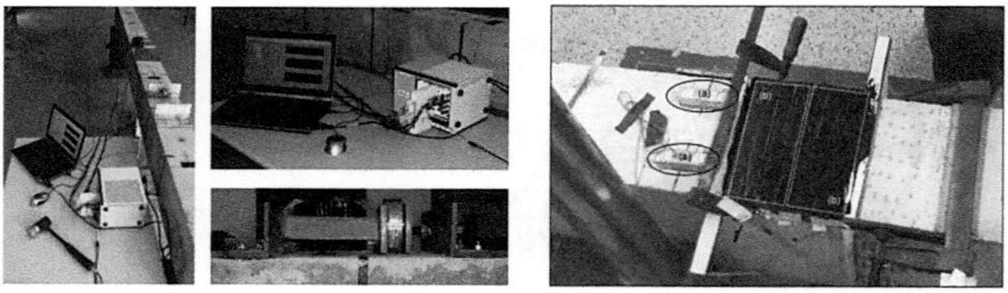

(a) (b)

Figure 10.16 Experimental setups for SHM using two cement-based sensors: (a) dynamic sensor in the RC beam and (b) damage sensor in RC beam. (a) Reprinted from Ref. [51], Copyright (2014), with permission from Elsevier. (b) https://creativecommons.org/licenses/by/3.0/.

moment of the RC beam. The CBS could detect strain even during RC beam failure in this study.

As shown in Fig. 10.17, Han et al. [53] investigated cement paste with nickel powder for a wireless stress/strain-sensing system. The wireless stress/strain measurement system can be used to achieve a sensitivity to stress/strain of a GF of 1336.5 and a stress/strain resolution of 150 Pa/0.02 $\mu\varepsilon$. The newly developed wireless stress/strain measurement system integrated with pressure-sensitive nickel powder-filled CBS has such advantages as high sensitivity to stress/strain, high stress/strain resolution, simple circuit, and low energy consumption.

Saafi et al. [58] also investigated cement paste with CNTs for damage sensing. The sensor could follow both monotonic and cyclic loading behavior and detect the initiation of damage at an early loading stage. The change in the electrical resistance CBS was measured by wireless networking.

In the literature several attempts to monitor traffic on the road are reported [53–55]. Shi et al. [54] found by simple testing that cement paste with a 0.5 or 1.0 wt% concentration of CFs was effective for traffic monitoring and weighing traffic in motion. Electrical resistance reversibly increased to 1 MPa stress and 55 mph speed. Han et al. [56, 57] tried to use a CBS embedded in pavement to detect vehicles. They constructed a self-sensing pavement embedded with nickel-particle-filled CBS, as shown in Fig. 10.18a. Since such CBS have high sensitivity, the self-sensing

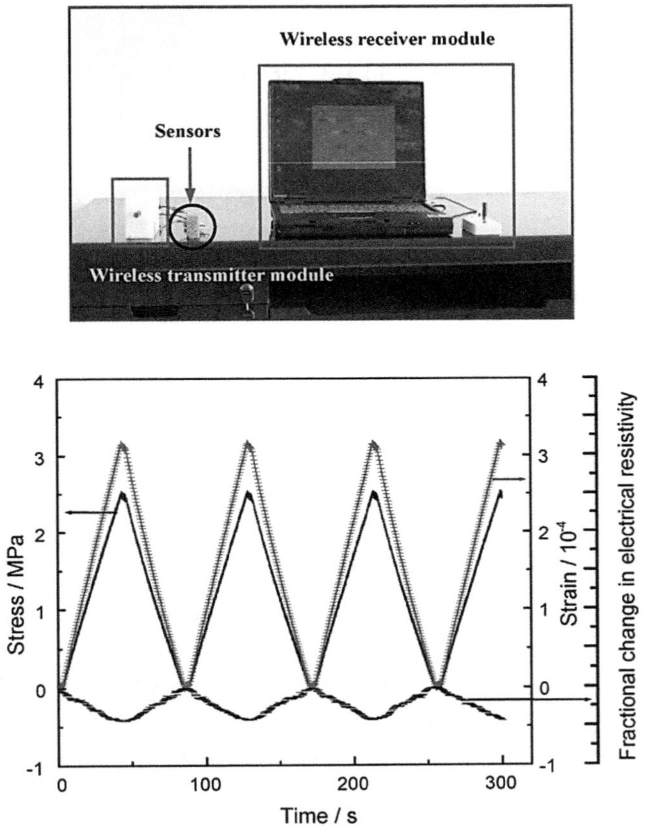

Figure 10.17 Experimental results of wireless stress/strain measurement system using CBS: (top) wireless acquisition system and (bottom) relationship between input and output of pressure-sensitive nickel powder-filled cement-based stress/strain sensors under uniaxial compression. Reprinted from Ref. [53], Copyright (2008), with permission from Elsevier.

pavement could accurately detect the passing vehicles, as shown in Fig. 10.18b. Similarly, the Minnesota Road Research Facility (MnROAD) used self-sensing pavement with CNT-filled CBS to detect passing vehicles [56]. Self-sensing pavement with CBS can also yield important traffic data such as flow rate, vehicle speed, and density [55].

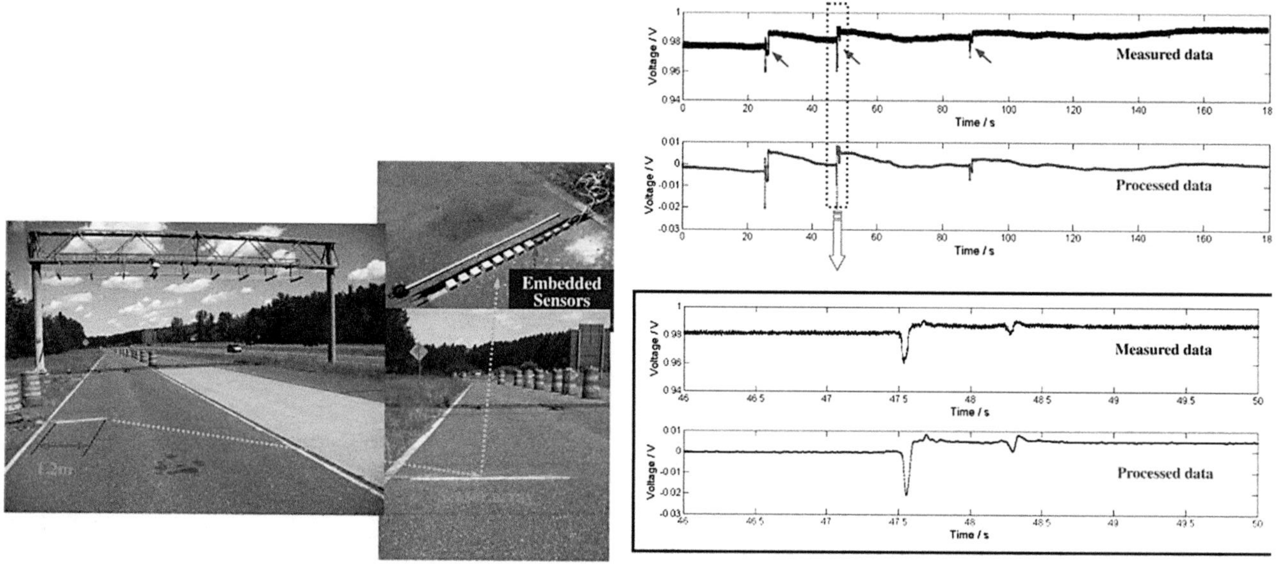

(a) (b)

Figure 10.18 Attempts to monitor traffic using cement-based sensors: (a) self-sensing pavement with embedded cement-based sensor and (b) test results to detect passing vehicles. Reprinted from Ref. [55], Copyright (2011), with permission from Elsevier.

10.4 Conclusions

The concept and state of the art of CBS incorporating carbon-based materials were described in this chapter. Incorporation of carbon-based materials into cement composites can allow self-sensing of mechanical strain. A number of studies of CBS with various carbon-based materials have been carried out over the past decade. At present, research on improving the performance and economic efficiency of CBS using hybrid fillers and on extending the self-sensing range through dynamic factors is ongoing and their feasibilities were verified.

Self-sensing performance is strongly determined by the type and concentration of carbon-based materials and especially the dispersion of nanoscale materials. By now, various methods for dispersing carbon-based materials have been introduced, including mechanical and chemical methods. However, there still remain challenges facing the manufacture of simple and large-sized CBS. These problems have to be solved and study is necessary for practical use in the field.

References

1. Chang, P. C., Flatau, A., Liu, S. C. (2003). Review paper: health monitoring of civil infrastructure, *Struct. Health Monit.*, **2**(3), pp. 257–267.
2. Yang, Y., Annamdas, V. G. M., Wang, C., Zhou, Y. (2008). Application of multiplexed FBG and PZT impedance sensors for health monitoring of rocks, *Sensors*, **8**(1), pp. 271–289.
3. Yoo, D., You, I., Zi, G., Lee, S. (2019). Effects of carbon nanomaterial type and amount on self-sensing capacity of cement paste, *Measurement*, **134**, pp. 750–761.
4. Banthia, N., Djeridane, S., Pigeon, M. (1992). Electrical resistivity of carbon and steel micro-fiber reinforced cements, *Cem. Concr. Res.*, **22**(5), pp. 804–814.
5. D'Alessandro, A., Rallini, M., Ubertini, F., Materazzi, A. L., Kenny, J. M. (2016). Investigations on scalable fabrication procedures for self-sensing carbon nanotube cement-matrix composites for SHM applications, *Cem. Concr. Compos.*, **65**, pp. 200–213.

6. Yu, X., Kwon, E. (2009). A carbon nanotube/cement composite with piezoresistive properties, *Smart Mater. Struct.*, **18**(5), p. 55010.
7. Yoo, D.-Y., You, I., Lee, S.-J. (2017). Electrical properties of cement-based composites with carbon nanotubes, graphene, and graphite nanofibers, *Sensors*, **17**(5), p. 1064.
8. Xu, J., Zhong, W., Yao, W. (2010). Modeling of conductivity in carbon fiber-reinforced cement-based composite, *J. Mater. Sci.*, **45**(13), pp. 3538–3546.
9. Han, B., Ding, S., Yu, X. (2015). Intrinsic self-sensing concrete and structures: a review, *Measurement*, **59**, pp. 110–128.
10. Lee, S. J., You, I., Zi, G., Yoo, D. Y. (2017). Experimental investigation of the piezoresistive property of cement composites with hybrid carbon fibers and nanotubes, *Sensors*, **17**(11), p. 2561.
11. Wen, S., Chung, D. D. L. (1999). Seebeck effect in carbon fiber-reinforced cement, *Cem. Concr. Res.*, **29**(12), pp. 1989–1993.
12. Li, H., Xiao, H. G., Ou, J. P. (2006). Effect of compressive strain on electrical resistivity of carbon black-filled cement-based composites, *Cem. Concr. Compos.*, **28**(9), pp. 824–828.
13. Le, J. L., Du, H., Pang, S. D. (2014). Use of 2-D Graphene Nanoplatelets (GNP) in cement composites for structural health evaluation, *Composites Part B*, **67**, pp. 555–563.
14. Yoo, D. Y., Zi, G., Kang, S. T., Yoon, Y. S. (2015). Biaxial flexural behavior of ultra-high-performance fiber-reinforced concrete with different fiber lengths and placement methods, *Cem. Concr. Compos.*, **63**, pp. 51–66.
15. Chen, B., Li, B., Gao, Y., Ling, T. C., Lu, Z., Li, Z. (2017). Investigation on electrically conductive aggregates produced by incorporating carbon fiber and carbon black, *Constr. Build. Mater.*, **144**, pp. 106–114.
16. Han, B., Yu, X., Ou, J. (2010). Effect of water content on the piezoresistivity of MWNT/cement composites, *J. Mater. Sci.*, **45**(14), pp. 3714–3719.
17. Li, G. Y., Wang, P. M., Zhao, X. (2007). Pressure-sensitive properties and microstructure of carbon nanotube reinforced cement composites, *Cem. Concr. Compos.*, **29**(5), pp. 377–382.
18. Han, B., Guan, X., Ou, J. (2007). Electrode design, measuring method and data acquisition system of carbon fiber cement paste piezoresistive sensors, *Sens. Actuators, A*, **135**(2), pp. 360–369.
19. Han, B., Ou, J. (2007). Embedded piezoresistive cement-based stress/strain sensor, *Sens. Actuators, A*, **138**(2), pp. 294–298.

20. Collins, F., Lambert, J., Duan, W. H. (2012). The influences of admixtures on the dispersion, workability, and strength of carbon nanotube-OPC paste mixtures, *Cem. Concr. Compos.*, **34**(2), pp. 201–207.
21. Luo, J., Duan, Z., Li, H. (2009). The influence of surfactants on the processing of multi-walled carbon nanotubes in reinforced cement matrix composites, *Phys. Status Solidi A*, **206**(12), pp. 2783–2790.
22. Parveen, S., Rana, S., Fangueiro, R. (2013). A review on nanomaterial dispersion, microstructure, and mechanical properties of carbon nanotube and nanofiber reinforced cementitious composites, *J. Nanomater.*, **2013**, p. 80.
23. Kim, H. K., Nam, I. W., Lee, H. K. (2014). Enhanced effect of carbon nanotube on mechanical and electrical properties of cement composites by incorporation of silica fume, *Compos. Struct.*, **107**, pp. 60–69.
24. Konsta-Gdoutos, M. S., Metaxa, Z. S., Shah, S. P. (2010). Highly dispersed carbon nanotube reinforced cement based materials, *Cem. Concr. Res.*, **40**(7), pp. 1052–1059.
25. Kharissova, O. V., Kharisov, B. I., de Casas Ortiz, E. G. (2013). Dispersion of carbon nanotubes in water and non-aqueous solvents, *RSC Adv.*, **3**(47), p. 24812.
26. Han, B., Sun, S., Ding, S., Zhang, L., Yu, X., Ou, J. (2015). Review of nanocarbon-engineered multifunctional cementitious composites, *Composites Part A*, **70**, pp. 69–81.
27. Sanchez, F., Ince, C. (2009). Microstructure and macroscopic properties of hybrid carbon nanofiber/silica fume cement composites, *Compos. Sci. Technol.*, **69**(7–8), pp. 1310–1318.
28. Fu, X., Lu, W., Chang, D. D. L. (1998). Ozone treatment of carbon fiber for reinforcing cement, *Carbon*, **36**(9), pp. 1337–1345.
29. Wen, S., Chung, D. D. L. (2006). The role of electronic and ionic conduction in the electrical conductivity of carbon fiber reinforced cement, *Carbon*, **44**(11), pp. 2130–2138.
30. Ibarra, L., Macias, A., Palma, E. (1996). Stress-strain and stress relaxation in oxidated short carbon fiber-thermoplastic elastomer composites, *J. Appl. Polym. Sci.*, **61**(13), pp. 2447–2454.
31. Tiwari, S., Bijwe, J. (2014). Surface treatment of carbon fibers: a review, *Procedia Technol.*, **14**, pp. 505–512.
32. Delamar, M., Désarmot, G., Fagebaume, O., Hitmi, R., Pinsonc, J., Savéant, J.-M. (1997). Modification of carbon fiber surfaces by electrochemical reduction of aryl diazonium salts: application to carbon epoxy composites, *Carbon*, **35**(6), pp. 801–807.

33. Pan, Z., He, L., Qiu, L., Korayem, A. H., Li, G., Zhu, J. W., et al. (2015). Mechanical properties and microstructure of a graphene oxide-cement composite, *Cem. Concr. Compos.*, **58**, pp. 140–147.
34. Azhari, F. (2008). Cement-based sensors for structural health monitoring, PhD Dissertation, University of British Columbia, Vancouver, p. 184.
35. Chiarello, M., Zinno, R. (2005). Electrical conductivity of self-monitoring CFRC, *Cem. Concr. Compos.*, **27**(4), pp. 463–469.
36. Hou, T.-C., Lynch, J. P. (2005). Conductivity-based strain monitoring and damage characterization of fiber reinforced cementitious structural components, *Smart Struct. Mater.*, pp. 419–429.
37. Fu, X., Ma, E., Chung, D. D. L., Anderson, W. A. (1997). Self-monitoring in carbon fiber reinforced mortar by reactance measurement, *Cem. Concr. Res.*, **27**(6), pp. 845–852.
38. Han, B., Zhang, K., Yu, X., Kwon, E., Ou, J. (2012). Electrical characteristics and pressure-sensitive response measurements of carboxyl MWNT/cement composites, *Cem. Concr. Compos.*, **34**(6), pp. 794–800.
39. Wen, S., Chung, D. D. L. (2008). Piezoresistivity-based strain sensing in carbon fiber-reinforced cement, *Mater. J.*, **104**(2), pp. 171–179.
40. Ou, J., Han, B. (2008). Piezoresistive cement-based strain sensors and self-sensing concrete components, *J. Intell. Mater. Syst. Struct.*, **20**(3), pp. 329–336.
41. Azhari, F., Banthia, N. (2012). Cement-based sensors with carbon fibers and carbon nanotubes for piezoresistive sensing, *Cem. Concr. Compos.*, **34**(7), pp. 866–873.
42. Sun, S., Han, B., Jiang, S., Yu, X., Wang, Y., Li, H., Ou, J. (2017). Nano graphite platelets-enabled piezoresistive cementitious composites for structural health monitoring, *Constr. Build. Mater.*, **136**, pp. 314–328.
43. Saafi, M., Tang, L., Fung, J., Rahman, M., Sillars, F., Liggat, J., Zhou, X. (2014). Graphene/fly ash geopolymeric composites as self-sensing structural materials, *Smart Mater. Struct.*, **23**(6), p. 65006.
44. Luo, J. L., Duan, Z. D., Zhao, T. J., Li, Q. Y. (2010). Self-sensing property of cementitious nanocomposites hybrid with nanophase carbon nanotube and carbon black, *Adv. Mater. Res.*, **143–144**, pp. 644–647.
45. Luo, J. L., Duan, Z. D., Zhao, T. J., Li, Q. Y. (2010). Hybrid effect of carbon fiber on piezoresistivity of carbon nanotube cement-based composite, *Adv. Mater. Res.*, **143–144**, pp. 639–643.

46. Lim, M. J., Lee, H. K., Nam, I. W., Kim, H. K. (2017). Carbon nanotube/cement composites for crack monitoring of concrete structures, *Compos. Struct.*, **180**, pp. 741–750.

47. Lee, H., Kang, D., Song, Y. M., Chung, W. (2017). Heating experiment of CNT cementitious composites with single-walled and multiwalled carbon nanotubes, *J. Nanomater.*, **2017**, p. 3691509.

48. Kim, G. M., Naeem, F., Kim, H. K., Lee, H. K. (2016). Heating and heat-dependent mechanical characteristics of CNT-embedded cementitious composites, *Compos. Struct.*, **136**, pp. 162–170.

49. Han, B., Yu, X., Kwon, E., Ou, J. (2010). Piezoresistive multi-walled carbon nanotubes filled cement-based composites, *Sens. Lett.*, **8**(2), pp. 344–348.

50. Materazzi, A. L., Ubertini, F., D'Alessandro, A. (2013). Carbon nanotube cement-based transducers for dynamic sensing of strain, *Cem. Concr. Compos.*, **37**(1), pp. 2–11.

51. Ubertini, F., Materazzi, A. L., D'Alessandro, A., Laflamme, S. (2014). Natural frequencies identification of a reinforced concrete beam using carbon nanotube cement-based sensors, *Eng. Struct.*, **60**, pp. 265–275.

52. Baeza, F. J., Galao, O., Zornoza, E., Garcés, P. (2013). Multifunctional cement composites strain and damage sensors applied on reinforced concrete (RC) structural elements, *Materials*, **6**(3), pp. 841–855.

53. Han, B., Yu, Y., Han, B. Z., Ou, J. P. (2008). Development of a wireless stress/strain measurement system integrated with pressure-sensitive nickel powder-filled cement-based sensors, *Sens. Actuators, A*, **147**(2), pp. 536–543.

54. Shi, Z. Q., Chung, D. D. L. (1999). Carbon fiber-reinforced concrete for traffic monitoring and weighing in motion, *Cem. Concr. Res.*, **29**(3), pp. 435–439.

55. Han, B., Zhang, K., Yu, X., Kwon, E., Ou, J. (2011). Nickel particle-based self-sensing pavement for vehicle detection, *Measurement*, **44**(9), pp. 1645–1650.

56. Han, B., Zhang, K., Burnham, T., Kwon, E., Yu, X. (2013). Integration and road tests of a self-sensing CNT concrete pavement system for traffic detection, *Smart Mater. Struct.*, **22**(1), p. 15020.

57. Fu, X., Lu, W., Chung, D. D. L. (1998). Ozone treatment of carbon fiber for reinforcing cement, *Carbon*, **36**, pp. 1337–1345.

58. Saafi, M. (2009). Wireless and embedded carbon nanotube networks for damage detection in concrete structures, *Nanotechnology*, **20**(39), p. 395502.

Chapter 11

Enhancing PCM Cement-Based Composites with Nanoparticles

Luisa F. Cabeza[a] **and Anna Laura Pisello**[b,c]
[a] *GREiA Research Group, INSPIRES Research Centre, Universitat de Lleida, Pere de Cabrera s/n, 25001-Lleida, Spain*
[b] *CIRIAF – Interuniversity Research Center on Pollution and Environment "Mauro Felli", Via Duranti 67, 06125 Perugia, Italy*
[c] *Department of Engineering, University of Perugia, Via Duranti 63, 06125 Perugia, Italy*
lcabeza@diei.udl.cat

11.1 Introduction

Phase change material (PCM) has been considered for thermal energy storage (TES) in buildings since before 1980 [1]. The use of PCM can overcome the lack of coincidence between energy supply and energy demand and can be applied in buildings as passive and active systems [2]. Passive TES systems can enhance effectively the naturally available heat energy sources in order to maintain the comfort conditions in buildings and minimize the use of mechanically assisted heating or cooling systems [3]. These systems include increased use of ventilated facades [4], thermal

Nanotechnology in Cement-Based Construction
Edited by Antonella D'Alessandro, Annibale Luigi Materazzi, and Filippo Ubertini

ISBN 978-981-4800-76-1 (Hardcover), 978-0-429-32849-7 (eBook)
www.jennystanford.com

mass [5, 6], shading effect using blinds [7], coated glazing elements [8], and solar heating and free cooling (night ventilation) techniques [9]. An up-to-date review of such systems can be found in de Gracia and Cabeza (2015) [10] and a review of building integrated TES in Navarro et al. (2016) [11].

PCM can be incorporated in construction materials using different methods, such as direct incorporation, immersion, encapsulation, microencapsulation, and shape stabilization [12]. In direct incorporation and immersion, potential leakage has to be assessed. When PCM is encapsulated or added in a shape-stabilized new material, a new layer appears in the construction system of the wall.

PCM can be encapsulated prior to incorporating it into building elements. There are four primary functions of the containment capsule: to meet the requirements of strength, flexibility, corrosion resistance, and thermal stability; to act as a barrier to protect the PCM from harmful interaction with the environment; to provide a sufficient surface for heat transfer; and to provide structural stability and easy handling.

11.2 Incorporation of PCM in Concrete, Mortar, or Cement

PCM can be impregnated or mixed with concrete [13] or mortar [14, 15]. One of the objectives pursued here is to preserve the concrete mechanical properties, while increasing its specific heat capacity. It is interesting to highlight that the addition of PCM in concrete decreases its density, which would have an interesting impact in the building structure weight. Fabiani and Pisello [15] also highlighted that the combination of PCM into a cement-based matrix may require dedicated thermal characterization methods because of the adaptive performance of such composite materials. Transient techniques may be recommended at a relatively larger scale than classic experimental techniques coming from chemical analysis, such as differential scanning calorimetry. In their work, they developed a brand new technique for detecting the activation of the transition phase of PCM within the specific concrete (Figs. 11.1 and 11.2).

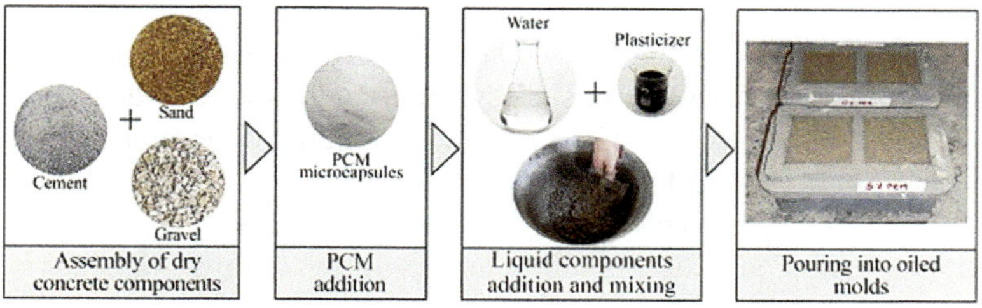

Figure 11.1 Preparation of concrete samples with PCM inclusion for dynamic tests of their thermal performance at an engineering scale. Reprinted from Ref. [15], Copyright (2018), with permission from Elsevier.

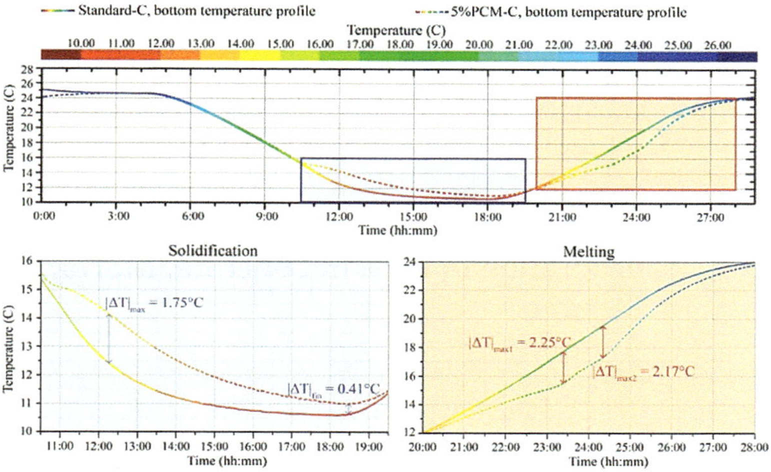

Figure 11.2 Thermal transient profiles of composite materials with PCM included into concrete, detecting transition-phase temperature ranges and duration. Reprinted from Ref. [15], Copyright (2018), with permission from Elsevier.

Cabeza et al. [16] (Fig. 11.3) showed the application of encapsulated PCM in two precast concrete house-like cubicles. One cubicle was made of conventional concrete and the other one of concrete with 5% of microencapsulated PCM. The objective here was to improve the thermal performance of buildings. Results presented a 2 h delay of the maximum peak temperature in summer because of

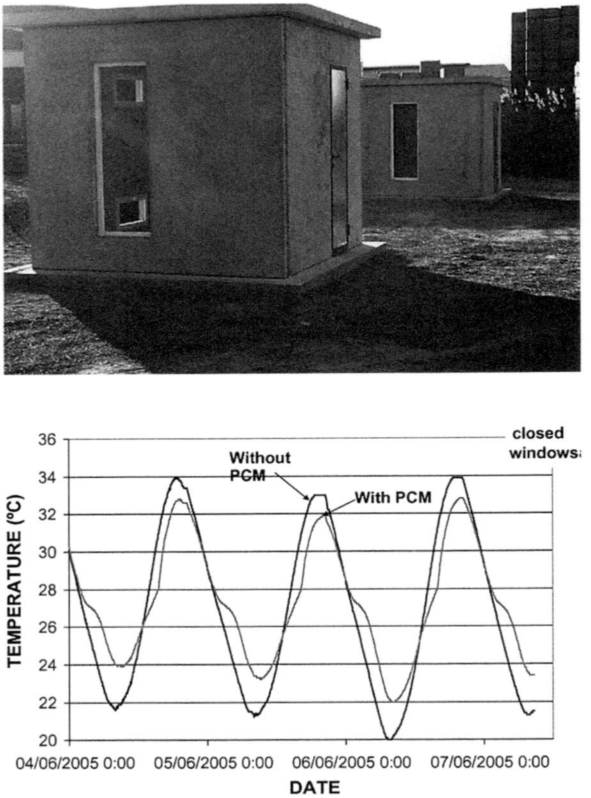

Figure 11.3 Incorporation of microencapsulated PCM in concrete-based walls. Left: Experimental setup. Right: Thermal performance. Reprinted from Ref. [16], Copyright (2007), with permission from Elsevier.

the PCM effect and also an internal temperature profile with lower fluctuations.

The most commonly used method for incorporating PCM into construction materials is microencapsulation, where small PCM particles (1–1000 μm) are encapsulated in a thin solid shell, which is made from natural and/or synthetic polymers. These microcapsules can then be directly added to the construction material, for example, concrete or plaster, during their mixing process. This method allows the possibility of providing a PCM composite over a large area, and hence it has the advantage of a high heat transfer rate per unit volume. Other advantages are that the capsules prevent leakage and resist volume change during phase change. They also have improved chemical stability and thermal reliability as phase separation during transition is limited to microscopic distances. However, the microcapsules can affect the mechanical properties of concrete and they are also relatively expensive.

A number of researchers have carried out studies using microencapsulated PCM in concrete. Hunger et al. [17] investigated the behavior and properties of self-compacting concrete mixes that contained 1%, 3%, and 5% of microencapsulated PCM by mass of concrete. A self-compacting concrete mix was used to mitigate damage to the microcapsules during compaction. Also the PCM was added to the mix at the latest possible moment to reduce its exposure to the mixing process. The authors recommended to use shells for the microencapsulated PCM in future applications to ensure that the capsules can withstand the high alkaline environment and mechanical impact during mixing. Another study in which microencapsulated high-purity paraffin PCM in an aqueous dispersion was added to self-compacting concrete was carried out by Fenollera et al. [18]. PCM dosages of 5%, 10%, 15%, 20%, and 25% of mass of cement were added to a self-compacting concrete mix, and slump, compressive strength, and density were compared in order to find an optimum combination. From the above two studies it can be concluded that 10% (by weight of cement) or 5% (by weight of concrete) is the maximum practical content of microencapsulated PCM to be used in a concrete mix application.

11.3 Enhancing PCM Microcapsules with Nanoparticles for Cement-Based Composites

If PCM is incorporated in cement-based composites in capsules with open pores, leakage will occur when the PCM melts. Therefore, Li et al. (2014) [19] evaluated the possibility of using three types of nano-SiO_2 powder (hydrophilic fumed silica [fs1], hydrophobic fumed silica [fs2], and precipitated silica [ps2]) to prevent such leakage in the commercial product PX25. PX25 is commercial form-stable PCM from Rubitherm, containing paraffin in a secondary supporting structure made of fs1. PX25 was incorporated into cement paste to fabricate TES composites. Previously, to prevent paraffin leakage, PX25 was surface-modified by mixing it with the three different nanosilicas separately. Because of the very high surface energy of the nanosilicas, they can easily deposit on the surfaces of PX25 particles. According to the authors, paraffin leakage was eliminated when fs2 or ps2 was used in a 2.3% and 9.0% mass fraction (relative to PX25), respectively. However, fs1 did not prevent leakage of paraffin. It should be highlighted that the authors reported that the compressive strength of cementitious composites decreased by incorporating hydrophobic nanosilica-modified PCM composites. A stated possible reason was slowdown of cement hydration due to the hydrophobic surface character of the PCM composites. In this view, PCM inclusion into concrete composites may also be used to partially absorb hydration heat during the curing process, as demonstrated by Fabiani et al. [19] (Fig. 11.4).

Another approach to the use of nanoparticles to decrease PCM leakage in cement-based composites is to increase the capsule strength of microencapsulated PCM. Authors found that it is important that the capsule itself be physically and chemically stable within the concrete matrix. The capsule needs to be hard and sustainable to avoid being damaged during the concrete-mixing and concrete-casting processes. Park et al. [20] discussed the use of zeolite and zeocarbon that some researchers have used for reinforcing microcapsules to enable them to withstand high friction and impact during the concrete-mixing process.

Yoo et al. (2017) [21] used cellulose nanocrystals (CNCs) to improve the mechanical and barrier properties of microcapsules. They used the advantageous characteristics of CNCs, such as high thermal conductivity, high stiffness, and low coefficient of thermal expansion with a high aspect ratio, to reinforce polymeric materials, forming a dense network resulting in sustainable, ecofriendly, and versatile polymer composites. In this study, PCM microcapsules were fabricated through an interfacial polymerization of poly(urea–urethane) (PU) as a shell component (Fig. 11.3)

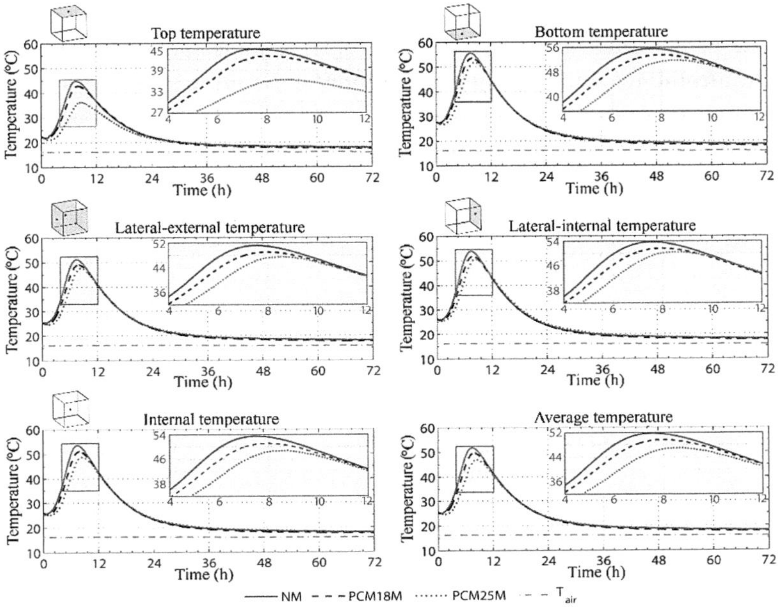

Figure 11.4 Thermal dynamics of concrete hydration process with PCM inclusion [19].

using modified cellulose nanocrystals (*h*CNCs). The strength of the microcapsules was tested physically breaking them applying a load to them. The authors claimed that these microcapsules would perform better in cement-based composites than those without nanocellulose.

Finally, carbon nanoparticles have been used to increase the thermal conductivity of a cement-based PCM composite [22]. In this study, octadecane is introduced to a cement-based composite inside diatomite particles. But before, diatomite was treated with carbon nanoparticles to increase its thermal conductivity. The results showed good results in microstructure and in mechanical and thermal properties of the composites. The compressive and flexural strength of the cement composites exhibited a decreasing tendency with an increase in the dosage of PCM. This phenomenon can be attributed to not only the poor mechanical strength and interface compatibility of the PCM–diatomite composite but also the increasing porosity of the prepared cement-based final composite. Nevertheless, the mechanical strength still satisfies the requirements for building applications. Although the properties of diatomite with PCM and carbon nanoparticle–enhanced diatomite with PCM were compared by the authors, this comparison was not done within the cement-based composite, not allowing us to see whether the addition of carbon nanoparticles would enhance the overall thermal performance of the final composite.

Acknowledgments

Prof. Luisa F. Cabeza would like to acknowledge the Spanish Government for funding (PRX17/00221), which allowed her to visit the University of Perugia for 6 months. Prof. Cabeza would like to thank the Catalan Government for the quality accreditation given to their research group (2017 SGR 1537). GREA is certified agent TECNIO in the category of technology developers from the Government of Catalonia. This work was partially funded by the Ministerio de Ciencia, Innovación y Universidades de España (RTI2018-093849-B-C31). This work is also partially supported by ICREA under the ICREA Academia program. Part of the findings also refer to SOS CITTA', supported by Fondazione Cassa di Risparmio di Perugia 2018.0499.026. Anna Laura Pisello thanks H2CU for supporting her international cooperation and research activities.

References

1. Mehling, H., Cabeza, L. F. (2008). *Heat and Cold Storage with PCM: An Up to Date Introduction Into Basics and Applications*, Springer-Verlag, Berlin Heidelberg, Germany.
2. Gil, A., Medrano, M., Martorell, I., Lázaro, A., Dolado, P., Zalba, B., Cabeza, L. F. (2010). State of the art on high temperature thermal energy storage for power generation. Part 1: Concepts, materials and modellization, *Renewable Sustainable Energy Rev.*, **14**, pp. 31–55.
3. Parameshwaran, R., Kalaiselvam, S., Harikrishnan, S., Elayaperumal, A. (2012). Sustainable thermal energy storage technologies for buildings: a review, *Renewable Sustainable Energy Rev.*, **16**, pp. 2394–2433.

4. Todorovic, B., Maric, B. (2002). The influence of double facades on building heat losses and cooling loads, Faculty of Mechanical Engineering, Belgrade University, Technical Report.
5. Castell, A., Martorell, I., Medrano, M., Pérez, G., Cabeza, L. F. (2010). Experimental study of using PCM in brick constructive solutions for passive cooling, *Energy Build.*, **42**, pp. 534–540.
6. de Gracia, A., Castell, A., Medrano, M., Cabeza, L. F. (2011). Dynamic thermal performance of alveolar brick construction system, *Energy Conserv. Manage.*, **52**, pp. 2495–2500.
7. Gratia, E., De Herde, A. (2007). The most efficient position of shading devices in a double-skin façade, *Energy Build.*, **39**, pp. 364–373.
8. Viljoen, A., Dubile, J., Wilson, M., Fontoynont, M. (1997). Investigations for improving the daylighting potential of double-skinned office buildings, *Sol. Energy*, **59**, pp. 179–194.
9. Gratia, E., De Herde, A. (2004). Natural cooling strategies efficiency in an office building with a double skin façades, *Energy Build.*, **36**, pp. 1139–1152.
10. de Gracia, A., Cabeza, L. F. (2015). Phase change materials and thermal energy storage for buildings, *Energy Build.*, **103**, pp. 414–419.
11. Navarro, L., de Gracia, A., Niall, D., Castell, A., Browne, M., McCormack, S. J., Griffiths, P., Cabeza, L. F. (2016). Thermal energy storage in building integrated termal systems: a review. Part 2: Integration as passive system, *Renewable Energy*, **85**, pp. 1334–1356.
12. Memon, S. A. (2014). Phase change materials integrated in building walls: a state of the art review, *Renewable Sustainable Energy Rev.*, **31**, pp. 870–906.
13. Lai, C., Hokoi, S. (2014). Thermal performance of an aluminum honeycomb wallboard incorporating microencapsulated PCM, *Energy Build.*, **73**, pp. 37–47.
14. Desai, D., Miller, M., Lynch, J. P., Li, V. C. (2014). Development of thermally adaptive engineered cementitious composite for passive heat storage, *Constr. Build. Mater.*, **67**, pp. 366–372.
15. Fabiani, C., Pisello, A. L. (2018). Coupling the transient plane source method with a dynamically controlled environment to study PCM-doped building materials, *Energy Build.*, **180**, pp. 122–134.
16. Cabeza, L. F., Castellón, C., M. Nogués, M., Medrano, M., Leppers, R., Zubillaga, O. (2007). Use of microencapsulated PCM in concrete walls for energy savings, *Energy Build.*, **39**, pp. 113–119.

17. Hunger, M., Entrop, A. G., Mandilaras, I., Brouwers, H. J. H., Founti, M. (2009). The behavior of self-compacting concrete containing micro-encapsulated phase change materials, *Cem. Concr. Compos.*, **31**(10), pp. 731–743.
18. Fenollera, M., Míguez, J. L., Goicoechea, I., Lorenzo, J., Alvarez, M. A. (2013). The influence of phase change materials on the properties of self-compacting concrete, *Materials*, **6**(8), pp. 3530–3546.
19. Fabiani, C., Pisello, A. L., D'Alessandro, A., Ubertini, F., Cabeza, L. F., Cotana, F. (2014). Effect of PCM on the hydration process of cement-based mixtures: a novel thermo-mechanical investigation, *Materials*, **11**(6), p. 871.
20. Park, S. K., Jay Kim, J. H., Nam, J. W., Hung, D. P., Kim, J. K. (2009). Development of antifungal mortar and concrete using zeolite and zeocarbon microcapsules, *Cem. Concr. Compos.*, **31**, pp. 447–453.
21. Yoo, Y., Martinez, C., Youngblood, J. P. (2017). Synthesis and characterization of microencapsulated phase change materials with poly(ure-uretane) shells containing cellulose nanocrystals, *ACS Appl. Mater. Interfaces*, **9**, pp. 31763–31776.
22. Qian, T., Li, J. (2018). Octadecane/C-decorated diatomite composite phase change material with enhanced thermal conductivity as aggregate for developing structuralefunctional integrated cement for thermal energy storage, *Energy*, **142**, pp. 234–249.

Chapter 12

Cement-Based Composites with PCMs and Nanoinclusions for Thermal Storage

Manila Chieruzzi and Luigi Torre
Civil and Environmental Engineering Department, University of Perugia, Strada di Pentima 4, 05100 Terni, Italy
manila.chieruzzi@unipg.it; luigi.torre@unipg.it

Carbon dioxide is one of the gases responsible for greenhouse gas emissions produced mainly by the burning of fossil fuels. Solid materials like concrete and structural cement can be used for thermal energy storage (TES).

Cement-based materials can store heat (as sensible heat) but this ability needs to be improved. To enhance the efficiency and thermal performance of cementitious materials (in particular to reduce the energy demands for heating and cooling), phase change materials (PCMs) can be used. They can store and release a large amount of thermal energy (in form of latent heat) during a phase change (usually melting/solidification). The introduction of PCMs into cement-based materials can increase the total energy stored, thus reducing energy consumption.

However, the most common PCMs suitable for cement-based materials for building applications have low thermal storage properties. Research is underway in order to develop other kinds

Nanotechnology in Cement-Based Construction
Edited by Antonella D'Alessandro, Annibale Luigi Materazzi, and Filippo Ubertini

ISBN 978-981-4800-76-1 (Hardcover), 978-0-429-32849-7 (eBook)
www.jennystanford.com

of PCMs to improve latent heat storage. In recent years, the addition of nanoparticles seems to be very promising to further enhance the thermal properties of PCMs.

This chapter reports the studies that have been carried out by many researchers on cement-based composites with PCMs for TES and the use of nanoinclusions.

12.1 Introduction

The increase in the global energy demand is an important issue in the past decades, with the consequent increase of greenhouse gas emissions.

The aim is to reduce the gas emissions by 20% by 2020.

Cement-based materials are greatly involved in this issue since the construction field represents one of the main contributors to energy consumption. The challenge is to reduce the energy demand of building constructions by enhancing their thermal energy storage (TES) capability. TES systems represent an addition of stored heat that can be used in later times.

The use of phase change materials (PCMs) as TES materials added into cement-based materials is becoming increasingly important. The energy stored by PCMs during their melting can be used later during their solidification; at this time they can release energy to the environment, thus reducing the need for electricity to maintain a certain comfort temperature.

The introduction of PCMs into cement-based materials can be done in several ways: They can be directly immersed into cements or directly added in the form of microcapsules. They can also be contained in a porous matrix able to avoid PCM leakage during the melting phase. Several researchers investigated the different methods of PCM incorporation into cements and the effect of the addition of these materials on the thermal performance of cement-based materials.

The purpose of this chapter is to report the improvements achieved in TES using PCMs in cement-based materials.

After a brief introduction on the types of TES, the chapter focuses on the experimental characterization of different PCMs and the mechanisms to realize cement-based composites with PCMs to be used as thermal storage media.

The next step of the actual research is to further improve the thermal storage behavior of cement-based materials with the use of nanoinclusions into PCMs. Thus, the effect of their addition into PCMs for cement-based materials will be reported here. These new composite materials made of cement-based materials, PCMs, and nanoinclusions may represent new solutions for the enhancement of the thermal storage ability of constructions materials.

12.2 Thermal Energy Storage

Energy storage is the temporary storage of energy that can be used at a later time, thus reducing the time or rate mismatch between energy supply and demand.

In TES systems, heat is transferred to a storage media and released at a later time [15, 20]. In particular, in solar energy systems the heat stored during the day can be preserved and released during the night. The use of heat storage in buildings can help to reduce the use of electricity, with consequent energy saving [17, 23, 45]. To make a TES system competitive with respect to other forms of energies the material used as storage media must be low cost, available in big quantities, and easy to introduce. Moreover, the storage material must have a high energy density, chemical stability, low heat losses, and low environmental impact.

Thermal storage can be sensible heat thermal energy storage (SHTES) or latent heat thermal energy storage (LHTES), with a phase transformation occurring. On the basis of storage temperature, TES can be divided into four categories: cold- ($>20^\circ$C), low- (20°C–150°C), medium- (150°C–300°C), and high- ($<300^\circ$C) temperature storage [26]. In particular, among other applications, low-temperature thermal energy storage (LTTES) is used in building constructions. Moreover, TES can use sensible heat, latent heat, and chemical heat (Fig. 12.1). The first two kinds are described in the next sections.

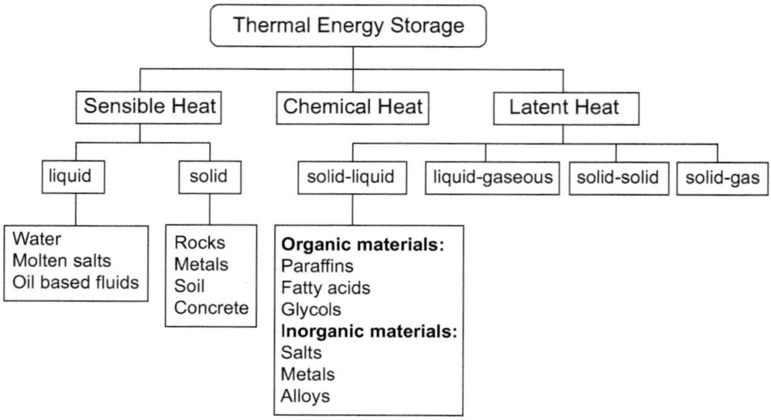

Figure 12.1 Thermal energy storage types.

12.2.1 Sensible Heat Thermal Storage

The thermal energy stored as sensible heat is due only to temperature change of a material without phase transformation. In this case the stored energy depends not only on the thermal and physical characteristics of the storage material (i.e., density, specific heat, and volume) but also on the temperature change. In particular the stored thermal energy can be expressed as follows:

$$Q = \int_{T_i}^{T_f} mC_p dT = m\overline{C}_p (T_f - T_i) = \rho\overline{C}_p V (T_f - T_i) , \qquad (12.1)$$

where Q is the stored thermal energy, C_p the specific heat at constant pressure, m is the mass of the storage medium, ρ the density of the storage medium, V the volume of the storage medium, and $(T_f - T_i)$ the temperature change (final minus initial temperature).

For example, concrete (with a thermal conductivity of 1.5 W/mK, a specific heat of 0.85 kJ/kgK, and a cost of \$0.05/kg) is a sensible heat storage medium. When it is used in buildings walls it stores sensible heat at low temperatures (from solar energy), but it could also be used as sensible heat storage medium at high temperature (200°C–400°C) in solar plants.

12.2.2 Latent Heat Thermal Storage

In latent heat thermal energy storage (LHTES) the heat is absorbed or released during a phase change of a storage medium at constant temperature [16]. These storage media are called PCMs. This TES type provides a high energy storage density and can absorb and release a large amount of heat at a constant temperature during the phase change (melting/solidification) [8].

The thermal energy stored as latent heat with solid–liquid phase change can be expressed as follows:

$$Q = m\lambda_{\mathrm{m}}, \tag{12.2}$$

where λ_{m} is the latent heat of fusion and m the PCM mass.

Usually, the melting point is not sharp (since the majority of substances are not pure), so there is a temperature range $(T_{\mathrm{i}} - T_{\mathrm{f}})$, including the melting temperature (T_{m}). Consequently, the total energy stored by a material is composed by latent heat due to phase change and also the sensible heat in both solid and liquid states. Thus, the total stored thermal energy is

$$\begin{aligned} Q &= \int_{T_{\mathrm{i}}}^{T_{\mathrm{m}}} mC_{\mathrm{pS}}dT + ma_{\mathrm{m}}\lambda_{\mathrm{m}} + \int_{T_{\mathrm{m}}}^{T_{\mathrm{f}}} mC_{\mathrm{pL}}dT \\ &= m\left[C_{\mathrm{pS}}\left(T_{\mathrm{m}} - T_{\mathrm{i}}\right) + a_{\mathrm{m}}\lambda_{\mathrm{m}} + C_{\mathrm{pL}}\left(T_{\mathrm{f}} - T_{\mathrm{m}}\right)\right], \end{aligned} \tag{12.3}$$

where a_{m} is the fused mass fraction, C_{pL} the mean specific heat at constant pressure of the liquid, and C_{pS} the mean specific heat at constant pressure of the solid.

12.3 Phase Change Materials

There are several PCMs available for the different applications. The most used PCMs are those characterized by a solid–liquid phase change, since they store a large amount of heat with a low volume change. There are three main classes of PCMs: organic compounds, inorganic compounds, and eutectics (mix of inorganics, organics, or organics/inorganics) (Fig. 1.2).

Organic PCMs can be paraffins, nonparaffins, or polyalcohols.

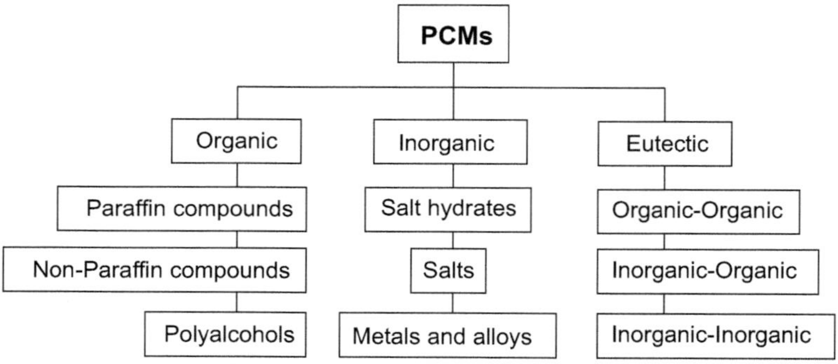

Figure 12.2 Types of PCMs.

In construction and building applications organic PCMs are the most suitable thermal storage materials to enhance the thermal performance of cement-based materials. In particular, paraffin wax is the most widely investigated in the literature, thanks to its melting point very close to thermal comfort conditions [40].

Paraffins are hydrocarbons (C_nH_{2n+2}) with the following advantages: high latent heat, chemical stability, and low price. Their melting points also range from 20°C to 70°C, making them very suitable for building applications (the higher the number of carbon atoms, the higher the melting temperature) [4, 67]. On the other hand, low thermal conductivity, large volume change during phase change, and high flammability (being hydrocarbons) represent the main limitations of this type of PCMs.

Among the nonparaffin PCMs, fatty acids ($C_nH_{2n}O_2$) have been widely investigated as energy storage material in buildings [46], thanks to their melting points being close to those of paraffins. However, their use is sometimes limited by their higher cost with respect to paraffins. The most used fatty acids are palmitic acid (or hexadecanoic acid), capric acid (CA, or decanoic acid), and lauric acid (or dodecanoic acid) [53]. Esters, glycols, and alcohols are also nonparaffin PCMs. The effect of the addition of some of these into concrete has been investigated, like 1-dodecanol (DD) [44], butyl stearate (BS) [68], and polyethylene glycol (PEG) [35].

In general, the main advantages of nonparaffins with respect to paraffins (besides their lower cost) are their smaller volume change and higher heat of fusion.

The advantages of inorganic PCMs (divided into salts, salt hydrates, and metals) include high specific heat and latent heat, sharp melting point, and nonflammability. Moreover, they show higher thermal conductivities, higher densities, and smaller volume changes compared to organic PCMs. The supercooling, the phase segregation, and instability after thermal cycles (especially for salts hydrates that tend to lose water during heating) represent the main disadvantages. The most common inorganic PCMs are salts hydrates that can be used in building applications, since they have a melting temperature ranging from −50°C to 120°C [29].

Metals and their alloys are also inorganic PCMs with a wide range of melting temperatures from low to high. The main characteristic is that metals show much higher thermal conductivities than paraffins (from 8 to 237 W/mK). Moreover, they show good physical and chemical stability. However, those characterized by low melting temperatures also show low heats of fusion; thus only metals with high melting points can compare with inorganic PCMs for the same application.

Another class of PCMs are eutectics, which are mixtures of different components, each one with a different melting temperature. Even these materials have been investigated as TES media

Table 12.1 Different type of PCMs for building applications

Class of PCM	PCM	T_m (°C)	Density (kg/m^3)	Latent heat (kJ/kg)	Thermal conductivity (W/mK)	Ref.
PA	Glycerin	17.9	-	198.7	-	[52]
NP	Butyl stearate	19	-	140	-	[67]
P	C_{16-18}	20–22	-	152	-	[67]
E	34%$C_{14}H_{28}O_2$+ 66%$C_{10}H_{20}O_2$	24	-	147.7	-	[52]
A	1-Dodecanol	26	-	200	-	[67]
FA	Capric acid	32	878 (l)	152.7	0.135 (l)	[34]
FA	Lauric acid	42–44	862 (l)	178	0.147(l)	[67]

PA, polyalcohol; NP, nonparaffin; P, paraffin; E, eutectic; A, alcohol; FA, fatty acid.

for building applications since they offer the possibility to obtain a wide range of PCMs by mixing different organic and/or inorganic compounds to obtain a particular melting temperature.

For example, eutectic mixtures of fatty acids can be used as good TES materials [24, 53]. The main physical and thermal properties of PCMs can be found in the literature [30, 52].

Some of them (focused on building applications) are reported in Table 12.1, considering that even if some have too high melting points, the combinations of some of them produce good PCMs with lower melting temperatures suitable for building applications.

12.4 Cement-Based Composites with PCMs

Cement-based materials consist mainly of mortar and concrete. They are the most common construction materials due to their large availability, low cost, and ease of production. Moreover, their large thermal mass makes them suitable as energy storage materials: during the day they can store the solar energy, while during the night they can release it to the surrounding environment.

The thermal storage ability of cement-based materials can be further increased by introducing PCMs into cement manufacturing or as part of the cement mixture, thus producing the so-called "thermocrete." Concrete, in fact, is made of cement, water, aggregates, and additives. Thus, a PCM can be easily considered as another additive or be introduced in other ways (as it will be seen later in this chapter). PCMs with low thermal conductivity integrated into the building envelopes of lightweight constructions can increase their thermal mass and reduce their thermal conductivity (passive systems). In this way, when the PCMs are introduced in wallboards they can absorb the solar energy (and melt) and release it during the nighttime (during solidification). As a consequence, summer overheats and winter heat losses can be avoided, thus achieving higher comfort.

Moreover, other studies showed that the addition of PCMs into concrete can reduce the indoor temperatures in summer or in warm climates and store energy in walls respect to conventional concrete walls, thus improving the thermal inertia [4].

One of the most important disadvantages of the introduction of PCMs into concrete is the reduction of mechanical strength and fire resistance [40]. Thus, a compromise between different final characteristics has to be done.

On the other hand, the reduction of the thermal conductivity of cement-based composite materials can be positive if they are used for thermal insulation of buildings.

In general, PCMs can be immersed or incorporated (macro-, micro- and nanoencapsulated or shape-stabilized) into cement-based materials [43, 54, 71]. The incorporation of PCMs can be done by different methods. The results obtained with different PCMs and incorporation methods are reported in the following sections.

12.4.1 Incorporation of PCMs in Cement-Based Materials Obtained with the Immersion Method

With this method, concrete is immersed in a liquid PCM bath for hours (depending on the degree of porosity of the concrete as well as on the melting temperature of the PCM). In this case, the concrete with the PCM shows similar mechanical behavior with respect to concrete without PCMs.

Hawes and Feldman [21] studied the thermal performance of several PCMs (paraffin, BS, dodecanol, tetradecanol) in concrete blocks. It was shown that autoclaved concrete blocks adsorb completely the paraffin into their pores in less time with respect to conventional concretes (less than 1 h with respect to 6 h) due to their higher porosity. Moreover, an increase of thermal storage up to 300% was recorded. However, one of the problems related to this type of method is the risk of PCM leakage during the heating cycles. For this reason PCMs with high chemical stability should be used.

12.4.2 Incorporation of PCMs in Cement-Based Materials Obtained with Direct Mixing

In this case the PCM is encapsulated so that it can be directly mixed into concrete. It is important to remember that PCMs must be able to cycle continuously for years without loss of their reacting volume. Thus encapsulation is a good way to avoid leakage of the material.

Different techniques are available to produce encapsulated PCMs. Among these, the most common are emulsion polymerization [50], interfacial polymerization [9], in situ polymerization [13, 55], and co-acervation and spray drying [22]. Moreover, encapsulation can be at the macro-, micro-, and nanoscale (up to 1 mm, from 1 μm to 1 mm, and below 1 μm, respectively) [57, 71].

In macroencapsulation the PCMs are encapsulated in a container (tubes, pouches, spheres, or panels) to be incorporated into building materials [31]. However, this solution presents the disadvantage of a possible solidification of the PCMs at the edge of the container, thus reducing the heat transfer (due to poor PCM conductivity). Thus, for macrocapsules, the large capsule size results in a temperature differential at the PCM core and boundary [4].

In micro-/nanoencapsulation of PCMs, the PCM material is surrounded by a shell (silica or organic polymer), thus forming capsules ranging from less than 1 μm up to about 300 μm [22]. In this way the PCMs can melt and solidify inside the capsule without volume change and leakage. In general, the shell protects the PCMs from leakage during the heating cycle (in particular at temperatures higher than the melting point) and must be chemically and physically stable. In fact, organic PCMs could change the concrete properties if they are free to fill the concrete, while inorganic PCMs could be corrosive if steel bars are present. Moreover, the shell must be hard enough to avoid any rupture during the concrete production. In this way microencapsulated PCMs can be introduced in concrete in liquid or powder form with no other protection. For example, it was shown that encapsulated nonflammable PCMs can be introduced into concrete, thus increasing the fire resistance of PCM-concrete. However, it should be noted that the presence of the shell material reduces the latent heat storage ability. In addition, attention should be paid since the presence of PCM capsules may decrease both the density and the compressive strengths of the resulting concrete. Figure 12.3 reports microencapsulated paraffin wax obtained with co-acervation and spray drying by Hawlader et al. [22] by using gelatin and acacia as coating materials.

The advantages of these microcapsules (50–100 μm) are mainly represented by the reduction of the reactivity of the paraffin wax with the outside environment, the increase of the heat transfer area,

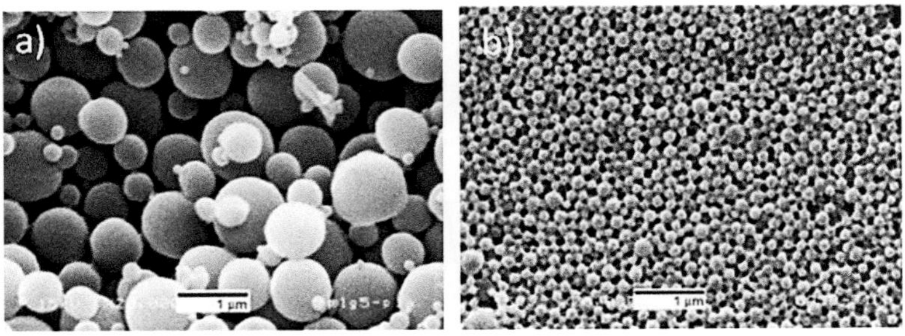

Figure 12.3 SEM images of (a) co-acervated microparticles and (b) spray-dried microparticles. Reprinted from Ref. [22], Copyright (2003), with permission from Elsevier.

and the capability to resist to several changes in volume due to the phase transformation. The authors found that that the encapsulation efficiency increases with the homogenizing time up to 10 min and the microcapsules obtained show high energy storage and release capacities (145–240 J/g).

Sari et al. [49] produced very small microcapsules (almost nanocapsules) with a size from 0.14 to 0.40 µm by using n-heptadecane (38 wt%) as core material and poly(methyl methacrylate) (PMMA) as shell material. They used the emulsion polymerization technique. They found a melting temperature of these capsules of 18.2°C, while the latent heat of melting was 81.5 J/g.

The two-step miniemulsion polymerization method was used by Li et al. [39] to produce nanocapsules with a diameter of 270 nm. In this work urea–formaldehyde resin was used for the shell and hexadecane as core material. The melting temperature of the nanocapsules produced was lower (18.2°C) with respect to pure PCM (20.3°C), as well as the melting heat (81.5 J/g with respect to 216.6 J/g of the pure PCM). The surfactant also played an important role: the higher the surfactant amount, the higher the enthalpy of the nanocapsules, and the lower the capsule size.

Several researchers investigated the effects of the addition of these capsules into cement-based materials.

Farid and Kong [14] introduced encapsulated $CaCl_2{\cdot}6H_2O$ (with 10% empty space for PCM volume expansion) into a concrete floor.

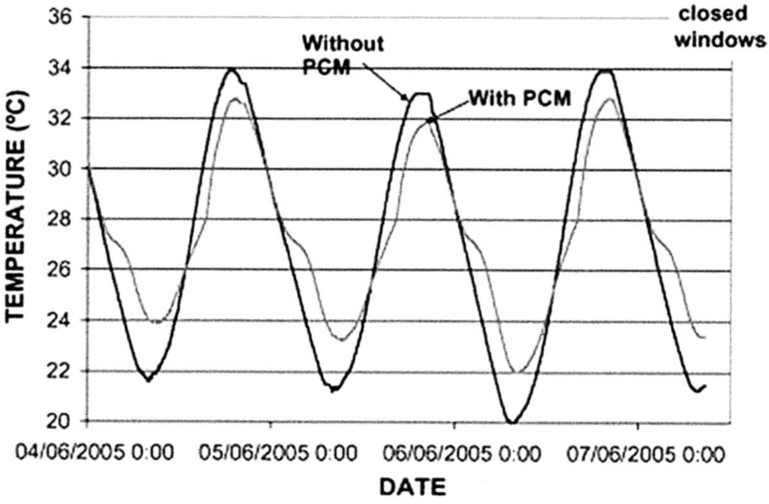

Figure 12.4 Temperature of the south wall of concrete cubicles with and without PCMs with closed windows for 2 days. Reprinted from Ref. [4], Copyright (2007), with permission from Elsevier.

This PCM showed a phase change temperature of 28°C. It resulted that the thermal mass of the concrete with the PCM increased significantly with respect to plain concrete, and it was able to keep the stored heat for more time, thus improving the existing underfloor heating system.

The energy-saving potential of microencapsulated PCMs into concrete walls was experimentally investigated by Cabeza et al. [4] by building two real concrete cubicles (2 m × 2 m × 3 m). One was made of conventional concrete and the other one of modified concrete with Micronal® PCM (from BASF) having a melting point of 26°C. In particular, they investigated the effect of PCMs on the temperatures in the cubicle and heat flux.

The cubicle without the PCM showed a maximum temperature 1°C higher than that with the PCM and a minimum temperature 2°C lower (Fig. 12.4). Moreover, the thermal inertia of the wall with the PCM increased (i.e., it reached the maximum temperature 2 h later than the wall without the PCM) with consequent lower energy consumption in the buildings.

Hunger et al. [25] analyzed the use of microcapsules (paraffin waxes encapsulated in PMMA) into concrete (1%, 3%, and 5% PCM by mass of concrete). They found that higher paraffin amounts could lead to higher heat capacity and lower thermal conductivity (up to about 3.5 times and 38%, respectively) with consequent enhancement of the thermal storage capacity of concrete.

The heat flux measurements showed a variation of the maximum and minimum peak temperatures up to 11% when 5% PCM was introduced into the concrete. Thus, even an energy saving up to 12% could be achieved to maintain the indoor temperature at 23.5°C.

In this work, however, many microcapsules were destroyed by the mixing process, thus releasing the PCM into the concrete. Moreover, it must be noted that the mechanical properties were lowered by the addition of PCM, even if they are still suitable for many structural applications.

Another study showed the possible introduction of PCMs into cement-based plastering (cement, lime, sand, and other fillers) to be used in interior coatings for buildings [47]. A microencapsulated paraffin powder based on PMMA, highly cross-linked paraffin mixture (Micronal DS5008x from BASF) was chosen as the PCM (T_m = 23°C; ΔH = 100 kJ/kg). The best weight fraction of the PCM that did not lead to cracks during the mixing procedure was found to be 25%. The cement plaster with the PCM then showed a latent heat of about 25 kJ/kg (as PCM percentage) since it is the only component in the plastering that undergoes phase change. The melting point ranged from 23°C to 25°C and thermal conductivity was 0.3 W/m°C. The tests showed a lower maximum temperature inside the environment during summer and spring due to the effect of the PCM.

Cui et al. [10] studied the incorporation of microencapsulated paraffin into cement paste produced by the in situ polymerization method.

The thermal characterization of microcapsules showed no variation of melting temperatures with respect to pure paraffin (49.9°C), while the latent heat was reduced as expected (202.71 J/g for the microcapsules with respect to 222 J/g for paraffin). The microcapsules in powder form were mixed with cement (the weight percentages used were 5%, 10%, 15%, 20%, and 25% by weight of cement). No breaking of the microcapsules during the mixing step occurred.

The work showed that the total energy storage capacity of 25% microencapsulated paraffin/cement increased by up to 3.9 times with respect to ordinary cement paste, while the thermal conductivity decreased with higher microcapsule contents. The thermal conductivity of paraffin in fact is less than that of cement. On the other hand, the latent heat of the modified cement increased with increasing content of the PCM (up to 20.47 J/kg with 25% of microcapsules).

However, a decrease of the mechanical strength of the microencapsulated paraffin/cement was found (with a maximum decrease of 67% with the addition of 20% paraffin) due to the presence of weak microcapsules.

12.4.3 Incorporation of PCMs in Cement-Based Materials Obtained with the Impregnation Method

The previous methods present some advantages and disadvantages. The immersion method, for example, is very simple and cheap, but leakage can occur into cement-based materials (with flammability problems).

The encapsulation method can resolve this issue, but it is an expensive procedure. For these reasons another method has been developed that exploits the different porosity of aggregates to absorb PCMs into them. By the impregnation method porous aggregates are firstly forced to absorb the PCM under vacuum, and then they are mixed with the other components to produce the final concrete. With this technique the degree of PCM absorption is higher than that obtained with the immersion method.

In this way, form-stable PCMs are produced by using porous supporting materials like vermiculite [27, 51], expanded perlite [51, 65], sepiolite [32], diatomite [33, 51, 63, 64], and expanded graphite (EG) [38, 69, 70].

Vermiculite is a hydrous silicate mineral (classified as a phyllosilicate) with a porous structure, nontoxic, and light (used in many commercial applications such as construction, thermal acoustic insulation, and agriculture). Perlite is a glassy amorphous

Table 12.2 Chemical compositions (wt%) of diatomite, perlite, and vermiculite [50]

Material	SiO_2	Al_2O_3	Fe_2O_3	CaO	MgO	K_2O	Other
Diatomite	92.8	4.2	1.5	0.6	0.3	0.67	0.5
Perlite	71.0–75.0	12.5–18.0	0.1–1.5	0.5–0.2	0.03–0.5	4.0–5.0	–
Vermiculite	38.0–46.0	10.0–16.0	6.0–13.0	1.0–5.0	16.0–35.0	1.0–6.0	0.2–1.2

volcanic rock with high porosity, a large surface area, very low density, high fire resistance, and low moisture retention. Sepiolite is an important natural clay material with a characteristic fibrous and tubular channel structure. The structural formula of sepiolite is $[(OH_2)_4Mg_8(OH)_4Si_{12}O_{30}]8H_2O$, which is responsible for its chemical and physical characteristics related to porosity and surface area [32]. Diatomite (or diatomaceous earth) is composed of fossil remains of diatoms, which are single-celled algae with silica cell walls with different shapes. It has low density, high porosity and purity, rigidity, and inertness [50]. An example of the chemical composition of some of these materials is shown in Table 12.2, as reported by Sari et al. [50].

Xu and Li prepared a paraffin/diatomite composite PCM and then added it into cement-based material, thus producing thermal energy storage cement-based composites (TESCs) [62]. The TESCs were then compared to normal cement (NC). The paraffin/diatomite composite PCM was fabricated by using direct impregnation method without vacuum treatment.

Paraffin and diatomite were dry-mixed, heated at 80°C ± 5°C for 4 h and remixed every 1.5 h. The mix proportion (paraffin:diatomite) of 0.9:1.0 resulted in powder form without aggregates with diatomite pores well impregnated by 47.4% of paraffin (Fig. 12.5). The latent heat and melting temperature of impregnated diatomite were found to be 70.51 J/g (with respect to 165.59 J/g of pure paraffin) and 41.1°C, respectively (just 1°C less than paraffin). The reduction of latent heat is obviously due to the presence of diatomite, and the experimental latent heat obtained is very close to the theoretical value calculated as 47.4% of paraffin latent heat. Thus, the impregnation is well done. After this, the authors analyzed the addition of 10%, 15%, 20%, and 30% of

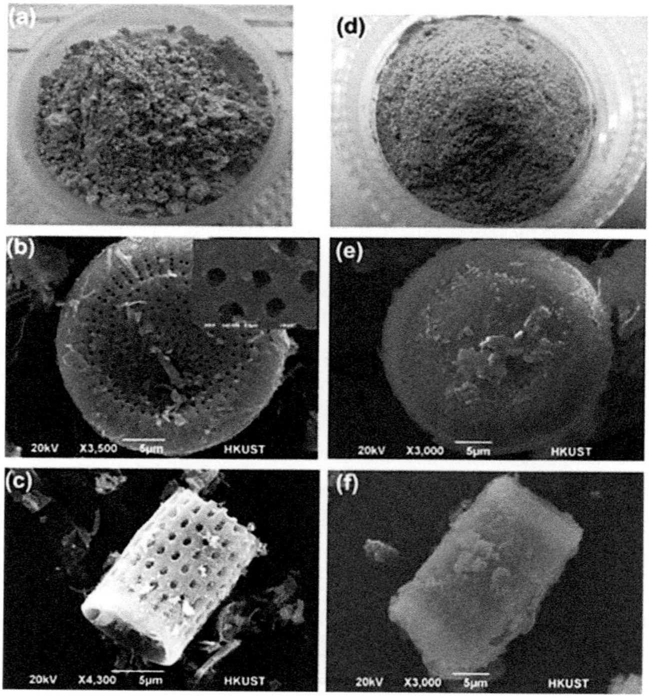

Figure 12.5 Morphology of diatomite and paraffin/diatomite composite PCM: (a) diatomite powder, (b, and c) SEM images of diatomite in disk-like and cylindrical shapes, (d) paraffin/diatomite composite PCM powder, and (e, and f) SEM images of paraffin/diatomite composite in disk-like and cylindrical shapes. Reprinted from Ref. [63], Copyright (2013), with permission from Elsevier.

paraffin/diatomite by weight of cement replacing part of the cement. Cement, paraffin/diatomite, and sand were dry-mixed at low speed for 1 min, followed by the addition of water and superplasticizer. The mixture was then mixed at low speed for 2 min and high speed for 3 min.

The compressive and flexural properties were reduced by the introduction of the paraffin/diatomite composite PCM by 48.7% to 33.6% corresponding to the minimum and the maximum percentage added. In any case this reduction is considered suitable for many applications.

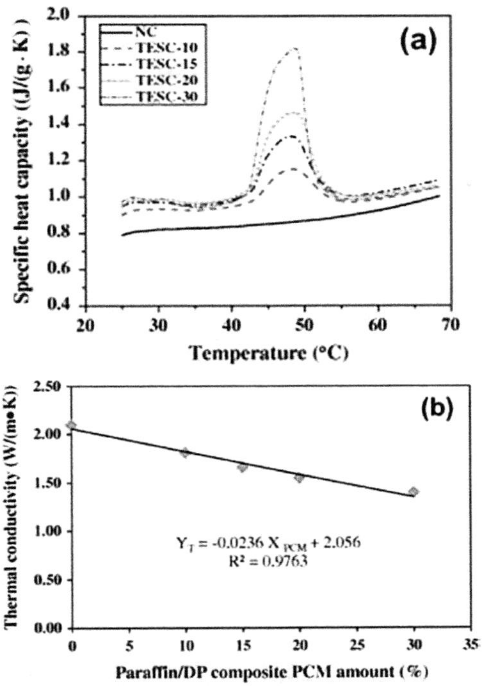

Figure 12.6 Thermal properties of paraffin/diatomite composite PCM and various TESCs: (a) specific heat capacity and (b) thermal conductivity. Reprinted from Ref. [63], Copyright (2013), with permission from Elsevier.

The authors also found that the specific heat of cement was significantly improved with the addition of paraffin/diatomite (Fig. 12.6a) due to the presence of the PCM, with a consequent increase of TES capacity from 0 (for only cement) to 5.438 J/g (for 30% of paraffin/diatomite added). Moreover, the thermal conductivity of this cement-based composite material was reduced with the incorporation of paraffin/diatomite with a positive effect if these materials are used for thermal insulation of buildings and need to be increased if they are used for thermal storage (Fig. 12.6b).

Sari et al. [50] used vermiculite, diatomite, and perlite to impregnate with fatty acid esters (galactitol hexa myristate [GHM] and galactitol hexa laurate [GHL]) esters to produce form-stable PCMs. The vacuum was kept for 90 min at 65 kPa. First, it was found that the maximum percentages of GHM confined into vermiculite, diatomite, and perlite were 52, 55, and 67 wt%, respectively, while for GHL the percentages were 39, 51, and 70 wt%, respectively.

These composite PCMs seem to have an important potential for TES applications in buildings. In fact, the melting temperatures and latent heats of the composite PCMs with GHM and GHL were reduced to about 39°C–46°C and 61–121 J/g with better results by using perlite as supporting material since it enabled a lower reduction of enthalpy. Moreover, they showed good thermal reliability, chemical stability, and thermal durability. The addition of 5 wt% of EG also enabled the increase of thermal conductivity values of the composite due to the high thermal conductivity of EG (4.26 W/mK) with respect to both PCMs and supporting materials values (ranging from 0.04 to 0.11 W/mK).

PCMs can also be confined into sepiolite, thus obtaining good TES materials for buildings applications, as showed by Konuklu and Ersoy [32]. They chose decanoic acid ($T_m = 29°C-32°C$) and paraffin ($T_m = 42°C-44°C$) as PCMs. To prepare a PCM/sepiolite composite, each PCM and sepiolite with a mass ratio of 50:50 were heated at 60°C. Heating and mechanical stirring on a hot plate was conducted for 30 min, followed by drying in an oven at 40°C for 48 h. Figure 12.7 shows the aspect of the single PCM and the produced PCM/sepiolite nanocomposite.

The authors used a simple formula to measure the real PCM content in the nanocomposite:

$$\text{PCM\%} = \frac{\Delta H_{\text{PCMcomposite}}}{\Delta H_{\text{PCM}}} 100, \tag{12.4}$$

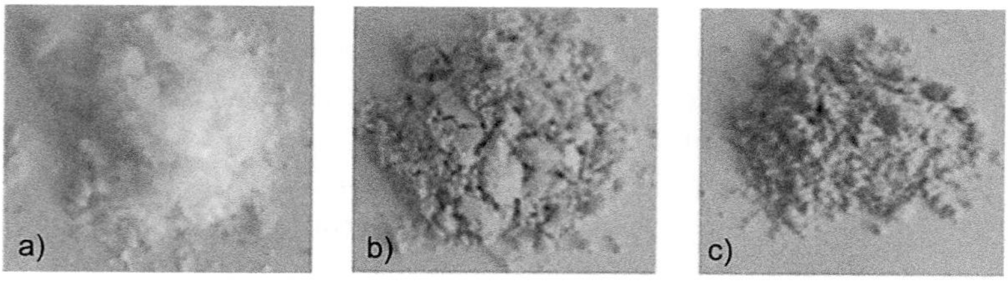

Figure 12.7 (a) Decanoic acid, (b) sepiolite, and (c) decanoic acid/sepiolite nanocomposite. Reprinted from Ref. [32], Copyright (2016), with permission from Elsevier.

where $\Delta H_{\text{PCMcomposite}}$ is the enthalpy of the composite and ΔH_{PCM} is the enthalpy of the pure PCM obtained by differential scanning calorimetry (DSC) analysis.

The latent heats of melting decreased as expected (due to the low amount of PCM in the composites) by about 63% for the paraffin/sepiolite nanocomposite (i.e., 37% of PCM content) and by about 78% for the decanoic acid/sepiolite nanocomposite (i.e., 22% of PCM content).

DCM analysis showed also melting temperature values of the composites of 35.70°C for the paraffin/sepiolite nanocomposite and 22.64°C for the decanoic acid/sepiolite nanocomposite. The composites showed good thermal stability and no leakage from sepiolite, thus obtaining good thermal storage materials.

Li et al. [38] investigated the heat storage properties of cement mortar with EG/paraffin. In this work 85 wt% of paraffin was impregnated into EG by the vacuum absorption method and then mixed with cement. The work showed that the melting temperature and latent heat of pure paraffin changed with the incorporation of EG/paraffin. In particular, the melting temperature slightly decreased (from 28.8°C to 28.5°C), while the latent heat was reduced (from 209.3 to 183.0 kJ/kg). This reduction is due to the reduced presence of paraffin (85% instead of 100%).

The heat storage of the cement mortar with EG/paraffin was also evaluated by recording the temperature inside a cube. In this way it was found that it can be decreased by 2.2°C, and also the indoor temperature fluctuation is reduced. The cement with EG/paraffin also showed a higher heat storage coefficient.

The authors calculated the heat storage coefficient (S) as follows:

$$S = \sqrt{\frac{2\pi kC\rho}{Z}}, \tag{12.5}$$

where S describes the ability to store and release heat and to resist the surface temperature variation under the heating effect of periodic fluctuation. A high value of S is correlated to a little surface temperature variation. They found that S for the composite cement is 1.74 times that of NCs. Moreover, the introduction of PCMs into cements decreases their mechanical properties. In this work 15%, 20%, and 25% of PCMs were studied, and it was found that 20% is

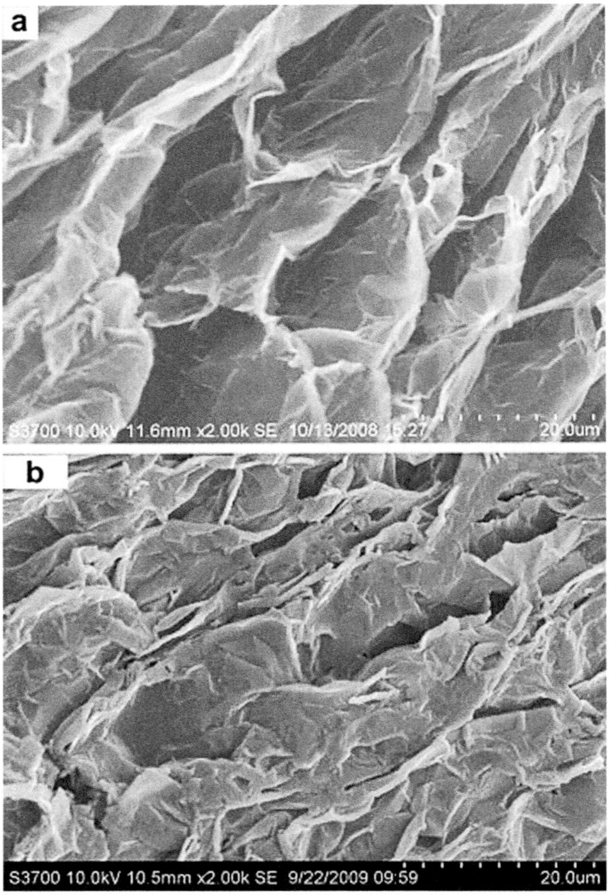

Figure 12.8 SEM images of (a) EG (×2000) and (b) the n-octadecane/EG composite PCM (×2000). Reprinted from Ref. [69], Copyright (2013), with permission from Elsevier.

the maximum percentage that can be added to increase the thermal storage properties and at the same time keep good mechanical performances.

Zhang at el. [68] also produced a cement mortar containing n-octadecane/EG for TES. First, it was observed that 90% of the PCM could be absorbed into the EG pores (Fig. 12.8), thus preventing PCM leakage, and the melting temperature of the composite PCM was close to pure n-octadecane (around 27°C). The latent heat was obviously decreased but still in agreement with the theoretical value. The thermal energy storage cement mortar (TESCM) was produced by adding the n-octadecane/EG composite PCM into the ingredients of the ordinary cement mortar and then mixing. Four different mass ratios of the composite PCM to the cement were tested: 0.02, 0.05, 0.07, and 0.1. In this way the percentages of the n-octadecane/EG composite PCM in the cement mortars were 0.5%, 1.2%, 1.7%, and 2.5%, respectively. A decrease of compressive strength of the mortars was highlighted with the increase of the composite PCM introduced. A similar behavior was shown by the thermal conductivity of the mortars (with a reduction up to 15.5%) with a positive effect if these mortars were used as insulation materials.

Moreover, the thermal performance of buildings with this composite PCM may take advantage from the presence of the PCM in the graphite pores, because the indoor temperature is kept constant by this material.

Mei et al. [41] prepared CA/halloysite nanotube (HNT) composite material for TES for building applications.

HNT is a two-layered aluminosilicate clay mineral similar to kaolin but with a hollow nanotubular structure and a large specific surface area. The idea was to absorb CA in these nanotubules. The composite with 60 wt% CA and 40 wt% HNT showed reduced latent heat (75.52 J/g compared to 139.77 J/g of pure CA) and a slightly changed melting temperature (around 29.34°C, which is only 1°C lower than that of pure CA). No leakage was detected during the thermal cycles. The most important result was the increase of thermal storage capacity by 1.8 times with the addition of graphite into the composite, as well as the increase of thermal conductivity of the CA/HNT with graphite.

12.5 PCMs and Nanoinclusions for Cement-Based Materials

The thermal properties of PCMs can be enhanced with the addition of nanoinclusions. It has been proved in fact that nanomaterials with dimensions in the range of 1–100 nm can be effective in improving the thermal conductivity, latent heat, and thermal storage capacity of a base PCM, thanks to their high surface-to-volume ratio.

The first step is the selection of the most suitable PCM, and the next step is the selection of the nanoinclusion to add. The materials obtained in this way are called nanoPCMs or nanoenhanced phase change materials (NEPCMs). These materials can then be added to cement-based materials in order to increase their overall thermal performances.

In the following sections, the production of PCMs with nanoinclusions suitable for cement-based materials and their thermal characteristics are reported. Finally, the characteristics of some cements produced with the addition of nanoPCMs are presented.

12.5.1 Selection of PCMs

The selection of a PCM is conditioned mainly by the melting temperature (which must be suitable for the application), latent heat, thermal conductivity, and specific heat. Other important characteristics are safety, cost, availability, and long-term durability and chemical stability. According to these properties, the most suitable PCM can be selected.

12.5.2 Selection of Nanoparticles

Nanoparticles are defined as particles with a size between 1 and 100 nm with a wide range of applications.

Nanoparticles are of great scientific interest since size-dependent properties can be observed on the nanometer scale due to their large surface area with respect to volume. In this way, they show a great driving force for diffusion. The addition of nanoparticles has been studied to improve the thermophysical properties of PCMs. In particular, in heat transfer and thermal storage heat applications, it has been shown that nanoparticles could enhance the specific heat of a base PCM since their specific heat is much higher than the bulk material [2, 6, 8, 57, 59]. The only disadvantage is their possible agglomeration into clusters.

There are different kinds of nanoinclusions. For cementitious materials the most used are oxide nanoparticles and carbon

Table 12.3 Types of nanoinclusions used in PCMs for cement-based materials

Type	Size (nm)	Density (g/cm^3)	Thermal conductivity (W/mK)
Cu	50–100	0.8	401
Graphite	1.5–4	1.2–2.8	1960
SiO_2	7	2.65	–
TiO_2	20	4.23	–
Al_2O_3	50	3.89	30
ZnO	15, 30, 50	0.15, 0.5, 0.4	–
CuOx	27	6.31	32.9
ZrO_2	40	5.89	–
SWCNT	di: 0.8–1.6 de: 1–2	2.1	50–200
MWCNT	di: 5–15 de: 3–5	2.1	3000
GNP	H: 420	–	3000
GNF	d: 2–100	–	3000

SWCNT, single-walled carbon nanotube; MWCNT, multiwalled carbon nanotube; GNP, graphene nanoplatelet; GNF, graphite nanofiber.

nanofillers. Table 12.3 shows some of the main nanoinclusions for cement-based materials. Recent studies showed that carbon-based nanostructures, in particular, may increase the thermal conductivity, heat storage ability, and thermal stability of PCMs.

Other researchers reported an increase of the thermal conductivity of PCMs with TiO_2, Al_2O_3, and ZnO nanoparticles. Some of these works mainly focused on PCMs and nanoinclusions for cement materials will be discussed in the following section.

12.5.3 PCMs and Nanoinclusions for Cement-Based Materials

The thermal conductivity and thermal storage of organic PCMs are low; thus even the heat storage/release rate of organic PCMs is low. This limits both the solar energy storage capability and the heat storage in a building (to overcome the heat dissipation speed to the environment and change the inner temperature).

Recent research tends to improve the thermal behavior of PCMs. The first idea was to introduce nanoinclusions with high thermal conductivity, such as carbon nanofillers: carbon nanotubes (CNTs), graphene nanoplatelets (GNPs), nanographite (NG), graphite and nanofibers (graphite nanofibers [GNFs], carbon nanofibers [CNFs]), and graphene oxide nanosheets (GONs). Graphene consists of a sheet of carbon atoms arranged in a honeycomb lattice with high thermal conductivity (in the range of 4840–5300 W/mK), while graphite is made of a sequence of graphene layers. CNTs can be considered a rolled sheet of graphene. According to the number of sheets, CNTs can be single-walled CNTs (SWCNTs) or multiwalled CNTs (MWCNTs). They are generally a few nanometers in diameter and several microns in length. In cement-based materials they have been considered mostly to enhance electrical and mechanical properties. NG consists of a 2D layer structure with a large length-to-diameter ratio. It has larger specific surface area than exfoliated graphite, and EG and forms a network structure in PCMs.

Recently, Babaei et al. [3] reported the improvement of thermal conductivity of paraffin, ethylene glycol, and palmitic acid obtained by different authors with the addition of several carbon nanofillers (Table 12.4). Other authors investigated this topic. Fan et al. [12] studied paraffin with different carbon nanofillers (short and long MWCNTs, CNFs, and GNPs) whose morphology was found with scanning electron microscopy (SEM) analysis (Fig. 12.9).

They showed a decrease of paraffin enthalpy due to the addition of these nanofillers and a strong dependence of thermal properties on the geometry of the carbon nanofillers. Moreover, it was shown that 5% of GNPs can increase the thermal conductivity of paraffin by

Table 12.4 Thermal conductivity enhancements with different carbon-based nanofillers [3]

Filler	PCM	Percentage	Phase	Thermal conductivity increase (%)
GNP	Paraffin	1 wt%	Solid	~40
GNP	Paraffin	2 vol%	Solid	~195
GNP	Paraffin	1 wt%	Liquid	20
GNP	Paraffin	1 wt%	Solid	~60
GNP	Paraffin	1 wt%	Liquid	~55
GON	Paraffin	5 vol%	Liquid	75.8

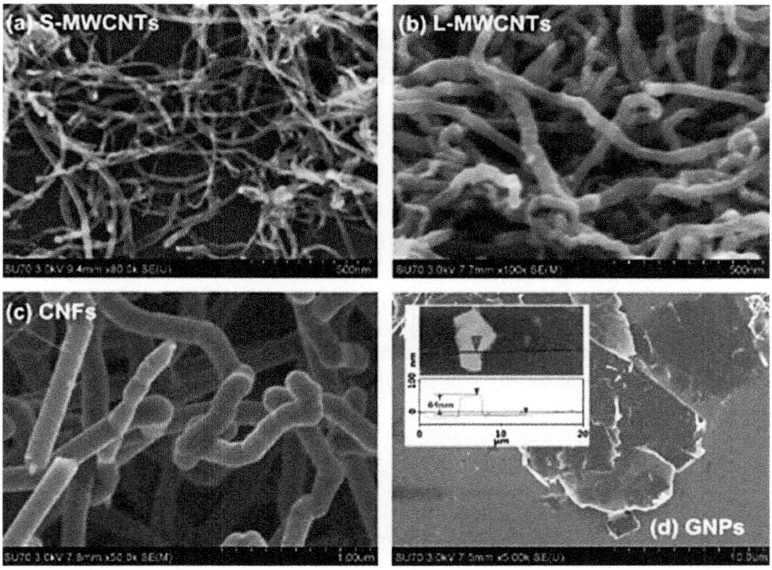

Figure 12.9 SEM images of (a) S-MWCNTs, (b) L-MWCNTs, (c) CNFs, and (d) GNPs. The inset in (d) is the atomic force microscopy (AFM) image of an individual GNP showing its thickness. Reprinted from Ref. [12], Copyright (2013), with permission from Elsevier.

170% due to the planar geometry of GNPs, inducing a lower thermal interface resistance.

Kaviarasu et al. [28] reviewed the use of nanoparticles in PCMs for building applications. They reported the use of exfoliated graphite nanoplatelets (xGnP), which is a porous nanosized carbon material useful as a supporting structure. In particular, the addition of 5 wt% xGnP in octadecane led to an increase of thermal conductivity by 101%. Organic fatty acid ester PCMs like coconut oil and palm oil (with T_m = 26.78°C and 17.26°C, respectively) were also incorporated into the porous structure of xGnP using the vacuum impregnation method [61].

An enhancement of thermal conductivity by 400% was shown. These materials can be used to reduce the building energy consumption and to improve the indoor comfort.

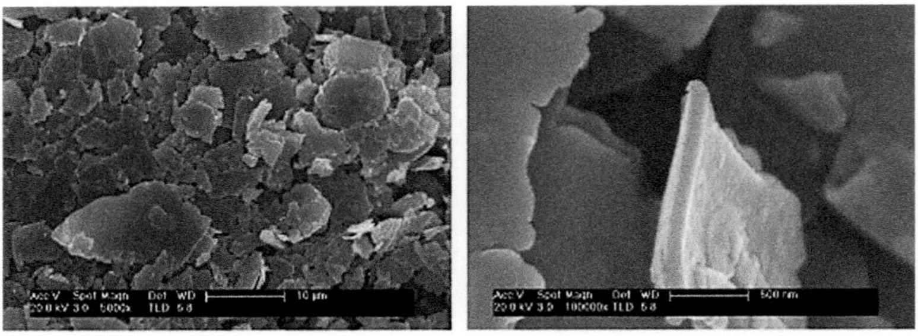

(a) x 5000 magnification (b) x 100000 magnification

Figure 12.10 Microstructure of nanographite at (a) 5000× and (b) 100000×. Reprinted from Ref. [37], Copyright (2013), with permission from Elsevier.

Li [37] studied the addition of NG to paraffin. Figures 12.10 and 12.11 show the microstructure of NG and the NG/paraffin composite. The study of Li aimed to accelerate the heat transmission and to improve the energy storage efficiency by using NG (1%, 4%, 7%, and 10%) with a diameter of 35 nm into paraffin. In particular, the paraffin/NG composite was produced by dispersing NG into the PCM by agitation and ultrasound at 60°C, cooling to 20°C, and grounding, obtaining a good distribution (Fig. 12.11).

It was found that the thermal conductivity of the composite PCM with 1% and 10% NG increased by 2.89 times and 7.41 times, respectively, with respect to pure paraffin.

On the contrary, the latent heat decreased with NG content increase as a consequence of the reduction in PCM content in the composite (Fig. 12.12).

The latent heat of the composite PCM ($\Delta H_{\text{PCMcomposite}}$) can be calculated by the following equation:

$$\Delta H_{\text{PCMcomposite}} = \eta \Delta H_{\text{PCM}}, \tag{12.6}$$

where ΔH_{PCM} is the latent heat of the PCM and η is the mass fraction of NG added to the PCM. Finally, the introduction of NG into paraffin doesn't seem to significantly change the melting temperature of the composites.

Sanusi et al. [48] studied the addition of GNFs into a paraffin (n-tricosane). GNFs have a diameter of 2–100 nm, length up to 100 μm, and a high surface area with thermal properties close to

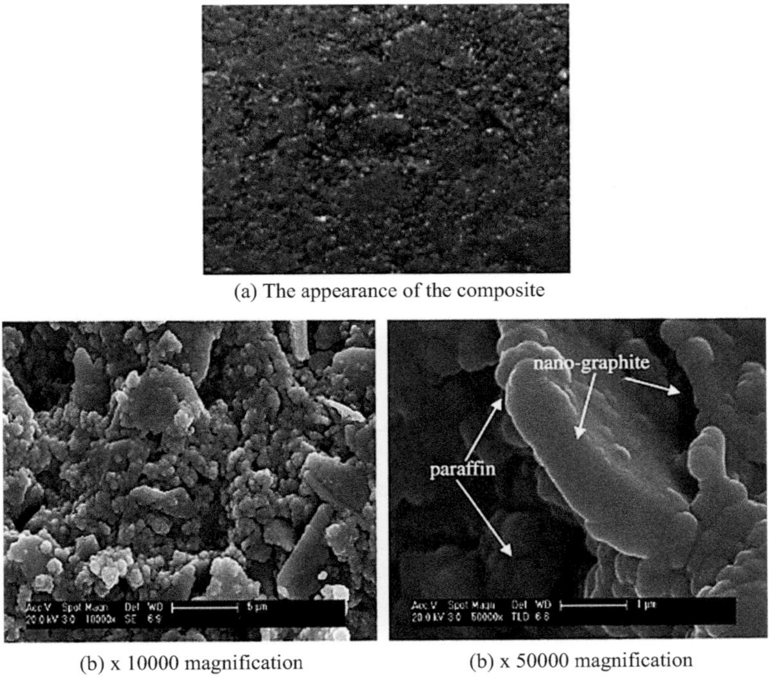

Figure 12.11 Morphology of paraffin/nanographite composite: (a) macromorphology and (b and c) microstructure at 10000× and 50000×, respectively. Reprinted from Ref. [37], Copyright (2013), with permission from Elsevier.

CNTs. The liquid paraffin was mixed with 10 wt% of GNFs using ultrasound. The study showed that paraffin/GNFs can be effective in increasing the TES of paraffin.

Moreover, a reduction in the solidification time by 61% was found with respect to pure paraffin, thus enhancing the diffusion rate.

Graphite nanosheets (GNs) with a sheet-like structure and a relatively high specific surface area were introduced into paraffin and the thermal properties investigated [5]. The thermal conductivities of the GN/paraffin composites with 5.0 wt% random and oriented GNs were 4.47 ± 0.15 and 1.68 ± 0.07 W/mK, respectively.

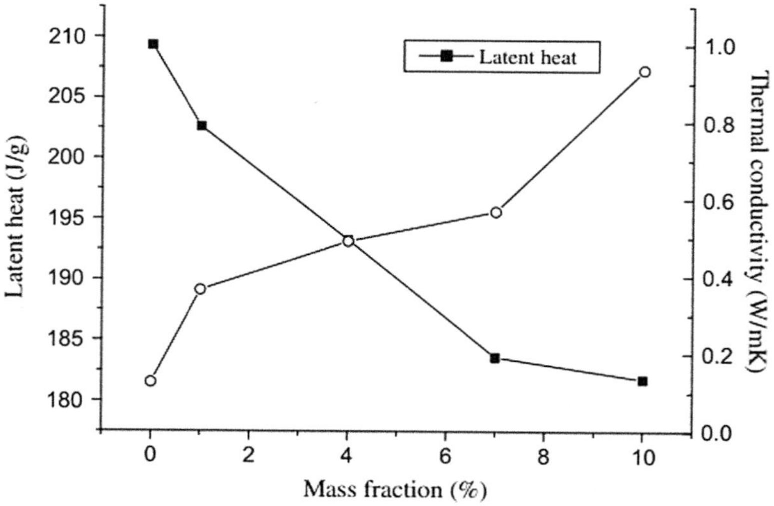

Figure 12.12 Latent heat and thermal conductivity of the composite as a function of NG mass fraction. Reprinted from Ref. [37], Copyright (2013), with permission from Elsevier.

The latent heats of the random GN/paraffin composites decreased with increased GN loading, while the melting point was not modified.

However, it was shown by Yu et al. [66] that GNPs are more effective in enhancing the thermal conductivity of paraffin with respect to CNTs and CNFs, thanks to their 2D planar structure. The final nanofluid viscosity also shows the lightest increase.

In the past decade oxide nanoparticles have also been added to PCMs for thermal storage. Teng et al. [56] added alumina (Al_2O_3), silica (SiO_2), titania (TiO_2), and zinc oxide (ZnO) to paraffin with a weight percentage of 1%, 2%, and 3%. They showed that titania was the most effective in enhancing heat conduction and thermal storage. Moreover, it was demonstrated that titania extends the temperature phase change range of paraffin by lowering the melting onset temperature and decreasing the solidification onset temperature. Finally, the latent heat is almost unaffected by the addition of titania.

New composite PCMs have been prepared with a mixture of lauric acid (LA) and stearic acid (SA) and TiO_2, ZnO, and CuO nanoparticles [19], and their TES behavior for building heating applications was investigated. It was shown that 1 wt% of CuO gives the best results in terms of thermal conductivity enhancement (62.12%) with respect to the other nanoparticles (34.85% and 46.97% with TiO_2 and ZnO, respectively) and in terms of thermal stability. The same kind of nanoparticle showed greater time savings for melting and solidification, making the composite LA/SA/CuO nanoparticles the best candidate for building heating applications. Negligible changes in phase change temperatures and latent heats indicate a strong physical interaction between the base PCM and the nanoparticles.

Ai et al. [1] investigated the performance of SA with the addition of zirconium oxide nanoparticles, while Wang et al. [59] focused on paraffin wax and alumina nanoparticles. In the first study the addition of 23 wt% of ZrO_2 nanoparticles to SA by using chloroform as a dispersant showed an increase of the heat capacity of the base PCM related also to the size of the PCM composite (with the best heat performance with 1.2 μm).

In the second study the Al_2O_3/paraffin composites decreased both the latent heat TES capacity and the melting point. On the contrary, the thermal conductivity was enhanced by the presence of alumina nanoparticles.

Besides oxide nanoparticles and carbon nanoinclusions that are most used, an example of the use of metallic nanoparticles can be found in the literature. In particular, a study concerning the use of 1.0 wt% of Cu, Al, and C/Cu nanoparticles in paraffin was conducted, and it was found that the nanoparticles have a different effect on the heat transfer rate. The heating time was reduced by 9.7%, 7.4%, and 4.2% for Cu/paraffin, Al/paraffin, and (C/Cu)/paraffin, respectively, while the cooling time was reduced by 6.7%, −0.4%, and −5%, respectively. Thus copper nanoparticles were more effective, and they also induced an enhancement of the thermal conductivity of paraffin as well as its melting rate and a reduction of latent heat [62].

12.5.4 NEPCM-Cement-Based Materials for Building and Construction Applications

It was shown that the mechanical properties of cement-based materials are reduced by the addition of microencapsulated PCMs. For example, the compressive strength of typical paraffin wax is about 4.5 MPa at room temperature, while the compressive strength of cement-based materials is higher than 30 MPa. Moreover, the thermal conductivity is also decreased. This can be either positive (when insulation is to be improved) or negative (when energy storage increase is the main scope).

To face these problems Han et al. [18] studied the performance of cementitious construction materials with microencapsulated paraffin wax ($T_m = 26°C$) and CNTs as nanoinclusions. CNTs, in fact, seem to be the best filler for cement-based materials to improve both mechanical and thermal behavior, thanks to their high aspect ratio, low density, and very good mechanical and thermal properties. Han et al. investigated in particular the use of CNTs and PCMs in cement mortars for house walls.

The PCM used was a commercial one (Micronal® DS 5000X provided by BASF), which is an aqueous microcapsule dispersion for construction materials containing PCMs. The microcapsules were made of PCMs in a core of a special wax mixture.

CNTs were added into an aqueous solution of superplasticizer and water and sonicated for 2 h to achieve a uniform dispersion. Afterward, a cement paste mixer was used to mix cement, 5% PCM, and the aqueous solution with superplasticizer and 1% CNTs for 3 min.

The modified cement mortar with PCM/CNTs showed improved heat insulation properties than plain cement mortar (6.8°C difference was observed among two building models made of plain cement mortar and cement mortar with PCM/CNTs). Moreover, the introduction of CNTs enhances the thermal conductivity of paraffin, which accelerates the heat absorption during the phase change. As a consequence buildings constructed with materials containing PCM/CNTs may ensure better thermal performance.

Xu et al. [64] also chose paraffin and MWCNTs for TESCs but with the use of diatomite. In particular, they used 47.37% paraffin,

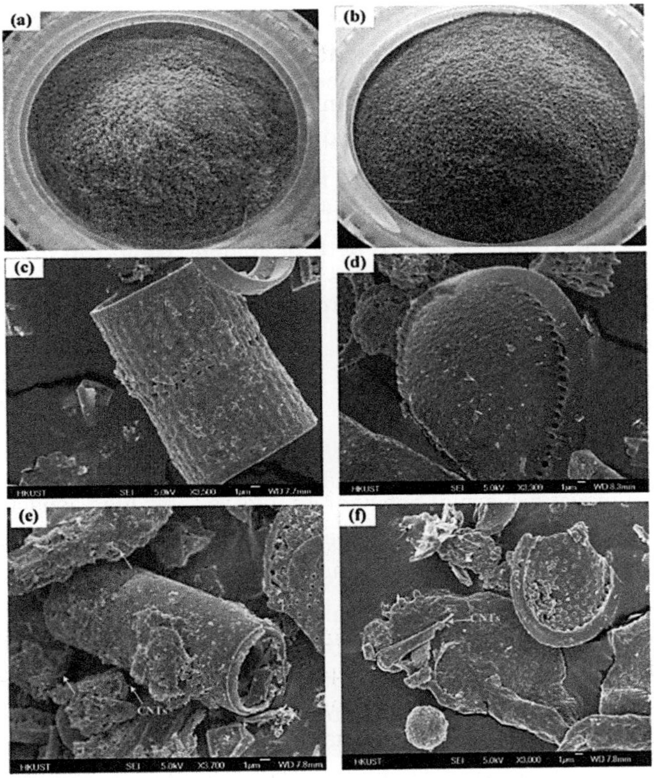

Figure 12.13 (a) PCM/diatomite composite, (b) PCM/diatomite/CNT composite, (c and d) SEM morphologies of PCM/diatomite, and (e and f) SEM morphologies of PCM/diatomite/CNTs. Reprinted from Ref. [64], Copyright (2014), with permission from Elsevier.

52.63% diatomite, and 0.26% MWCNTs. They used the direct impregnation method without vacuum treatment. The composite was prepared by first mixing nanotubes and diatomite (in acetone by magnetic stirring and evaporation) and dry paraffin, followed by heating at 80°C in an oven for 4 h (Fig. 12.13). The paraffin actually was found well confined into the diatomite pores with CNTs in the structure (Fig. 12.13c–f). They found that diatomite calcined at 600°C for 2 h is optimal to produce an innovative PCM composite (with a melting temperature of 27.12°C and a latent heat of 89.40 J/g, which is almost half of pure paraffin).

The composite also showed good thermal stability and chemical compatibility. An enhancement of the thermal conductivity by about 42.45% was due to the presence of CNTs in low amount (0.26%). Moreover, heat/storage release is fastened by the addition of CNTs.

Finally, the work also demonstrated that the use of diatomite can enhance the compressive strength of cement when the diatomite replaces part of the cement (i.e., 10% of the cement). Another study conducted by Cui et al. [11] focused on the performance of cement paste with microencapsulated paraffin and MWCNTs as nanoinclusions with a phase change temperature of 28°C for structural applications.

This research follows a previous one where the thermal effect of the introduction of microcapsules into cement was verified [10] (Section 12.4.2).

In this work the CNTs were mixed with PCM/cement in dry state for 5 min at high speed. Then, water and superplasticizer were added and mixed for 10 min to achieve a uniform dispersion of CNTs.

The introduction of MWCNTs in microencapsulated PCM cement paste accelerated the cement hydration reaction and also enhanced the mechanical performance after 28 days (+41% for flexural strength and +5% for compressive strength). The best CNT percentage was found to be 0.5 wt%. Moreover, it was showed that the temperature inside the room decreased by 4.6°C with increasing PCM nanoinclusions. Thus it was possible to realize cement with good thermal behavior without lowering the mechanical properties.

Lucas [41] proposed a new multifunctional mortar combining energy storage, self-cleaning, and air-depolluting capabilities by using TiO_2 nanoparticles and an encapsulated PCM (paraffin wax into capsule of PMMA) with a melting point of 23°C and latent heat of 135 J/g. The author used 20%–30% PCM and 2.5%–5% TiO_2.

The heat storage capability of these modified mortars was experimentally investigated by submitting test cells to a temperature cycle (held at 10°C for 10 min, heating up to 40°C at 0.5°C/min, and held at 40°C for 10 min) inside a climatic chamber. The cycle was repeated for 24 h.

Interesting results showed that the internal pore distribution plays an important role in heat transfer. It was found that mortars with 20%PCM-2.5%TiO_2 and 30%PCM-2.5%TiO_2 led to similar temperature reductions with respect to the outside temperature, while the mortar with 20%PCM-5%TiO_2 nanoparticles was more efficient than the one with 30% PCM. The latter in fact showed a strong reduction of macropores (above 5 μm) that seem to be the only responsible for the heat efficiency.

Another application of MWCNT/cement-based material was reported recently by Li et al. [36] for a novel self-deicing road system. In this case, the MWCNT/cement-based layer was proposed as a thermal conduction layer among others. It was shown that this layer with 3 wt% of CNTs had higher thermal conductivity (2.83 W/mK) than common cement-based composites (1.58 W/mK), thus contributing to the deicing efficiency and low energy consumption of the proposed self-deicing road system.

12.5.5 Recent Developments in NEPCM-Cement-Based Materials for High-Temperature Thermal Storage

In the past few years a new application of cement-based materials as thermal storage materials was studied and proposed. Besides building applications, cement materials can, in fact, be used to store heat at medium/high temperatures in solar heat plants.

In this way the cement-based materials could be used as TES materials in solar power plants by absorbing the latent heat during sunny days and release it during the night or on cloudy days. An attempt to produce such cementitious materials was made by Chieruzzi et al. [7]. They produced a modified cement mortar by using a nitrate salt mixture ($NaNO_3/KNO_3$, 60/40) with a phase change temperature of 220°C to cement, sand, and water. They tried different type of additions from the direct mixing of the salt into the cement mixture to the impregnation of the salt into diatomite and subsequent mixing with the other additives. Cement mortar samples with the PCM were then subjected to mechanical and thermal characterization. A good impregnation of the PCM into the diatomite pores was the first result. Then, the study showed an enhancement of compressive and flexural strength for cement with PCM/diatomite. The thermal conductivity was also enhanced. The materials also showed good thermal stability after several heating cycles with no leakage.

12.6 Conclusions

In this chapter, an overview was provided on the cement-based materials with PCMs and with nanoinclusions for TES. In particular, after a brief introduction of the different types of TES systems and the main PCMs used for cementitious materials, the effect of PCMs and nanoPCM on cement mortar, concrete, and plastering has been reported and discussed. It has been shown that the addition of PCMs into cement-based materials can lower the temperature peak variation inside buildings and constructions, thus enhancing the thermal comfort and saving energy for heating and cooling. Research has been conducted on the effect of nanoparticles into PCMs. It was proved that nanoparticles can increase the thermal storage of PCMs, promoting their use as potential additives for cementitious materials. However, little investigation has been made on cement-based materials incorporating nanoPCM. Thus the future challenge is to characterize concrete with nanoPCM to evaluate their thermal and mechanical behavior.

References

1. Ai, D. S., Su, L. Z., Gao, Z., Deng, C. S., Dai, X. M. (2010). Study of ZrO_2 nanopowders based stearic acid phase change materials, *Particuology*, **8**(4), pp. 394–397.
2. Avramov, I., Michailov, M. (2008). Specific heat of nanocrystals, *J. Phys.: Condens. Matter*, **20**(29), p. 295224.
3. Babaei, H., Keblinski, P., Khodadadi, J. M. (2013). Improvement in thermal conductivity of paraffin by adding high aspect-ratio carbon-based nano-fillers, *Phys. Lett. A*, **377**(19–20), pp. 1358–1361.
4. Cabeza, L. F., Castellon, C., Nogues, M., Medrano, M., Leppers, R., Zubillaga, O. (2007). Use of microencapsulated PCM in concrete walls for energy savings, *Energy Build.*, **39**(2), pp. 113–119.

5. Chen, Y. J., Nguyen, D. D., Shen, M. Y., Yip, M. C., Tai, N. H. (2013). Thermal characterizations of the graphite nanosheets reinforced paraffin phase-change composites, *Composites Part A*, **44**, pp. 40–46.
6. Chieruzzi, M., Cerritelli, G. F., Miliozzi, A., Kenny, J. M. (2013). Effect of nanoparticles on heat capacity of nanofluids based on molten salts as PCM for thermal energy storage, *Nanoscale Res. Lett.*, **8**(1), pp. 448–448.
7. Chieruzzi, M., Kenny, J., Torre, L. (2015). Studio di nuovi materiali avanzati di accumulo termico a cambiamento di fase con proprietà termiche incrementate.
8. Chieruzzi, M., Miliozzi, A., Torre, L., Kenny, J. M. (2017). Nanofluids with enhanced heat transfer properties for thermal energy storage, in Tiwari, A., Mishra, Y. K., Kobayashi, H., Turner, A. P. F. (eds.) *Intelligent Nanomaterials*, Wiley, pp. 295–360.
9. Cho, J. S., Kwon, A., Cho, C. G. (2002). Microencapsulation of octadecane as a phase-change material by interfacial polymerization in an emulsion system, *Colloid. Polym. Sci.*, **280**(3), pp. 260–266.
10. Cui, H. Z., Liao, W. Y., Memon, S. A., Dong, B. Q., Tang, W. C. (2014). Thermophysical and mechanical properties of hardened cement paste with microencapsulated phase change materials for energy storage, *Materials*, **7**(12), pp. 8070–8087.
11. Cui, H. Z., Yang, S. Q., Memon, S. A. (2015). Development of carbon nanotube modified cement paste with microencapsulated phase-change material for structural-functional integrated application, *Int. J. Mol. Sci.*, **16**(4), pp. 8027–8039.
12. Fan, L. W., Fang, X., Wang, X., Zeng, Y., Xiao, Y. Q., Yu, Z. T., Xu, X., Hu, Y. C., Cen, K. F. (2013). Effects of various carbon nanofillers on the thermal conductivity and energy storage properties of paraffin-based nanocomposite phase change materials, *Appl. Energy*, **110**, pp. 163–172.
13. Fang, Y. T., Kuang, S. Y., Gao, X. N., Mang, Z. G. (2008). Preparation and characterization of novel nanoencapsulated phase change materials, *Energy Convers. Manage.*, **49**(12), pp. 3704–3707.
14. Farid, M., Kong, W. J. (2001). Underfloor heating with latent heat storage, *Proc. Inst. Mech. Eng. Part A*, **215**(A5), pp. 601–609.
15. Fernandes, D., Pitie, F., Caceres, G., Baeyens, J. (2012). Thermal energy storage: "how previous findings determine current research priorities", *Energy*, **39**(1), pp. 246–257.
16. Fleischer, A. S. (2015). *Thermal Energy Storage Using Phase Change Materials: Fundamentals and Applications*, Springer.

17. Guarino, F., Dermardiros, V., Chen, Y., Rao, J., Athienitis, A., Cellura, M., Mistretta, M. (2015). PCM thermal energy storage in buildings: experimental study and applications, *International Conference on Solar Heating and Cooling for Buildings and Industry*, SHC 2014, **70**, pp. 219–228.
18. Han, B. G., Zhang, K., Yu, X. (2013). Enhance the thermal storage of cement-based composites with phase change materials and carbon nanotubes, *J. Sol. Energy Eng.*, **135**(2).
19. Harikrishnan, S., Deenadhayalan, M., Kalaiselvam, S. (2014). Experimental investigation of solidification and melting characteristics of composite PCMs for building heating application, *Energy Convers. Manage.*, **86**, pp. 864–872.
20. Hasnain, S. M. (1998). Review on sustainable thermal energy storage technologies. Part I: Heat storage materials and techniques, *Energy Convers. Manage.*, **39**(11), pp. 1127–1138.
21. Hawes, D. W., Feldman, D. (1992). Absorption of phase-change materials in concrete, *Sol. Energy Mater. Sol. Cells*, **27**(2), pp. 91–101.
22. Hawlader, M. N. A., Uddin, M. S., Khin, M. M. (2003). Microencapsulated PCM thermal-energy storage system, *Appl. Energy*, **74**(1–2), pp. 195–202.
23. Heier, J., Bales, C., Martin, V. (2015). Combining thermal energy storage with buildings: a review, *Renewable Sustainable Energy Rev.*, **42**, pp. 1305–1325.
24. Huang, J. Y., Lu, S. L., Kong, X. F., Liu, S. B., Li, Y. R. (2013). Form-stable phase change materials based on eutectic mixture of tetradecanol and fatty acids for building energy storage: preparation and performance analysis, *Materials*, **6**(10), pp. 4758–4775.
25. Hunger, M., Entrop, A. G., Mandilaras, I., Brouwers, H. J. H., Founti, M. (2009). The behavior of self-compacting concrete containing micro-encapsulated phase change materials, *Cem. Concr. Compos.*, **31**(10), pp. 731–743.
26. IEA (2014). *Technology Roadmap: Energy Storage*, OECD/IEA, Paris.
27. Karaipekli, A., Sari, A. (2010). Preparation, thermal properties and thermal reliability of eutectic mixtures of fatty acids/expanded vermiculite as novel form-stable composites for energy storage, *J. Ind. Eng. Chem.*, **16**(5), pp. 767–773.
28. Kaviarasu, C., Prakash, D. (2016). Review on phase change materials with nanoparticle in engineering applications, *J. Eng. Sci. Technol. Rev.*, **9**(4), pp. 26–36.

29. Kenisarin, M., Mahkamov, K. (2016). Salt hydrates as latent heat storage materials: thermophysical properties and costs, *Sol. Energy Mater. Sol. Cells*, **145**, pp. 255–286.
30. Kenisarin, M. M. (2014). Thermophysical properties of some organic phase change materials for latent heat storage: a review, *Sol. Energy*, **107**, pp. 553–575.
31. Khudhair, A. M., Farid, M. M. (2004). A review on energy conservation in building applications with thermal storage by latent heat using phase change materials, *Energy Convers. Manage.*, **45**(2), pp. 263–275.
32. Konuklu, Y., Ersoy, O. (2016). Preparation and characterization of sepiolite-based phase change material nanocomposites for thermal energy storage, *Appl. Therm. Eng.*, **107**, pp. 575–582.
33. Konuklu, Y., Ersoy, O., Gokce, O. (2015). Easy and industrially applicable impregnation process for preparation of diatomite-based phase change material nanocomposites for thermal energy storage, *Appl. Therm. Eng.*, **91**, pp. 759–766.
34. Lane, G. A. (1980). Low temperature heat storage with phase change materials, *Int. J. Ambient Energy*, **1**(3), pp. 155–168.
35. Li, H., Fang, G. Y. (2010). Experimental investigation on the characteristics of polyethylene glycol/cement composites as thermal energy storage materials, *Chem. Eng. Technol.*, **33**(10), pp. 1650–1654.
36. Li, H., Zhang, Q. Q., Xiao, H. G. (2013a). Self-deicing road system with a CNFP high-efficiency thermal source and MWCNT/cement-based high-thermal conductive composites, *Cold Reg. Sci. Technol.*, **86**, pp. 22–35.
37. Li, M. (2013). A nano-graphite/paraffin phase change material with high thermal conductivity, *Appl. Energy*, **106**, pp. 25–30.
38. Li, M., Wu, Z. S., Tan, J. M. (2013b). Heat storage properties of the cement mortar incorporated with composite phase change material, *Appl. Energy*, **103**, pp. 393–399.
39. Li, M. G., Zhang, Y., Xu, Y. H., Zhang, D. (2011). Effect of different amounts of surfactant on characteristics of nanoencapsulated phase-change materials, *Polym. Bull.*, **67**(3), pp. 541–552.
40. Ling, T. C., Poon, C. S. (2013). Use of phase change materials for thermal energy storage in concrete: an overview, *Constr. Build. Mater.*, **46**, pp. 55–62.
41. Lucas, S. (2016). New construction materials combining self-cleaning and heat storage properties, in Jankovic, L. (ed.) *Zero Carbon Buildings: Today and in the Future 2016*, Birmingham Birmingham City University, pp. 125–130.

42. Mei, D. D., Zhang, B., Liu, R. C., Zhang, Y. T., Liu, J. D. (2011). Preparation of capric acid/halloysite nanotube composite as form-stable phase change material for thermal energy storage, *Sol. Energy Mater. Sol. Cells*, **95**(10), pp. 2772–2777.
43. Memon, S. A. (2014). Phase change materials integrated in building walls: a state of the art review, *Renewable Sustainable Energy Rev.*, **31**, pp. 870–906.
44. Memon, S. A., Lo, T. Y., Cui, H. Z., Barbhuiya, S. (2013). Preparation, characterization and thermal properties of dodecanol/cement as novel form-stable composite phase change material, *Energy Build.*, **66**, pp. 697–705.
45. Parameshwaran, R., Kalaiselvam, S., Harikrishnan, S., Elayaperumal, A. (2012). Sustainable thermal energy storage technologies for buildings: a review, *Renewable Sustainable Energy Rev.*, **16**(5), pp. 2394–2433.
46. Rozanna, D., Chuah, T. G., Salmiah, A., Choong, T. S. Y., Sa'ari, M. (2004). Fatty acids as phase change materials (PCMs) for thermal energy storage: a review, *Int. J. Green Energy*, **1**(4), pp. 495–513.
47. Sa, A. V., Azenha, M., de Sousa, H., Samagaio, A. (2012). Thermal enhancement of plastering mortars with phase change materials: experimental and numerical approach, *Energy Build.*, **49**, pp. 16–27.
48. Sanusi, O., Warzoha, R., Fleischer, A. S. (2011). Energy storage and solidification of paraffin phase change material embedded with graphite nanofibers, *Int. J. Heat Mass Transfer*, **54**(19–20), pp. 4429–4436.
49. Sari, A., Alkan, C., Karaipekli, A. (2010). Preparation, characterization and thermal properties of PMMA/n-heptadecane microcapsules as novel solid-liquid microPCM for thermal energy storage, *Appl. Energy*, **87**(5), pp. 1529–1534.
50. Sari, A., Alkan, C., Karaipekli, A., Uzun, O. (2009). Microencapsulated n-octacosane as phase change material for thermal energy storage, *Sol. Energy*, **83**(10), pp. 1757–1763.
51. Sari, A., Bicer, A. (2012). Thermal energy storage properties and thermal reliability of some fatty acid esters/building material composites as novel form-stable PCMs, *Sol. Energy Mater. Sol. Cells*, **101**, pp. 114–122.
52. Sharma, A., Tyagi, V. V., Chen, C. R., Buddhi, D. (2009). Review on thermal energy storage with phase change materials and applications, *Renewable Sustainable Energy Rev.*, **13**(2), pp. 318–345.
53. Shilei, L., Neng, Z., Feng, G. H. (2006). Eutectic mixtures of capric acid and lauric acid applied in building wallboards for heat energy storage, *Energy Build.*, **38**(6), pp. 708–711.

54. Soares, N., Costa, J. J., Gaspar, A. R., Santos, P. (2013). Review of passive PCM latent heat thermal energy storage systems towards buildings' energy efficiency, *Energy Build.*, **59**, pp. 82–103.

55. Song, Q. W., Li, Y., Xing, J. W., Hu, J. Y., Yuen, C. W. M. (2007). Thermal stability of composite phase change material microcapsules incorporated with silver nano-particles, *Polymer*, **48**(11), pp. 3317–3323.

56. Teng, T. P., Yu, C. C. (2012). Characteristics of phase-change materials containing oxide nano-additives for thermal storage, *Nanoscale Res. Lett.*, **7**, pp. 1–10.

57. Tyagi, V. V., Kaushik, S. C., Tyagi, S. K., Akiyama, T. (2011). Development of phase change materials based microencapsulated technology for buildings: a review, *Renewable Sustainable Energy Rev.*, **15**(2), pp. 1373–1391.

58. Wang, B. X., Zhou, L. P., Peng, X. F. (2006). Surface and size effects on the specific heat capacity of nanoparticles, *Int. J. Thermophys.*, **27**(1), pp. 139–151.

59. Wang, J. F., Xie, H. Q., Li, Y., Xin, Z. (2010). PW based phase change nanocomposites containing gamma-Al_2O_3, *J. Therm. Anal. Calorim.*, **102**(2), pp. 709–713.

60. Wang, L., Tan, Z. C., Meng, S. G., Liang, D. B., Li, G. G. (2001). Enhancement of molar heat capacity of nanostructured Al_2O_3, *J. Nanopart. Res.*, **3**(5–6), pp. 483–487.

61. Wi, S., Seo, J., Jeong, S. G., Chang, S. J., Kang, Y., Kim, S. (2015). Thermal properties of shape-stabilized phase change materials using fatty acid ester and exfoliated graphite nanoplatelets for saving energy in buildings, *Sol. Energy Mater. Sol. Cells*, **143**, pp. 168–173.

62. Wu, S. Y., Wang, H., Xiao, S., Zhu, D. S. (2012). Numerical simulation on thermal energy storage behavior of Cu/paraffin nanofluids PCMs, *International Conference on Advances in Computational Modeling and Simulation*, **31**, pp. 240–244.

63. Xu, B. W., Li, Z. J. (2013). Paraffin/diatomite composite phase change material incorporated cement-based composite for thermal energy storage, *Appl. Energy*, **105**, pp. 229–237.

64. Xu, B. W., Li, Z. J. (2014). Paraffin/diatomite/multi-wall carbon nanotubes composite phase change material tailor-made for thermal energy storage cement-based composites, *Energy*, **72**, pp. 371–380.

65. Ye, R. D., Fang, X. M., Zhang, Z. G., Gao, X. N. (2015). Preparation, mechanical and thermal properties of cement board with expanded

perlite based composite phase change material for improving buildings thermal behavior, *Materials*, **8**(11), pp. 7702–7713.

66. Yu, Z. T., Fang, X., Fan, L. W., Wang, X., Xiao, Y. Q., Zeng, Y., Xu, X., Hu, Y. C., Cen, K. F. (2013). Increased thermal conductivity of liquid paraffin-based suspensions in the presence of carbon nano-additives of various sizes and shapes, *Carbon*, **53**, pp. 277–285.
67. Zalba, B., Marin, J. M., Cabeza, L. F., Mehling, H. (2003). Review on thermal energy storage with phase change: materials, heat transfer analysis and applications, *Appl. Therm. Eng.*, **23**(3), pp. 251–283.
68. Zhang, D., Li, Z. J., Zhou, H. M., Wu, K. (2004). Development of thermal energy storage concrete, *Cem. Concr. Res.*, **34**(6), pp. 927–934.
69. Zhang, Z. G., Shi, G. Q., Wang, S. P., Fang, X. M., Liu, X. H. (2013). Thermal energy storage cement mortar containing n-octadecane/expanded graphite composite phase change material, *Renewable Energy*, **50**, pp. 670–675.
70. Zhang, Z. G., Zhang, N., Peng, J., Fang, X. M., Gao, X. N., Fang, Y. T. (2012). Preparation and thermal energy storage properties of paraffin/expanded graphite composite phase change material, *Appl. Energy*, **91**(1), pp. 426–431.
71. Zhou, D., Zhao, C. Y., Tian, Y. (2012). Review on thermal energy storage with phase change materials (PCMs) in building applications, *Appl. Energy*, **92**, pp. 593–605.

Chapter 13

Self-Heating Conductive Cement-Based Nanomaterials

E. Seva, O. Galao, F. J. Baeza, E. Zornoza, R. Navarro, and P. Garcés
Department of Civil Engineering, University of Alicante, Alicante, Spain
pedro.garces@ua.es

13.1 Introduction

In cement composites, multifunctionality consists of taking advantage of the structural material itself to develop nonstructural functions, without the need of any type of external device. It can be achieved by combining a cementitious material with different additions that provide the resulting material with a new range of applications [1–11], keeping or even improving its structural characteristics [12–17]. Thus, cost is reduced, design is simplified, and the use of embedded devices is minimized. Functional properties include anode for electrochemical chloride extraction [2, 4], electromagnetic wave shielding [6], strain/stress sensor [7–9], dynamic monitoring and damage detection [9, 11, 18], temperature sensor [19], and heating and thermal control [20–26], among others.

Nanotechnology in Cement-Based Construction
Edited by Antonella D'Alessandro, Annibale Luigi Materazzi, and Filippo Ubertini

ISBN 978-981-4800-76-1 (Hardcover), 978-0-429-32849-7 (eBook)
www.jennystanford.com

With its application as an ice controller on different transportation infrastructures—highways, interchanges, bridges, airport runways, for example—safety for the drivers would be increased, while not compromising the durability of the structures with the use of substances that can damage it.

Carbonaceous materials have a high thermal conductivity (although not as high as metals), a low coefficient of thermal expansion (lower than metals), and high resistance to corrosion [14–16], which makes them good candidates for thermal applications in multifunctional cementitious composites such as heating of buildings or pavement de-icing. Carbon fiber is one these materials, which when added to concrete can transform its high resistivity to become an electrically conductive cement composite, that is, carbon fiber–reinforced concrete (CFRC) [1].

Thermal engineering deals with heat transfer processes and the methodology for calculating the speed produced by them, and the purpose of this field is to design applied components and systems. Its application covers areas including temperature measurement, heating, refrigeration, and heat retention. In addition, theses aspects are especially relevant to structures and other elements of concrete where thermal phenomena relating to heating and cooling impact durability and performance.

Thermal engineering in concrete structures is normally applied through the use of nonstructural materials or devices such as thermometers, objects of high thermal mass for heat retention, embedded resistance, or hot water pipes for heating. However, little attention has been given to the use of the structural material itself and its application to thermal engineering, which could reduce or eliminate the need for nonstructural peripheral materials. Given that structural materials are relatively economical and durable, its application in this field will reduce costs and maintenance. Additionally, the elimination of embedded objects will reduce the risk of the degradation of mechanical properties resulting from the required openings in the structure [22].

As previously mentioned, cementitious materials are materials with an elevated level of electrical resistivity that can be transformed into compound multifunctional cementitious materials with a variety of practical applications through the addition of another conductive material. Its function as a heating element is directly related to the improvement in electrical and thermal conductivity of the compound.

Electrical heating involves passing an electrical current through a resistance as a heating element. With respect to the heating of buildings and other structures, heating involves embedding heating elements in the materials used in the construction, such as concrete or gypsum. Materials used as heating elements are typically metal alloys, such as nickel-chrome. The metallic fibers are embedded in the structural material in order to attain heating through resistance. However, this procedure degrades the mechanical properties of the structural component, rendering difficult the repairing of the heating elements. Moreover, embedding these elements in the structural components results in non-uniform heating as it is limited to selected areas. The non-uniform heating worsens due to the poor thermal conductivity of structural materials. A conductive cementitious material, by the addition of a carbon-based material, can be used as a resistor—the Joule effect—and, therefore, as a heating element. There is no need to embed fibers in a structural element, which, in turn, minimizes the aforementioned problems. The resistivity of conventional concrete is too high to provide effective electric resistance heating; however, by the addition of carbon-based materials, such as carbon fibers, it is possible to reduce the electrical resistivity to adequate levels.

To ensure good heating a high electrical current is necessary if the electric resistivity of the material is very low. By contrast if the electric resistivity of the material is very high, the application of a high electrical voltage is necessary, and at the same time a low electric current is applied. Up to now, the study of self-heating concrete has been based on compounds of midrange electrical resistivity.

So far, the study of self-heating concrete has been mainly based on composites with medium electrical resistivity, as a high current would be needed if it is too low in electrical resistivity or a high voltage would be needed if it is too high in electrical resistivity—the current in the heating element would be very low [22]. Nevertheless, other researchers investigated the de-icing performance of CFRC with an electrical resistivity of only 10 $\Omega \cdot$cm at 30°C [27] with promising results.

Many methods and techniques have been investigated for pavement anti-icing and de-icing as snow and ice on roads cause enormous loss of human lives, infrastructure, and materials [1, 20, 28, 29]. The mechanical snow removal technique is the most used, but it is intensive, expensive, and time-consuming labor. Furthermore, not all the snow is completely cleaned from the driveway by the snowplows, leaving a small layer to be eliminated. Many of the methods currently used to remove ice from roads are based on the use of chemicals. Most of them are harmful to both reinforced concrete and steel structures—viaducts, tunnels, airport runways—and for the environment.

Several researchers have analyzed the feasibility of using conductive multifunctional concrete, with different additions, for pavement de-icing [29–33]. Since the late 1990s, steel fibers, steel shavings, carbon fibers, and graphite products have been added to concrete as conductive materials to greatly improve the electrical conductivity. Some drawbacks about using steel shavings in the mixtures were noticed, and thus carbon products were used to replace the steel shavings in the conductive concrete mixture design.

A conductive concrete deck using carbon products of different particle sizes was implemented for de-icing on a 36 m long and 8.5 m wide highway bridge at Roca, Nebraska [31]. The de-icing system worked well in four major snowstorms in the winter of 2003, delivering an average power density of 452 W/m^2 to melt snow and ice. Also a method of de-icing with carbon fiber heating wires (CFHWs) embedded inside concrete slabs has also been verified, showing that with an input power of 1134 W/m^2, the temperature on the slab surface rises from -25°C to above 0°C in 2.5 h at an approximate rate of 0.17°C/min [33].

Previous research [26] studied the heating caused by the flow of electric current through a cement paste containing carbonaceous materials such as nanofibers, nanotubes, carbon fiber powder, and graphite powder, which reduce the electrical resistance of the resulting composite. An experimental study and a mathematical model were developed in it. That research showed the best results in the composites with high electrical conductivity.

The following sections briefly describe the various models.

13.2 Heating/Cooling Model

The results of an experiment of heating/cooling, in which the water evaporation is not important, are showed in Fig. 13.1. It concerns a specimen of 1% carbon nanotube (CNT) and a water/cement ratio (w/c) of 0.50 that was subjected to 100 V. This potential difference produced an average electric current of 42 mA, which means a power consumption of 4.17 W that remains constant while the source is connected (up to 18,000 s). At that moment (t_{off}) the source is disconnected, and consequently, intensity and electrical power fall to 0.

To model the process, it is assumed that the changes in the electric resistance of the specimen are insignificant and all the points of the specimen are at the same temperature T, due to the thickness of the specimens and the surface exposed to the air is large.

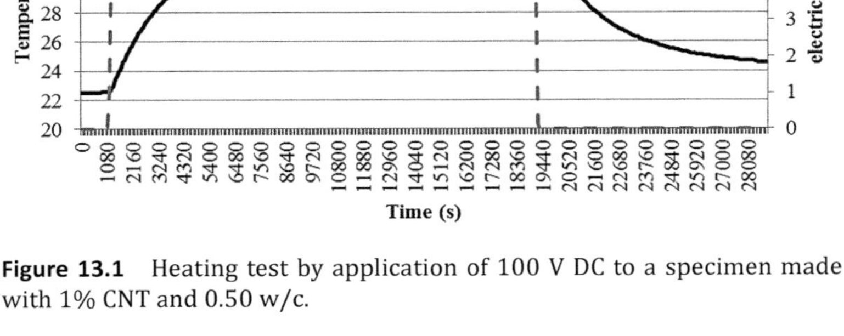

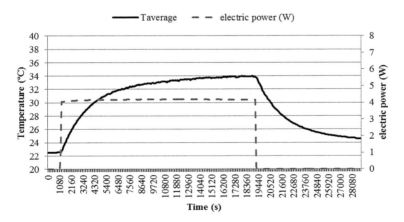

Figure 13.1 Heating test by application of 100 V DC to a specimen made with 1% CNT and 0.50 w/c.

The process of modeling will be divided in two stages, where the respective mathematical equations will be defined.

13.3 Stage of Heating Produced by the Application of Electric Current

In this period, the specimen is heated by the Joule effect in a nonsteady process from the initial temperature T_r at time t_1 to temperature T at time t.

The variations of thermal energy are due to the applied electrical power (P) minus the rate of energy loss by cooling, which is established by Newton's law of cooling.

$$mc_{\mathrm{p}}\frac{\mathrm{d}T}{\mathrm{d}t} = P - hA\left(T - T_{\mathrm{r}}\right) \tag{13.1}$$

where m (kg) is the mass of the specimen, c_{p} (J/kg°C) the specific heat of the material of the specimen, T (°C) the average temperature of the specimen, t (s) the time, h (W/m^{2}°C) a parameter that measures the rate of energy transport toward the outside, P (W) the applied power, A (m^2) the surface of the specimen exposed, and T_{r}(°C) the room temperature.

If this equation is integrated with the initial condition: for $t = t_1$ (the time at which the application of the voltage starts) the specimen temperature is T_{r}, an expression is obtained that calculates the temperature of the specimen with respect to time:

$$T = T_{\mathrm{r}} + \frac{P}{hA}\left[1 - e^{-\frac{hA}{mc_{\mathrm{p}}}(t-t_1)}\right] \tag{13.2}$$

If the time passed is long enough (in the example given, longer than 5000 s), the steady state is reached and the temperature remains constant. When $t \to \infty$ Eq. 13.2 turns into

$$T = T_{\mathrm{r}} + \frac{P}{hA} \tag{13.3}$$

which says that in the steady state the electric power supplied is equal to the loss of energy calculated using Newton's law of cooling:

$$P = hA(T - T_{\mathrm{r}}) \tag{13.4}$$

The electric power against the maximum temperature rise for each test ought to be in a line of slope hA, as the geometry of all the specimens is the same. With sufficient tests, the electric power should arrive to a constant value of hA.

The heating data of each test can be correlated representing $-\ln(P - hA(T - T_r)/P)$ versus time using the hA value obtained. The experimental data should perfectly fit to straight lines whose slopes should be hA/mc_p.

13.4 Stage of Cooling

The specimen, which is not electrically heated at this stage and has a temperature above the ambient one, will release energy to the outside and reduce its temperature.

The thermal energy of the specimen is released to the outside with a rate that must also be calculated with Newton's law of cooling. The mathematical equation that rules the process is a differential equation

$$mc_p \frac{dT}{dt} = -hA\,(T - T_r) \qquad (13.5)$$

which is integrated to the limit condition. For $t = t_{off}$ (time when the electric power is disconnected) the specimen temperature is T_{off} (temperature at that moment) and gives

$$T = T_r + (T_{off} - T_r)\,e^{-\frac{hA}{mc_p}(t-t_{off})}. \qquad (13.6)$$

As the heating data of each test, the cooling data has been correlated representing $-\ln[(T - T_r)/(T_{off} - T_r)]$ versus time. The experimental data fits perfectly to straight lines whose slopes should be hA/mc_p.

Equations 13.2, 13.6, and 13.7 can be used to model the heating test. For that, the necessary parameters are time when the voltage is connected, t_1 (s); time when the voltage is disconnected, t_{off} (s); electrical resistance of the test specimen, R (Ω); ratio hA/mc_p (s^{-1}); applied voltage V (V); and room temperature T_r (°C).

Moreover, Ohm's and Joule's laws, which relate the electric power P (W) with the applied voltage V (V) and current I (A) through the specimen of electrical resistance R (Ω), should be taken into account:

$$P = VI = \frac{V^2}{R} \qquad (13.7)$$

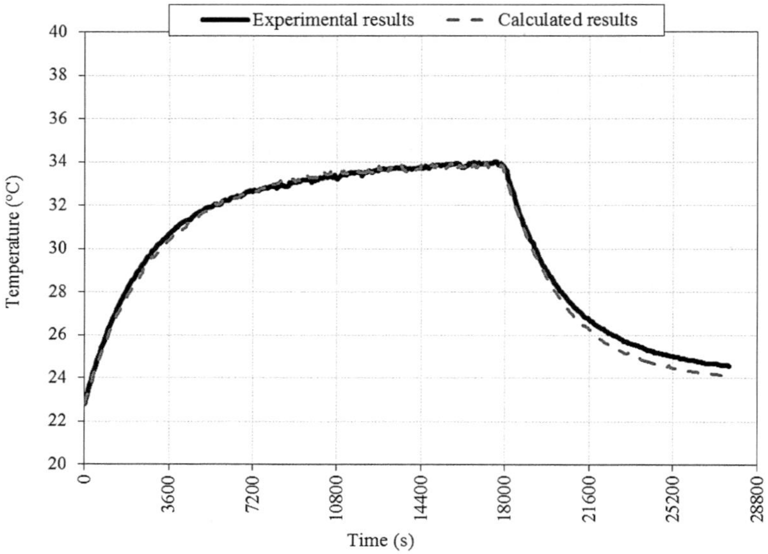

Figure 13.2 Experimental results (continuous line) and calculated results (dashed line) for the heating test on the specimen with 1% CNT, w/c of 0.50, and applied voltage of 100 V DC.

All the experimental tests have been simulated by using these equations and parameters. As an example, Fig. 13.2 shows the results for the test on the specimen with 1% CNT, w/c of 0.50, and applied voltage of 30 V. As can be observed, the differences between experimental and model data are small and always smaller than 2°C. In the rest of the tests, similar small differences have been found, and consequently, the model can be used to study theoretically the behavior of specimens when a voltage is applied. But the model is only solid when there is no great loss by evaporation of moisture from the specimens.

The simplified model is an excellent tool to study the influence of the different parameters in the heating of specimens. For example, the model allows calculating that when the applied voltage is duplicated the rise in temperature is four times greater and that the specific heat of the specimen has no influence on the maximum temperature rise or that the specimen's electrical resistance is inversely proportional to the temperature rise. As a result, the cement material with the lowest electrical resistivity is the most useful to carry out the heating feature, requiring heating adjustable with the applied voltage.

A recent study, with the objective of studying the thermal function, focused on the heating effect produced by an electric current passing through concrete specimens with the addition of CNTs, which transform ordinary Portland cement into a conductive paste. An AC power source at different fixed voltages was used to study two different performances in this research: anti-icing (prevention) and de-icing (curing). For the prevention test, specimens at room temperature were introduced into a freezer with a temperature of $-15^\circ C$. When the specimens reached approximately $+5^\circ C$, a power supply was connected to allow maintaining the surface specimen temperature above $0^\circ C$, preventing it from freezing. For the curing test, specimens were kept inside the freezer for 24 h, reaching the same temperature, that is, $-15^\circ C$. After this, the power supply was connected in order to increase the specimen temperature above $0^\circ C$. CNTs and cement dosage are shown in Tables 13.1 and 13.2, respectively.

Figures 13.3 and 13.4 show the specimens' geometry and location of temperature sensors (Pt100), and the experimental setup for self-heating tests, respectively.

Table 13.1 Carbon nanotube properties

Property	**Value and unit**
Average diameter	9.5 nm
Average length	1.5 μm
Carbon purity	90 wt%
Volume resistivity	$10^{-4} \Omega \cdot cm$
Surface area	250–300 m^2/g

Table 13.2 Paste dosage (g)

Cement	1800
Water	900
Superplasticizer	7.2
Carbon nanotubes	18

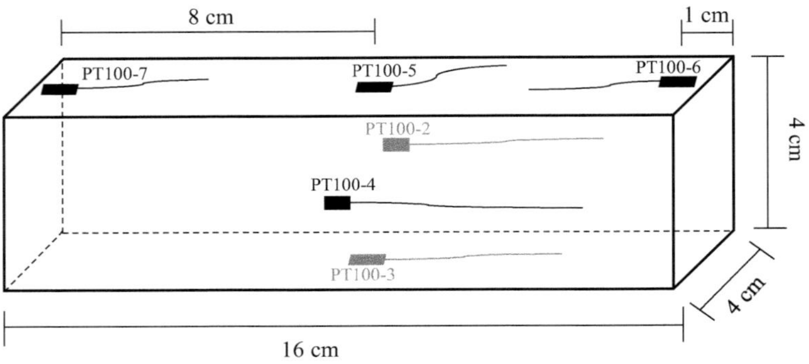

Figure 13.3 Specimen geometry and location of temperature sensors (PT100).

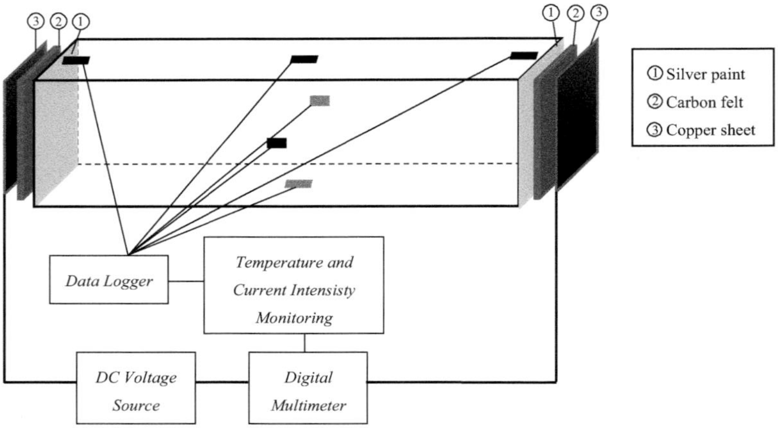

Figure 13.4 Experimental setup for self-heating test.

Fixed voltages of 150 V or 140 V prevented both specimens from freezing (Fig. 13.5) and increased the temperature above 0°C in an initial environment of −15°C (Fig. 13.6). Also, the mathematical model published in Ref. [26] was checked in laboratory conditions, showing very good correlation, being, therefore, applicable for this type of concrete with high conductivity in the former conditions.

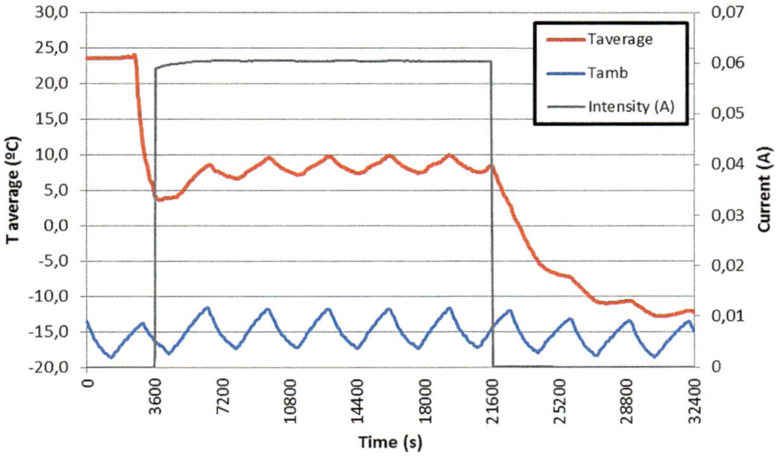

Figure 13.5 Prevention test performed at a fixed voltage of 150 V DC.

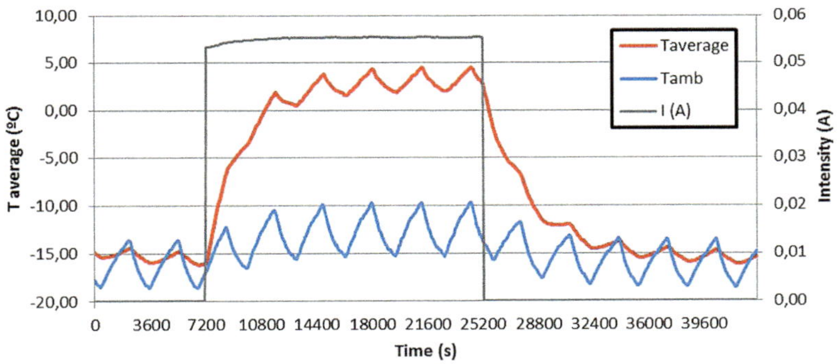

Figure 13.6 Curing test performed at a fixed voltage of 140 V DC.

References

1. Chung, D. D. L. (2003). *Multifunctional Cement-Based Materials*, Buffalo, New York, USA.
2. Pérez, A., Climent, M. A., Garcés, P. (2010). Electrochemical extraction of chlorides from reinforced concrete using a conductive cement paste as the anode, *Corros. Sci.*, **52**(5), pp. 1576–1581.

3. D'Alessandro, A., Fabiani, C., Pisello, A. L., Ubertini, F., Materazzi, A. L., Cotana, F. (2016). Innovative concretes for low-carbon constructions: a review, *Int. J. Low-Carbon Technol.*, **12**(3), pp. 289–309.
4. del Moral, B., Galao, O., Anton, C., Climent, M. A. A., Garcés, P. (2013). Usability of cement paste containing carbon nanofibres as an anode in electrochemical chloride extraction from concrete, *Mater. Constr.*, **63**(309), pp. 39–48.
5. Lu, S.-N., Xie, N., Feng, L.-C., Zhong, J. (2015). Applications of nanostructured carbon materials in constructions: the state of the art, *J. Nanomater.*, **2015**, pp. 1–10.
6. Wang, B., Guo, Z., Han, Y., Zhang, T. (2013). Electromagnetic wave absorbing properties of multi-walled carbon nanotube/cement composites, *Constr. Build. Mater.*, **46**, pp. 98–103.
7. Sun, M., Liew, R. J. Y., Zhang, M.-H., Li, W. (2014). Development of cement-based strain sensor for health monitoring of ultra high strength concrete, *Constr. Build. Mater.*, **65**, pp. 630–637.
8. Baeza, F. J., Zornoza, E., Andión, L. G., Ivorra, S., Garcés, P. (2011). Variables affecting strain sensing function in cementitious composites with carbon fibers, *Comput. Concr.*, **8**(2), pp. 229–241.
9. Baeza, F. J., Galao, O., Zornoza, E., Garcés, P. (2013). Multifunctional cement composites strain and damage sensors applied on reinforced concrete (RC) structural elements, *Materials (Basel)*, **6**(3), pp. 841–855.
10. Materazzi, A. L., Ubertini, F., D'Alessandro, A. (2013). Carbon nanotube cement-based transducers for dynamic sensing of strain, *Cem. Concr. Compos.*, **37**(1), pp. 2–11.
11. Ubertini, F., et al. (2014). Novel nanocomposite technologies for dynamic monitoring of structures: a comparison between cement-based embeddable and soft elastomeric surface sensors, *Smart Mater. Struct.*, **23**(4), p. 45023.
12. Sanchez, F., Ince, C. (2009). Microstructure and macroscopic properties of hybrid carbon nanofiber/silica fume cement composites, *Compos. Sci. Technol.*, **69**(7–8), pp. 1310–1318.
13. Catalá, G., Ramos-Fernández, E. V., Zornoza, E., Andión, L. G., Garcés, P. (2011). Influence of the oxidation process of carbon material on the mechanical properties of cement mortars, *J. Mater. Civ. Eng.*, **23**(3), pp. 321–329.
14. Garcés, P., Fraile, J., Vilaplana-Ortego, E., Cazorla-Amorós, D., Alcocel, E. G. G., Andión, L. G. G. (2005). Effect of carbon fibres on the mechanical

properties and corrosion levels of reinforced Portland cement mortars, *Cem. Concr. Res.*, **35**(2), pp. 324–331.

15. Garcés, P., Zornoza, E., Alcocel, E. G. G., Galao, O., Andión, L. G. G. (2012). Mechanical properties and corrosion of CAC mortars with carbon fibers, *Constr. Build. Mater.*, **34**, pp. 91–96.
16. Garcés, P., Andión, L. G., De la Varga, I., Catalá, G., Zornoza, E. (2007). Corrosion of steel reinforcement in structural concrete with carbon material addition, *Corros. Sci.*, **49**(6), pp. 2557–2566.
17. Galao, O., Zornoza, E., Baeza, F. J., Bernabeu, A., Garcés, P. (2012). Effect of carbon nanofiber addition in the mechanical properties and durability of cementitious materials, *Mater. Constr.*, **62**(307), pp. 343–357.
18. Giner, V. T., Baeza, F. J., Ivorra, S., Zornoza, E., Galao, O. (2012). Effect of steel and carbon fiber additions on the dynamic properties of concrete containing silica fume, *Mater. Des.*, **34**, pp. 332–339.
19. Wen, S., Chung, D. D. L. (1999). Carbon fiber-reinforced cement as a thermistor, *Cem. Concr. Res.*, **29**(6), pp. 961–965.
20. Chung, D. D. L. (2001). Materials for thermal conduction, *Appl. Therm. Eng.*, **21**(16), pp. 1593–1605.
21. Yehia, S., Tuan, C. Y. (1999). Conductive concrete overlay for bridge deck deicing, *ACI Mater. J.*, **96**(3), pp. 382–390.
22. Chung, D. D. L. (2004). Self-heating structural materials, *Smart Mater. Struct.*, **13**(3), pp. 562–565.
23. Yehia, S., Tuan, C. Y., Ferdon, D., Chen, B. (2000). Conductive concrete overlay for bridge deck deicing: mixture proportioning, optimization, and properties, *ACI Struct. J.*, **97**(2), pp. 172–181.
24. Tuan, C. Y. (2004). Electrical resistance heating of conductive concrete containing steel fibers and shavings, *ACI Mater. J.*, **101**(1), pp. 65–71.
25. Galao, O., Baeza, F. J., Zornoza, E., Garcés, P. (2014). Self-heating function of carbon nanofiber cement pastes, *Mater. Construcción*, **64**(314), p. e015.
26. Gomis, J., Galao, O., Gomis, V., Zornoza, E., Garcés, P. (2015). Self-heating and deicing conductive cement: experimental study and modeling, *Constr. Build. Mater.*, **75**, pp. 442–449.
27. Tang, Z., Li, Z., Qian, J., Wang, K. (2005). Experimental study on deicing performance of carbon fiber reinforced conductive concrete, *J. Mater. Sci. Technol.*, **21**(1), pp. 113–117.
28. Vitaliano, D. F. (1992). An economic assessment of the social costs of highway salting and the efficiency of substituting a new deicing material, *J. Policy Anal. Manag.*, **11**(3), pp. 397–418.

29. Yu, W., Yi, X., Guo, M., Chen, L. (2014). State of the art and practice of pavement anti-icing and de-icing techniques, *Sci. Cold Arid Reg.*, **6**(1), pp. 14–21.

30. Zhang, Q., Li, H. (2011). Experimental investigation of road snow-melting based on CNFP self-heating concrete, in *Proceedings of SPIE - The International Society for Optical Engineering*, **7978**, p. 797825.

31. Tuan, C. Y., Yehia, S. (2004). Evaluation of electrically conductive concrete containing carbon products for deicing, *ACI Mater. J.*, **101**(4), pp. 287–293.

32. Chang, C., Ho, M., Song, G., Mo, Y.-L., Li, H. (2009). A feasibility study of self-heating concrete utilizing carbon nanofiber heating elements, *Smart Mater. Struct.*, **18**(12), p. 127001.

33. Zhao, H., Wu, Z., Wang, S., Zheng, J., Che, G. (2011). Concrete pavement deicing with carbon fiber heating wires, *Cold Reg. Sci. Technol.*, **65**(3), pp. 413–420.

Chapter 14

Functional Cementitious Composites for Energy Harvesting and Civil Engineering Applications: An Overview

Ashok Batra,[a] **Aschalew Kassu,**[b] **Bir Bohara,**[a] **Timir B. Roy,**[c] **and Antonella D'Alessandro**[d]

[a] *Department of Physics, Chemistry and Mathematics (Materials Science Group), Alabama A&M University, PO Box 4900 Normal (Huntsville), AL 35762, USA*

[b] *Department of Mechanical, Civil Engineering and Construction Management, Alabama A&M University, Normal (Huntsville), AL 35762, USA*

[c] *Department of Building Civil and Environmental Engineering, Concordia University, Montreal, Quebec, Canada*

[d] *Department of Civil and Environmental Engineering, University of Perugia, Perugia, Italy*

ashok.batra@aamu.edu

This chapter describes the recent findings on cement-based modified nanocomposite materials with applications in ambient energy harvesting, sensor and actuator technology, and structural health monitoring (SHM). The synergies that nanomaterials can provide in structural materials with new and enhanced properties make them "smart" because of their multifunctionalities. A brief introduction of new composite materials is also provided. Finally, a review and future outlook on the possible applications of nanotechnology to cement-based composites are presented.

Nanotechnology in Cement-Based Construction
Edited by Antonella D'Alessandro, Annibale Luigi Materazzi, and Filippo Ubertini

ISBN 978-981-4800-76-1 (Hardcover), 978-0-429-32849-7 (eBook)
www.jennystanford.com

14.1 Introduction

Engineers consistently demand new material systems, vital in emerging novel technological applications. Thus, this demand dictates that material scientists develop new material systems. Modern electronic devices and systems require various and specific functional properties in materials that single-phase materials don't possess [1]. Composite technology, where a novel functional material is fabricated by combining two or more chemically different materials or phases (e.g., ceramics and polymers) in an ordered manner or just mixing, is playing an important role. In recent decades, a large number of ceramic-polymer electronic composites have been introduced for medical, telecommunication, and microelectronics applications and for devices as microelectromechanical systems (MEMS; bio-MEMS) or sensors and actuators [2]. The composites have a unique blend of polymeric properties such as mechanical flexibility, high strength, design flexibility and formability, and low cost, with the high electroactive functional properties of ceramic materials. In these materials, it is, thus, possible to tailor physical, electronic, and mechanical properties catering to a variety of applications. As a result, the composite can be designed using a set of microstructural characteristics: for example, connectivity, volume fractions of each component, spatial distribution of the components, percolation threshold, and other parameters. Thus, the response of an electronic composite (electro-ceramic/polymer) to an external excitation (electric field, temperature, stress, etc.) depends upon the response of individual phases, their interfaces, as well as the type of connectivity. As a result, an electronic composite can broadly be described as exhibiting electromagnetic, thermal, and/or mechanical capabilities, while maintaining structural integrity [2]. Engineering materials' classification, including composites, are presented in Fig. 14.1.

With the recent progress in nanoscience and nanotechnology, there is an increasing interest in polymer nanocomposites, both in scientific and in engineering applications [3–8]. However, recently the demand for smart structures, capable of being sensitive and responding properly to certain external stimuli, created a necessity

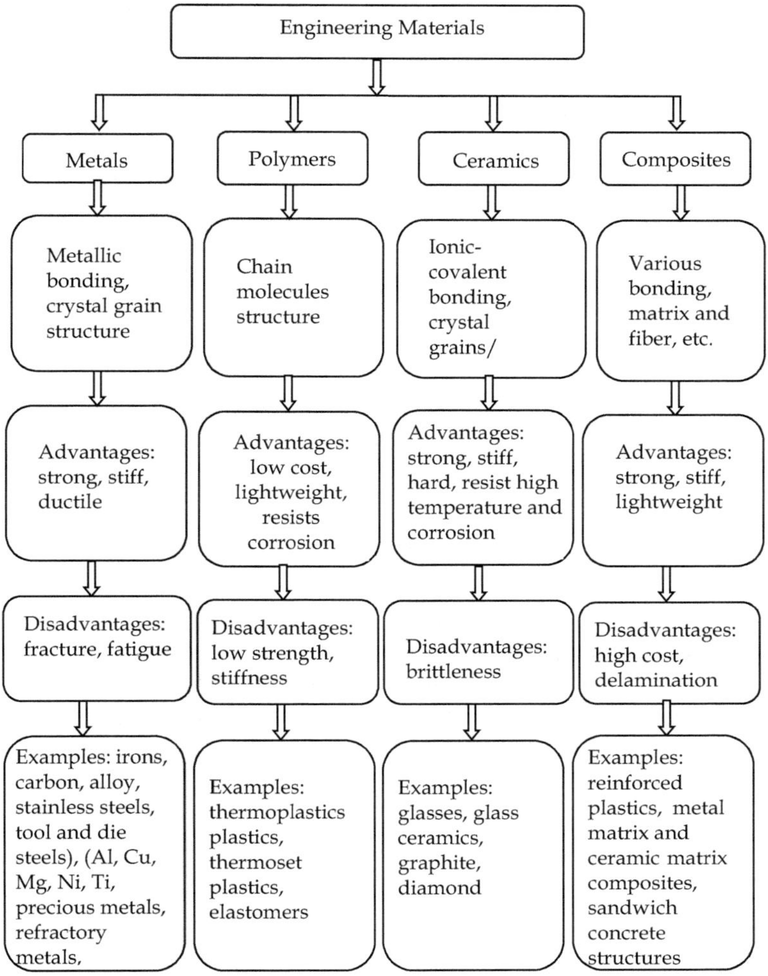

Figure 14.1 General characteristics of the major classes of engineering materials.

of new construction materials (cement-based functional nanocomposites) with additional functionalities such as strain or damage sensing and energy harvesters. In the light of many technologically important applications in this field of cement-based engineering materials, the proposed chapter will be a significant contribution.

14.2 Composite Materials and Their Constituents

14.2.1 Major Phases

Composite materials are materials made up of two or more materials combined together in such a way that the constituent materials are easily distinguished. The composite materials consist of basically two major phases, as presented in Fig. 14.2.

14.2.1.1 Matrix phase

The matrix phase is a continuous phase or the primary phase. It holds the dispersed phase and shares a load with it. It is made up of metals, ceramics such as cement, or polymers, depending on the different applications or requested characteristics.

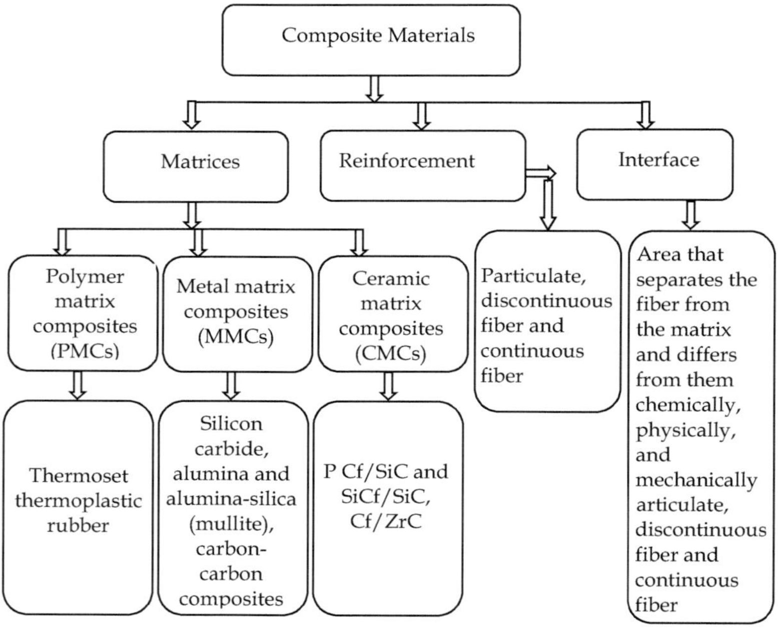

Figure 14.2 Composite materials and types of constituents.

14.2.1.2 Dispersed (reinforcing) phase

This is the second phase (or phases) that is imbedded in the matrix in a continuous/discontinuous form. The dispersed phase is usually stronger than the matrix; therefore it is sometimes called the reinforcing phase in the case of structural composites. This reinforcement is a strong, stiff (functional) integral component that is incorporated into the matrix to achieve desired properties or functionalities—basically it means desired property enhancement. It can be constituted by fibers or particles of any shape and size, including nanoparticles as well.

14.2.1.3 Interface in the composite structure

This is the zone across where the matrix phase and reinforcing phases interact (chemical, physical, mechanical, and electrical effects and others). This region in most composite materials has a finite thickness because of diffusion and/or chemical reactions between the fibers or particles and the matrix, as illustrated in Fig. 14.3.

A multiphase material is formed by a combination of materials that differ in composition or form, remain bonded together, and retain their identities and properties. Composites maintain an interface between components and act in concert to provide

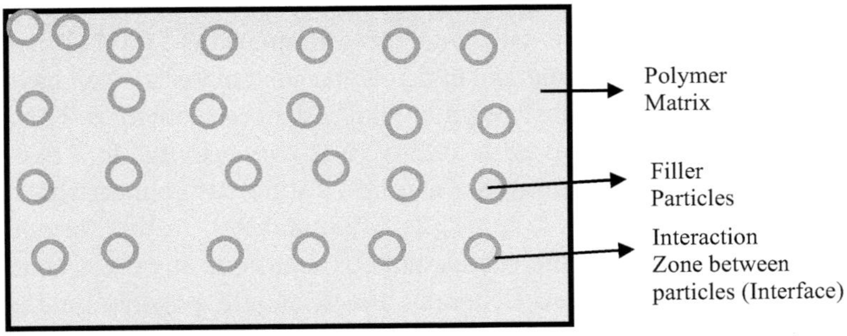

Figure 14.3 Presentation of the interface in a composite with filler particles.

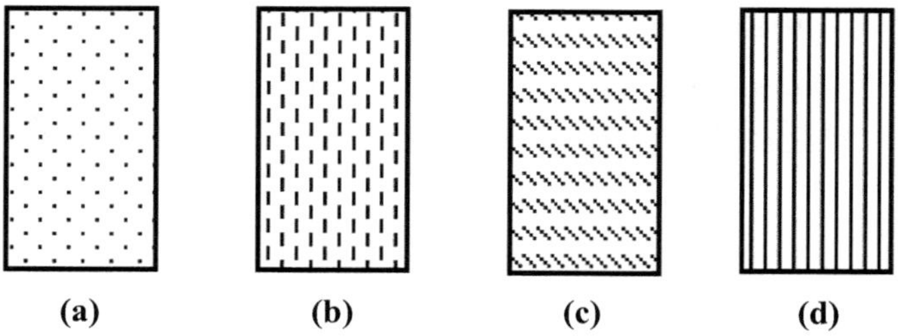

Figure 14.4 (a) Particulate and random, (b) discontinuous fibers and unidirectional, (c) discontinuous fibers and random, and (d) continuous fibers and unidirectional.

improved specific or synergistic characteristics, different from those of the original components acting alone. Composites include (i) particulate and random, (ii) discontinuous fibers and unidirectional, (iii) discontinuous fibers and random, (iv) continuous fibers and unidirectional, and (v) hybrid (combinations of any of the above) [9]. A few common types of composites as listed before are presented in Fig. 14.4.

14.2.2 Design of Composites: Connectivity Models

For fabrication of a composite, properties of the components, amount of each phase present, and how they are interconnected viz. connectivity are important. Newnham et al. proposed the concept of connectivity. Any phase in a mixture can be self-connected in zero, one, two, and three dimensions [10, 11]. For example, inclusions dispersed in a polymer host material shall have connectivity 0, while host polymer shall have connectivity 3. Thus, we can say "composites with '0-3' or '3-0' connectivity." In a two-phase composite system, there can be 10 different connectivities: 0-0, 0-1, 0-2, 0-3, 1-1, 1-2, 1-3, 2-2, 2-3, and 3-3. In this format, the first digit denotes the connectivity of inclusions and the second digit indicates the host. Generally, the host is a polymer in the case of polymer composites. A few connectivities are presented in Fig. 14.5 using a cube as a building block, along with real examples. On the basis of the above concept, in a 0-3 connectivity composite,

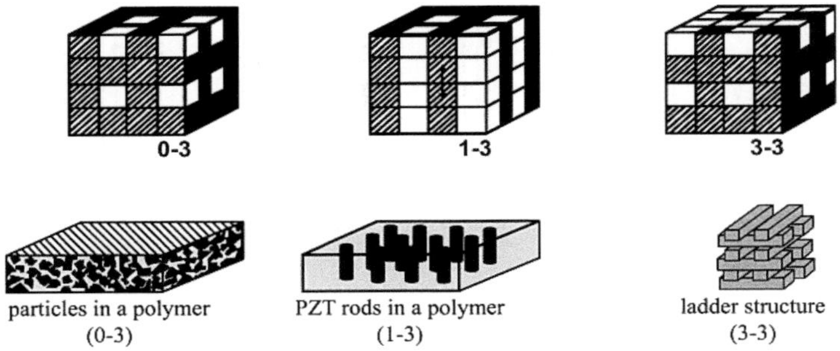

Figure 14.5 Connectivity models.

there is random distribution of active particulates in a 3D host polymer matrix.

14.3 Composite Materials with Piezoelectric, Ferroelectric, and Pyroelectric Functionalities

14.3.1 Classification

All crystals can be categorized into 32 different classes. These classes are point groups divided by using the following symmetry elements: (i) center of symmetry, (ii) axes of rotation, (iii) mirror planes, and (iv) several combinations of them. The 32 point groups are subdivisions of 7 basic crystal systems that are, in order of ascending symmetry, triclinic, monoclinic, orthorhombic, tetragonal, rhombohedral (trigonal), hexagonal, and cubic. Of the 21 classes of the 32 point groups that are noncentrosymmetric, which is a necessary condition for piezoelectricity to exist, 20 of them are piezoelectric. Among the 20 piezoelectric crystal classes, 10 crystals have pyroelectric properties [12]. Within a given temperature range, this group of materials is permanently polarized. Compared to the general piezoelectric polarization produced under stress, the pyroelectric polarization is developed spontaneously and kept as permanent dipoles in the structure. As this polarization varies

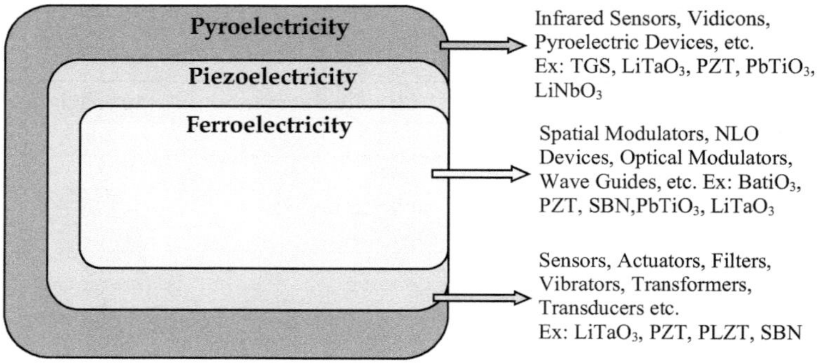

Figure 14.6 Some important ferroelectric materials along with important uses.

with temperature, the response is termed "pyroelectricity." The ferroelectric group is a subgroup of spontaneously polarized pyroelectric crystals. Though the polarization of ferroelectric is similar to the polarization of pyroelectric materials, there is a difference in the two polarizations because the ferroelectric polarization is reversible by an external applied electric field, provided that the applied field is less than the dielectric breakdown of the materials. Therefore, materials that can be defined as ferroelectrics must have two characteristics: the presence of spontaneous polarization and the reversibility of the polarization under an electric field. A ferroelectric material is therefore pyroelectric, piezoelectric, and noncentrosymmetric. It follows that not all piezoelectric materials are pyroelectric and not all pyroelectric materials are ferroelectric material, as depicted in Fig. 14.6. Table 14.1 lists the important ferroelectric/pyroelectric materials being investigated and/or are being used in the various devices [12]. Figure 14.6 shows some important ferroelectric materials and their uses in various devices based on their unique piezoelectric, pyroelectric, ferroelectric, electro-optic, acousto-optic, and dielectric properties. The most commonly used materials for pyroelectric and piezoelectric applications include triglycine sulfate (TGS); lead (Pb) zirconate titanate, $PbZrTiO_3$ (PZT); lead titanate, $PbTiO_3$ (PT); barium titanate, $BaTiO_3$(BT); lithium (Li) tantalate, $LiTaO_3$ (LT); and polymers: poly(vinylidene difluoride) (PVDF) and copolymers,

Table 14.1 Pyroelectric/Piezoelectric materials investigated/being investigated for use in various devices and structures

Ferroelectrics' family	Ferroelectric material	Chemical formula	Abbreviation
Perovskite type	Barium titanate	$BaTiO_3$	BT
	Lead zirconate titanate (ceramics)	$PbZr_xTi_{1-x}O_3$	PZT
	Barium strontium titanate	$Ba_xSr_{1-x}TiO_3$	BST
Lithium niobate	Lithium niobate	$LiNbO_3$	LN
	Lithium tantalate	$LiTaO_3$	LT
Barium type	Barium strontium niobate	$Ba_{5x}Sr_{5(1-x)}Nb_{10}O_{30}$	BSN
TGS type	Triglycine sulfate	$(NH_2CH_2COOH)_3{\cdot}H_2SO_4$	TGS
Polymer	Poly(vinyldene fluoride)	-	PVDF
	Poly(vinylidene fluoride)-trifluoroethylene	-	PVDF-TrFE

including poly(vinylidene difluoride)-trifluoroethylene (PVDF-TrFE) and poly(vinylidene difluoride)-tetrafluoroethylene (PVDF-TFE).

14.3.2 Physics and Chemistry of Composite Materials

There are thousands of materials available for use in engineering applications. Most materials fall into one of three classes that are based on the atomic bonding forces (such as metallic, ionic, covalent, and van der Walls) of a particular material: metallic, ceramic, and polymeric. Additionally, different materials can be combined to create a composite material. Within each of these classifications, materials are often further organized into groups on the basis of their chemical composition or certain physical or mechanical properties. Composite materials are often grouped by the types of materials combined or the way the materials are arranged together. Figure 14.1 clearly shows a list of some of the common classifications of materials within these four general groups of materials [2].

14.4 Fabrication of Composites

14.4.1 Fabrication of Polymer-Ceramic Composites

The 0-3 connectivity composites are easy to fabricate, which allows for commercial production of these composites in a cost-effective manner [13]. The composites that have shown some promising results and are prepared by simple composite fabrication routes are described in this section. As a result, most composites are fabricated as active pyroelectric and polymer-based diphasic samples. The polymer component can be polar or nonpolar. The second, more popular approach is using the sol-gel synthesis route. Though glass-ceramics have also shown noticeable pyroelectric sensing elements, this section doesn't include specific descriptions of such materials.

A 0-3 connectivity composite may be prepared by mixing ceramic particles in a hot rolling mill with a softened thermoplastic polymer, and thin films of composites then are produced by high-pressure casting at the softening temperature of the polymer. With a thermoset polymer such as epoxy, the mixing can be made at room temperature with the right proportion of the resin, hardener, and ceramic powder [13, 14].

14.4.2 Fabrication of Cement-Ceramic Composites

The 0-3 connectivity composites based on cement with functional components and carbon nanoparticles are easy to fabricate as polymer-based diphasic samples, which allows for commercial production of these composites in a cost-effective manner. Generally Portland cement (PC) paste is made with distilled water and other component constituents to form composites of desired chemical characteristics. To attain good dispersion and workability, it can be mechanically mixed and/or ultrasonicated. The fresh mix is then poured into a cylindrical steel mold of the desired thickness and diameter and then demolded for curing at room temperature. After the curing period elapses and the samples are externally dried, the surfaces is coated with electrically conductive silver paint to form a full-face capacitor under a high electric field and to create peculiar functionality in the presence of piezoelectric/pyroelectric components (if present). The fabricated composite materials are characterized for desired functionality and applications.

14.5 Ambient Energy Harvesting and Structural Health Monitoring of Civil Structures via Cement Nanocomposites

14.5.1 Energy Harvesting via Cement Nanocomposites

Energy harvesting (also known as power harvesting or energy scavenging) technology allows capturing unused ambient energy such as vibration, strain, light, temperature gradients, and/or changes in the energy of gas and liquid flows and converting it into usable electrical energy, which is stored for performing sensing and actuation and other applications [15]. Energy harvesting is a perfect match for wireless devices and wireless networks, which otherwise depend on battery power and other maintenance-free systems, thus providing cost-effective and environment-friendlier solutions for various applications. With the high price of fossil fuels, there is a renewed emphasis on energy conservation and development of alternative energy resources and systems. As a result, there is emphasis on low-cost energy conversion materials. Due to recent advances in low-power portable electronics and the fact that batteries in general provide a finite amount of power, attention has been given to explore methods for energy harvesting and scavenging. Energy can be recovered from mechanical vibration, light, and spatial and temporal temperature variations [15]. Thermal energy in the environment is a potential source of energy for low-power electronics. The pyroelectric effect converts a temperature change into electrical current or voltage. In pyroelectric materials, the thermal energy is converted to electrical energy. Electric dipoles are present in the materials due to the presence of ionic bonding. Pyroelectric materials are mainly used as sensors. One popular application of pyroelectric sensors is in the regular household motion sensors, where a slight change in temperature from infrared radiation from human bodies can produce electrical charges that can trigger an alarm [12]. In particular, charge is produced when the material's temperature is altered as a function of time. Several factors must be considered to optimize the performance of such materials for a given application, such as the material's geometry, boundary conditions, and even the circuitry used to harvest power. Recent work was presented on using lead zirconate titanate (PZT), lead magnesium niobate–lead titanate (PMN-PT), PVDF, and other pyroelectric materials for energy harvesting and storage [15]. Recently, focus has been put on pyroelectric materials as energy harvesters. If such electric energy can be trapped, it can be used for potential energy harvesting purposes. There have been several studies performed in this emerging field recently. In a recent paper Cuadras et al. [16] have shown the potential of using pyroelectric materials as energy harvesters [16]. The authors tested several PZTs and PVDFs as potential pyroelectric materials under various temperature conditions. The research results indicated that both PZT and PVDF can be used as pyroelectric materials. Novel applications require diverse and specific properties in materials that cannot be found in single-phase materials. Composites contain two or more chemically different materials or phases. In these materials, it is possible to tailor electrical, piezoelectric, pyroelectric, optical, and mechanical properties catering to a variety of applications in civil engineering. The composites show properties derived from individual components. Recent studies on new cement-based composites show potential usefulness in structural health monitoring (SHM) and other applications. Several studies have shown the potential of using nanocomposites. Han et al. fabricated nickel powder-filled cement-based composite piezoresistive sensors [17]. The gauge factor of piezoresistive sensors is higher than 895.45 and goes up to 1929.5 maximum with the compressive strain in the range from 0 to 125,689 $\mu\varepsilon$. Chaipnaich investigated the effect of PZT particle size on dielectric and piezoelectric properties of PZT-cement composites [18]. The author proposed that the enhancement in dielectric and piezoelectric properties was due to the lesser contacting surfaces between the cement matrix and the PZT particles. Wen and Chung presented damage monitoring of cement paste by electrical resistance measurement [19]. Gong et al. exploited the piezoelectric and dielectric behavior of 0-3 cement–based composites mixed with carbon black for civil engineering applications. The authors showed that piezoelectric sensitivities of the composites can be enhanced dramatically with incorporation of a small amount of carbon black [20]. Wen and Chung investigated the pyroelectric behavior of cement-based materials [21]. They showed that the steel/carbon nanofibers (CNFs) increase the dielectric properties of cement composites. However, there are limited studies on the pyroelectric and dielectric properties of cement-based nanocomposites for infrared sensing and energy harvesting and other civil engineering applications. Therefore, further electrical and pyroelectric investigations of cement-based composites, including nanocomposites, are warranted for the purpose of a fundamental understanding of their behavior. Piezoelectric materials have high potential of being used in the field for energy harvesting, especially under high-volume traffic conditions. However, limited studies and applications can be found in energy harvesting from pavements. In the United Kingdom, PaveGen Systems, Inc., has developed a piezoelectric material–based energy harvesting system from people walking on pavements [22]. The most serious attempt to generate electricity from vehicle movements has come from Innowattech in Israel, where they have installed piezoelectric devices under 1 km length of pavement lanes. The maximum generated power is 250 kWh from 600 v/h traffic flow [23]. Similar devices have been installed also in the Tokyo Railway Station to harness 1400 kW/s energy from human movements [24]. Specially, the use of pyroelectric materials in energy harvesting from pavements is important as well. If the pavement temperature can be converted into electrical current, it would be an ideal choice for using an alternative sustained power source to wireless sensors and other devices, which require low but sustained power. This would ensure sustained and uninterrupted power supply to pavement management system hardware and will contribute to energy conservation. In this section, the pyroelectric properties of several single and polycrystal materials and cement composites consisting of ordinary PC as a matrix are reported. The influence of carbon section fibers on the dielectric and pyroelectric properties of the composite has been examined for admixtures prepared under identical temperature and humidity conditions. The possibility of harvesting electrical energy from pavement temperature variation is described. Single-crystal- and polycrystal-based materials, and a polycrystalline composite material based on ordinary PC with CNFs, have been used as pyroelectric smart materials. Smart materials based on regular available PC can capture ambient thermal energy available from pavements. When CNFs are added, the cement-based composite acts as a pyroelectric material. Addition of CNFs increases the pyroelectric properties of cement. The simulation of the energy harvesting capacity of the materials with available pavement temperature data indicates that pyroelectric energy harvesting is an attractive and feasible method. The laboratory tests indicate that smart materials can produce enough electricity from traffic vibrations to power wireless bridge sensors. Simple material such as PC can be used to capture heat energy from pavements, which can be stored in a capacitor for use as a power source for other sensor electronics. It can be concluded that the power extracted from a pyroelectric element is typically low but can guarantee continuous operation for low-power autonomous sensors. It has also been proposed that piezoelectric converted energy can be accumulated into a storage capacitor by proper power management techniques and then transferred to the load during short time intervals [15]. Different pyroelectric materials have been investigated for possible use as infrared detectors [25]. Some of the materials have possible uses in energy harvesting from pavements. As indicated in Section 14.3.1, all crystalline materials can be classified into 32 different classes. Some of these classes of materials have pyroelectric properties. Within a given temperature range, this group of materials is permanently polarized. Compared to the general piezoelectric polarization produced under stress, pyroelectric polarization is developed spontaneously and kept as permanent dipoles in the structure. Because this polarization varies with temperature, the response is termed "pyroelectricity." The ferroelectric group is a subgroup of the spontaneously polarized pyroelectric crystals. On the one hand, the polarization of ferro-electric is similar to the polarization of pyroelectric. On the other hand, there is a difference between the two polarizations because the ferroelectric polarization is reversible by an external applied electric field, provided that the applied field is less than the dielectric breakdown of the materials. Therefore, materials that can be

Table 14.2 Properties and constituents of pyroelectric materials [26]

Material properties of pyroelectric materials			
Material	**Pyroelectric coefficient (p, nano C/cm^2/°C)**	**Dielectric constant (ε')**	
Doped PZT	40.8	768	
PLZT	45.0	700	
PZT Multilayer	34.0	845	
TGSe	420	420	
LT	12	-	
Carbon nanofiber properties			
Fiber diameter, nm (average)		150	
CVD carbon overcoat present on fiber		No	
Surface area, m^2/gm		20–30	
Dispersive surface energy, mJ/m^2		20–140	
Moisture, wt%		<5	
Iron, ppm		<14,000	
Composition of cement nanocomposites fabricated			
Sample ID	**Nanocarbon (% by mass)**	**PVA (% by mass)**	**Cement (% by mass)**
SSS3	0.070	0.513	99.417
SSS5	0.246	0.500	99.254
SSS7	0.490	0.508	99.002
SSS8	0.629	0.508	98.863
SSS9	1.173	0.509	98.318

defined as ferroelectrics must have two characteristics: the presence of spontaneous polarization and reversibility of the polarization under an electric field. A ferroelectric material is therefore both pyroelectric and piezoelectric.

Two different pyroelectric materials have been discussed in this section: single-crystal- and polycrystal-based materials. Their material properties are listed in Table 14.2.

14.5.1.1 Single-crystal-based materials

Triglycine selenate (TGSe) is a well-known ferroelectric material used for infrared detection devices at room temperature. It has a Curie temperature in the vicinity of 60°C, which makes its pyroelectric coefficient very high.

14.5.1.2 Polycrystalline-based materials

A wide range of polycrystalline materials are available for pyroelectric device applications, such as doped-PZT, lead lanthanum zirconate titanate (PLZT), multilayer PZT, lithium tantalate (LT; $LiTaO_3$), etc. Doped-PZT and PLZT are PZT-based ferroelectric oxide ceramics that are used for infrared detection devices. Ferroelectric oxide ceramics for use in detectors are relatively low cost to manufacture by using standard mixed-ceramic oxide processing steps and they are both mechanically and chemically robust. Their good mechanical strength allows large-area wafers to be made, which can be easily machined into thin sections [25]. These properties are suitable for use in highway pavement applications. LT is a crystalline solid that possesses both pyroelectric and piezoelectric properties and is a common pyroelectric material used as ordinary household motion detectors. It has excellent mechanical stability, making it a good candidate for piezoelectric materials as well, which can be used under heavy loading. The doped-PZT ceramics and LT crystals were grown in the laboratory using the Czochralski technique and other materials were procured from outside. The materials were tested for electrical and pyroelectric properties for applications in highway pavements [26].

Another pyroelectric material chosen was a cement-based composite. To prepare the composite specimens, ordinary PC, polyvinyl alcohol (PVA), and CNFs were mixed thoroughly. PVA acts as a binder within the cement. No aggregate was added. The CNF (PR-19-XT-LHT) was obtained from Pyrograph Products (OH, USA) and the material properties are listed in Table 14.2 [26].

Two electrical parameters were calculated by the authors that were relevant to the materials' potential application of charge storage capacity: the real (ε') and imaginary (ε'') parts of dielectric constant and dielectric loss ($\tan\delta$) and they are defined as [26]

$$\varepsilon' = \frac{C_p d}{\varepsilon_0 A} \quad \varepsilon'' = \varepsilon' \tan\delta, \tag{14.1}$$

where A is the electrode area (identical areas for the opposite electrodes were used in each sample), d is the thickness of the sample, $\omega = 2\pi f$ is the frequency of AC measurement, and $\varepsilon_0 = 8.854 \times 10^{-12}$ F/m is the permittivity of vacuum. A series of AC frequencies was selected from very low (10 Hz) to as high as 1 MHz to obtain a wide spectrum. The detailed measurement procedures of these parameters are described in Batra et al. [26]. The pyroelectric current I_p was also measured at various temperatures and the pyroelectric coefficient (p) was calculated using the relationship

$$p = \left(\frac{I_p}{A}\right) / \left(\frac{dT}{dt}\right), \tag{14.2}$$

where A is the electrode area and dT/dt is the rate of change of temperature, which was kept constant throughout the measurement. The change in the pyroelectric coefficient will indicate the change in dipole orientation inside the material; the higher the coefficient, the better the material for converting the temperature change in electrical charge. The additional charge generated via heating or cooling within a temperature change dT can be calculated as

$$dQ = dI_p dt = pA\frac{dT}{dt}dt = pAdT, \tag{14.3}$$

where the pyroelectric coefficient p could be constant or as a function of temperature.

Recently, Batra et al. numerically simulated the energy harvesting capacity of laboratory-fabricated and commercially available pyroelectric elements/transducers by capturing the thermal energy of pavements [15] and the environment [27, 28]. The single- and polycrystalline elements characterized for applicable performance parameters were triglycine selenate, LT, modified PZT, modified PT, modified lead meta-niobate, pyroelectric polymer nanocomposites (such as PC), nanocarbon fibers, polymer-LT embedded with silver nanoparticles, and others [27]. The modeling and numerical simulation of the energy harvesting capacity of these samples with the available pavement's temperature profile data over an extended period of time were investigated. The results indicated that electrical energy harvesting via pyroelectricity is a feasible technique for powering autonomous low-duty electric devices. On the basis of the analysis of a single electric energy harvesting unit, Batra et al. proposed that the triglycine selenate pyroelements will perform better than others with regard to the amount of voltage and energy densities extracted with respect to time [27].

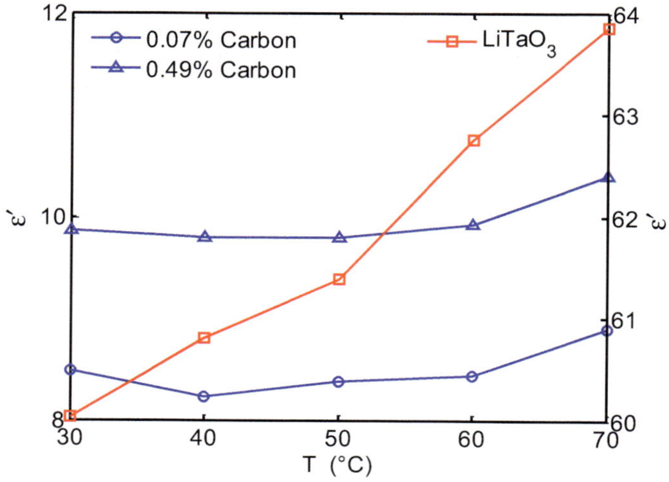

Figure 14.7 Typical dependence of dielectric constant (ε') of $LiTaO_3$ (red line) and cement-carbon nanocomposites (blue line) on temperature.

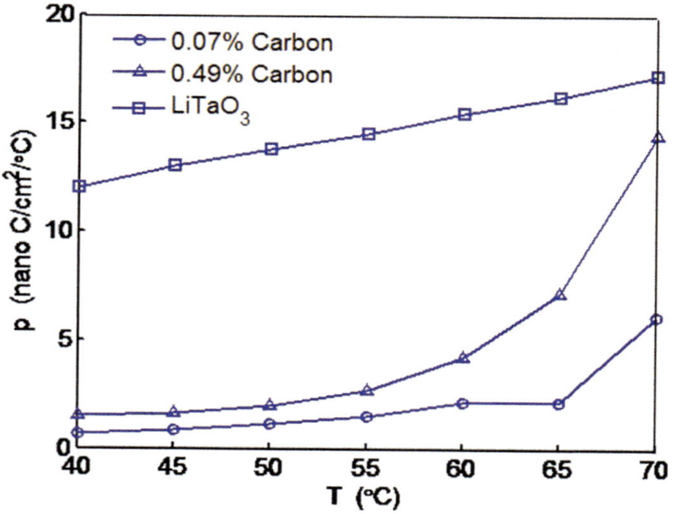

Figure 14.8 Typical dependence of pyroelectric coefficient (p) of $LiTaO_3$ and cement-carbon nanocomposites on temperature.

Figure 14.7 shows the dependence of dielectric constant (ε') of cement composites and LT on temperature. Figure 14.8 shows the characteristics of temperature dependence of the pyroelectric coefficient (p) of these two materials. As the figures indicate, these parameters increase with an increase in temperature. The observed increase in dielectric constant in cement composites can be attributed to the increase in mobility of ions as the temperature increases, whereas in the case of LT, the increase of p is due to the increase in dipolar moments in the material. The figures also indicate that the inclusion of CNFs increases the dielectric constant and the pyroelectric coefficient of the cement composite. One major conclusion from these figures is that ordinary PC can act as a pyroelectric material. The influence of carbon fiber content on the dielectric constant and pyroelectric coefficient at 1 kHz frequency and 40°C are shown in Figs. 14.9 and 14.10, respectively. It can be seen that the dielectric constant values increase with the increase in CNF content.

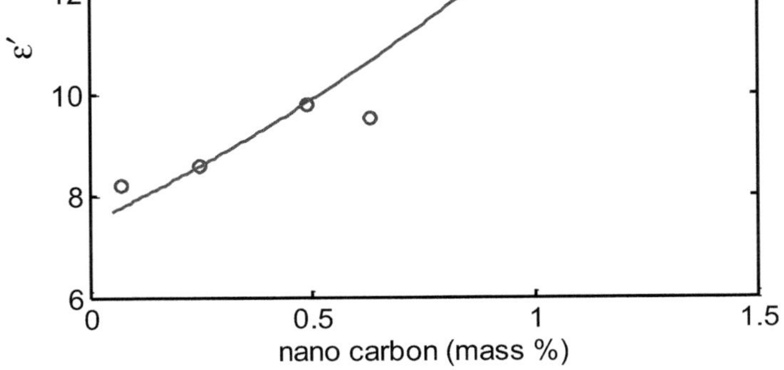

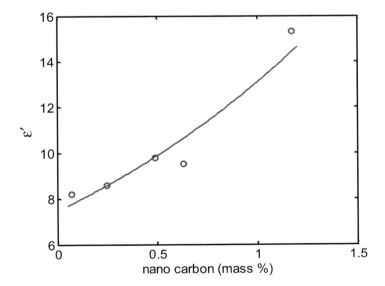

Figure 14.9 Dependence of dielectric constant (ε') of cement-carbon nanocomposites on carbon content (40°C and 1 kHz).

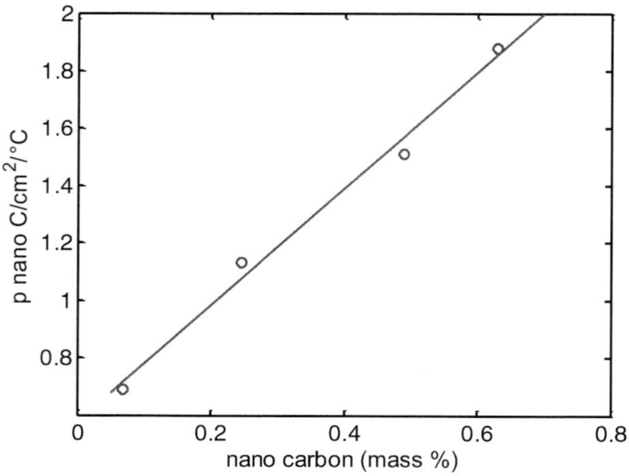

Figure 14.10 Dependence of pyroelectric coefficient of cement-carbon nanocomposites on carbon content (40°C and 1 kHz).

14.5.1.3 Charge storage via the pyroelectric effect

The conversion of thermal energy into electrical energy is essential in pyroelectric sensors. A large amount of heat absorbed by pavements is radiated to the atmosphere. The feasibility of capturing the thermal energy from pavements via the pyroelectric effect is reported in the next section [29].

14.5.1.4 Thermal energy harvesting from pavements via modeling and simulation

The idea explored in this section was proposed by Cuadras et al. [16] for pyroelectric energy harvesting. Figure 14.11 shows the circuit used to model pyroelectric sensors used for energy conversion and charge storage [16]. The pyroelectric sensor is modeled as a capacitor (C) and resistor (R) in parallel with a current source (i_{p}). The current is generated within the cell with the change in temperature. Equation 14.2 indicates that the pyroelectric current is directly proportional to the rate of change of temperature. However, one problem with this behavior is that the current will flow in yjr opposite direction when the rate changes from positive

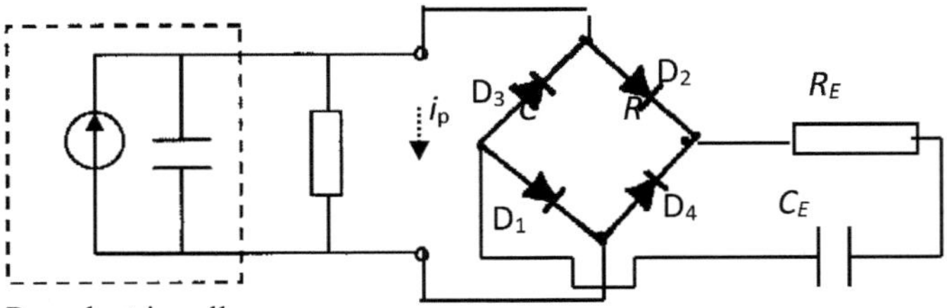

Figure 14.11 Pyroelectric cell with full bridge rectifier circuit for charge storage; D_1–D_2 are used in one direction of current flow (solid arrow), and D_3–D_4 are used in the other direction of current flow (dotted arrow).

to negative or from negative to positive. In other words, heating followed by cooling or cooling followed by heating will produce charge accumulation in a different direction. However, to charge an external capacitor it is essential that the capacitor be charged continuously.

To mitigate the problem, a full bridge diode rectifier circuit, as shown in Fig. 14.11, can be used. The charge pyroelectric source is connected to an external capacitor (C_E) and an external resistor (R_E). There are two pairs of diodes: one pair is used for each direction of current flow. Diodes D_1–D_2 are used when current is flowing in one direction, and diodes D_3–D_4 are used when current flows in another direction. At each time only the forward-biased diodes work, the other pair blocks current flow under a reverse-biased condition. As seen, in both cases, the external capacitor (C_E) is charged via charge flow in one direction, and that causes the voltage to increase across the external storage capacitor.

The charge flow in the two-capacitor system can be modeled as follows: When the new charge is accumulated, it is distributed in both capacitors and the charge balance equation can be written as [16]

$$\Delta Q_n = Q_{\mathrm{E},n} - Q_{\mathrm{E},n-1} + Q_{\mathrm{P},n} \pm Q_{\mathrm{P},n-1} \tag{14.4}$$

$$\begin{aligned} Q_{\mathrm{P},n} &= V_n C_\mathrm{P} \\ Q_{\mathrm{P},n-1} &= V_{n-1} C_\mathrm{P}, \\ Q_{\mathrm{E},n} &= V_n C_\mathrm{E} \\ Q_{\mathrm{E},n-1} &= V_{n-1} C_\mathrm{E} \end{aligned} \tag{14.5}$$

where C_p and C_E are pyroelectric cell capacitance and external charging capacitance, respectively; V_n and V_{n-1} are voltage at temperature data points n and $n-1$, respectively; Q_p and Q_E are charge accumulated in the pyroelectric cell and external capacitance, respectively; and ΔQ_n is the additional charge generated at the nth data point (from heating or cooling). The $\pm$ sign in front of the right-hand term indicates that charge stored in the pyroelectric cell can be in the opposite direction if the sign of the rate of change of temperature changes from the $(n-1)^{\mathrm{th}}$ data point to the nth data point.

Substitution of Eq. 14.5 into Eq. 14.4 results in the following recurrence equation, from which the voltage across the external capacitance can be calculated at a given temperature data point:

$$V_n = \frac{\Delta Q}{C_\mathrm{P} + C_\mathrm{E}} + \frac{C_\mathrm{E} \pm C_\mathrm{P}}{C_\mathrm{E} + C_\mathrm{P}} V_{n-1} = \frac{pA\Delta T}{C_\mathrm{P} + C_\mathrm{E}} + \left(\frac{C_\mathrm{E} \pm C_\mathrm{P}}{C_\mathrm{E} + C_\mathrm{P}}\right) V_{n-1} \tag{14.6}$$

Once the voltage is determined, the energy stored at the nth data point can be calculated from the following equation:

$$E_n = 0.5 C_\mathrm{E} V_n^2 \tag{14.7}$$

Equation 14.9 has been used to simulate the voltage produced from a measured temperature profile of actual pavement temperature (shown in Fig. 14.12). Figure 14.12 shows the temperature profile between May and October at a station located in Huntsville, Alabama. The temperature profile was obtained from the Environmental and Climatic Database of Mechanistic Empirical Pavement Design Guide (MEPDG) [30]. The simulated voltage and energy stored are shown in Figs. 14.13 and 14.14 for cement composites and other pyroelectric materials. The circuit is simulated with the following values: $C_\mathrm{P} = 3.761 \times 10^{-12}$ F, $C_\mathrm{E} = 10 \times 10^{-6}$ F, $p = 1.5 \times 10^{-9}$ C/cm^2/^{o}C, and $A = 0.907$ cm^2. When $C_\mathrm{E} \gg C_\mathrm{p}$, Eq. 14.6 reduces to $V_n \approx (pA\Delta T)/C_\mathrm{E} + V_{n-1}$. This is the simplified version of Eq. 14.6; when storing energy from temperature fluctuations, the values of C_p and C_E should be optimized for the highest energy storage. Furthermore, the external resistor, R_E, and the external capacitor, C_E, can be adjusted to obtain fast charging. Figure 14.13 shows that the accumulated voltage increases as the summer months go by.

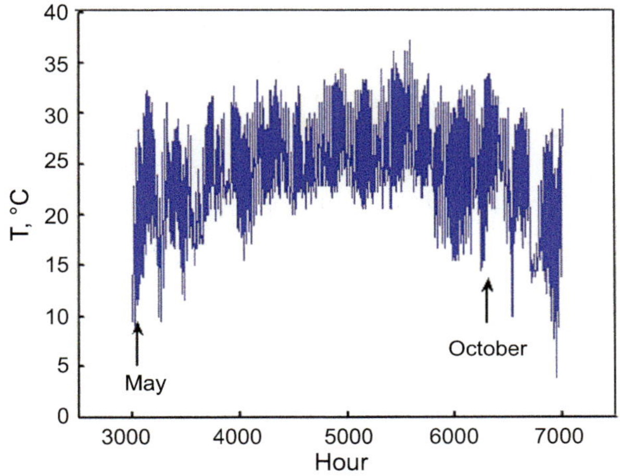

Figure 14.12 Temperature profile at Huntsville, AL, May–October (Station ID = 03856).

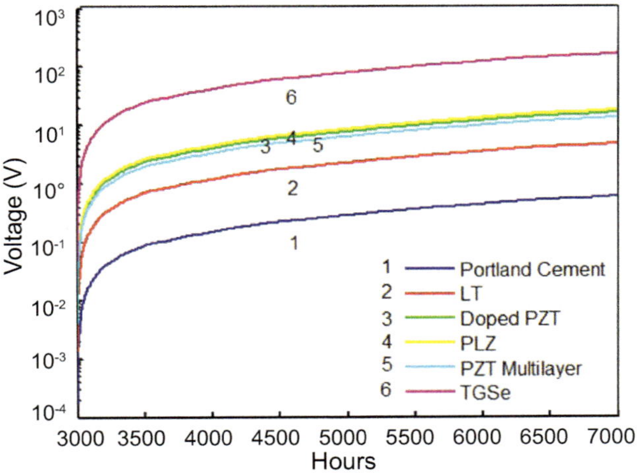

Figure 14.13 Generated voltage across the external storage capacitor from a single pyroelectric device.

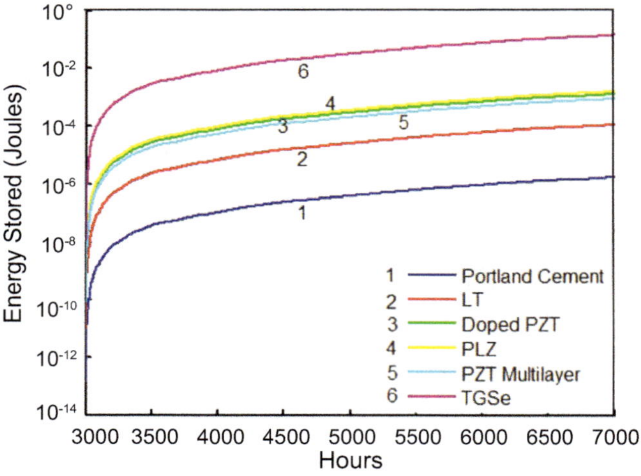

Figure 14.14 Energy (in joules) stored from a single pyroelectric device.

Figure 14.14 shows the accumulated energy (joules). The highest maximum voltage accumulated at the end of October is around 100 V. It should be noted that this maximum voltage can be controlled by choosing a suitable value for the external storage capacitor, C_E.

Assuming that one single device will produce 10 $\mu W/cm^2$ of energy, this converts to $10 \times (12 \times 2.54)^2\ \mu W/ft^2 \approx 10\ mW/ft^2$. Most of the wireless devices use energy in the range of several mW; therefore, even one single device will be sufficient to produce power for the wireless sensors. To check how many devices will be required to produce a power of 10 W, which is typical for a street intersection light-emitting diode (LED) signal, the area required $= 10/(10 \times 10^{-3})\ ft^2 = 1000\ ft^2$, that is, for a 10 ft wide lane, 100 ft length of the lane needs to be covered by the sensors.

It is possible to increase the voltage by placing several of the pyroelectric cells in a series combination. Batra et al. [32] performed modeling and simulation of energy harvesting capacity with temperature variations. Their results show that harvested energy can be compatible with use in autonomous sensors modules working in low-duty cycle switched supply mode [31, 32]. Hadi et al. studied the waste heat energy harvesting method from roads by the pyroelectric effect [33]. Their main focus was to develop an energy harvesting system that can modulate high-intensity solar radiation using a rotating cardboard to produce a higher rate of temperature variation in pyroelectric materials. In their investigation, PZT material was used as a transducer to capture heat and convert it into electric energy for ultralow-power device usage. The maximum positive and negative voltages produced from the prototype are 133.9 mV and −146.4 mV, respectively. Authors found that using the developed energy harvesting system, the output voltage produced is much higher compared to directly attaching the pyroelectric material onto the road, which is 23 mV.

Tao and Hu proposed using PVDF as a hybrid piezo-pyroelectric energy harvester via controlled experiments and found that there exists an interesting coupling phenomenon between pyroelectric and piezoelectric effects of PVDF: the voltage generated by simultaneous mechanical and thermal stimulations is an algebraic sum of voltages generated by separate stimulations [34]. They concluded that energy stored in a day is more than sufficient to power commonly used wireless sensor nodes and even LED lights. For more details about pyroelectric materials and devices for energy harvesting applications, readers can refer to an excellent review by Bowen et al. [35].

14.5.1.5 Waste heat harvesting via thermoelectric cement composites

In the United States, 60% of energy produced is wasted as heat, which can be directly converted into electric energy using thermoelectric (TE) materials through Seebeck effects. The Seebeck effect is a TE effect that is the basis for thermocouples for temperature measurement. This effect involves charge carriers moving from a hot point to a cold point within a TE material, thereby resulting in a voltage difference between two points. The Seebeck coefficient is the voltage difference per unit temperature between two points. TE technology can be explored in civil structures where waste heat is available, such as thermal storage in concrete structures, building envelopes, and heating, ventilation, and air-conditioning systems. It is also possible to harvest thermal energy for power generation in concrete and concrete asphalt pavement systems. Ghahari et al. [36] investigated by incorporating nano-zinc oxide (ZnO) particles in cement paste to form ZnO-cement composites that had an increased Seebeck coefficient of cement by 17% [36]. Thermal conductivity decreased to 9%, and electrical conductivity was found to have decreased to 37% compared to plain cement paste, thereby increasing the Z-factor of TE. Thus, the authors suggested that the resulting materials can potentially be used for energy harvesting to convert stored energy in concrete infrastructures for electric power. Sun et al. demonstrated the enhancement by 4–5-folds of the TE effect of carbon fiber–reinforced cement compositions by a metallic oxide/cement interface [37]. The largest Seebeck coefficient was found to be $+100$ μV/°C in carbon fiber–reinforced cement with 5.0 wt% Bi_2O_3. Thus, significant effort has to be applied in this field of energy harvesting to make the TE nanocomposite cement-based pavement and building roof come to fruition. The TE effect in cement-based materials was first reported by Ji et al. [38]. Authors fabricated a composite by incorporating short carbon fibers in cement, which contributed to the majority of the total TE effect [38].

Wei et al. [39] incorporated metallic oxide nanopowders in a cement matrix to increase the TE effect of cement composites [39]. The enhanced Seebeck coefficient of cement composites by adding metallic oxide nanopowders was found to be higher than 1000 μV/°C. [39] It is worth mentioning that a high Seebeck coefficient of this kind of a cement composite is important for the potential application in energy harvesting and air-conditioning systems of future buildings. Lee et al. [40] developed a system that can collect energy from the temperature difference between surface and inside of concrete structure, which presents a feasible energy harvesting using TE technology on concrete structure in roads [40].

14.5.1.6 Electric power harvesting via application of piezoelectric transducers in pavements

Mechanical energy appears everywhere in industry; road vibration energy created by vehicles will not only damage the pavement structure but also be difficult to collect. However, utilizing electromechanical conversion characteristics of piezoelectric materials (discussed in the next section), directly converting mechanical energy to electric energy, collects vibration energy when vehicles pass on the pavements by designing an appropriate transducer package box based on cantilevers, for traffic lights along roads, signs, and so on. Thus, saving the cost of laying long-distance transmission lines also take advantage of the loss of mechanical energy. Various efforts have been made to enhance the piezoelectric coefficients of cement by incorporating piezoelectric particles and other variations [41–45]. To mention a few, Li et al. [46] fabricated cement-based 0-3 piezoelectric composites with enhanced values of k_t, ε_r, and d_{33}: 18.1%, 130.5, and 53.7 pC/N, respectively [46]. These highest values of the performance parameters were obtained by incorporating PZT nanoparticles due to good crystallinity of PZT grains and the network-like distribution of PZT particles in the cement. Jaitanong et al. [47] investigated PC-based piezoelectric composites using PZT (xPZT(1–x)PC; where $x = 0.3$, 0.6, and 0.9) and found values of dielectric constant (ε_r) and piezoelectric coefficient (d_{33}) up to 536 and 87 pC/N, respectively [47]. Thus, there is good potential for application in civil engineering. Gong et. al. [48] fabricated a 0-3 composite from white cement, PZT powder, and a small amount of carbon black with d_{33}, ε_r, tan δ, and k_t values of 28.5 pC/N, 202.6, 0.19, and 12.2%, respectively, with potential applications in civil engineering [48]. PC with addition of 70 vol% PZT powders and various amount of carbon nanotubes (CNTs) from 1 to 1.0 vol% was also exploited for its potential in smart civil engineering applications, and it was found that the 0.3 vol% CNT composite showed the highest piezoelectric strain factors (d_{33}) of 62 pC/N and piezoelectric voltage factors (g_{33}) 60×10^{-3} Vm/N. Rianyoi et al. [49] investigated the 0-3 lead-free piezoelectric PC composites (PC/barium titanate) and obtained an electromechanical coupling coefficient (k_t) value 16.6% when the content of BT reaches 70% with a particle size of 425 μm [49]. Piezoelectric properties of 0-3 PZT/sulfoaluminate cement composites were investigated for the first time by Huang et al. [50], and results showed that with 85% PZT, k_p and k_t are 28.54% and 28.19%, respectively [50]. Jaitanong et al. [51] fabricated new and novel cement-based piezoelectric composites using PZT, PVDF, and PC. The authors found that PZT-PVDF-PC composites with 50 vol% PZT and 5 vol%

PVDF have g_{33} and d_{33} values 25.7×10^{-3} Vm/N and 24pC/N, respectively [51].

14.5.2 Functional Cement-Based Nanocomposites for Structural Health Monitoring in Civil Engineering and Sensor Applications

Civil engineering infrastructure is an important and most expensive national investment and asset of any country. Furthermore, civil engineering structures have long operational life along with being costly to maintain and replace once they are erected [52]. The most important civil engineering structures include bridges, multistory high-rise buildings, bridge piers, power utilities, nuclear power plants, and dams. The deterioration of all civil engineering structures is due to aging of materials, overloading, aggressive and continuous exposure conditions, lack of sufficient maintenance, and sometimes difficulties encountered in proper inspection methods. These factors contribute to material and structure degradation as internal and external damages emerge and coalesce, and then evolve and progress. Thus, failure of the structural integrity occurs, including possible loss of life and huge investment. To safeguard structural integrity and safety, major civil structures have to be equipped with a structural health monitoring (SHM) system for continuous monitoring, visual inspection, and damage detection of civil structures with minimum personnel involvement. An effective SHM system can detect various defects or imperfections in real time and online. It can also monitor strain, stress, temperature, and humidity so that the optimum maintenance of structures can be carried out to ensure safety and normal service life. The SHM system has three major components: a sensor system, a data processor, and a health evaluation system. In this section, only a stable and reliable structural sensing capability component is described, viz. a sensing system fabricated via smart materials. It is worth mentioning that in civil engineering, concrete is the most popularly used smart structural material.

Candidate piezoelectric composites, such as polymer-based piezoelectric composites, are not suitable for civil engineering applications because of their distinct difference in volume stability from concrete. To meet the requirement of civil engineering structures, a 0-3 cement-based piezoelectric ceramic composite has been developed in the recent past [45]. Smart materials/sensors, such as fiber optic sensors, piezoelectric sensors, magnetostrictive sensors, and self-diagnosing fiber-reinforced structural composites, possess very important capabilities of sensing various physical and chemical parameters related to the health of a structure [53]. Sensors are a fundamental component and the heart of SHM systems. Sensors allow one to monitor structural damage caused by fatigue load, environment corrosion, or natural disasters.

Chen et al. [54] investigated mechanical and smart properties, such as piezoresistivity of carbon fiber and graphite conductive concrete, for internal damage monitoring of the structure [54]. The strain-dependent resistivity characteristic of modified concrete was subject to a quadratic and fitted correlation coefficient of over 0.95. A strong correlation with strain was found with 0.6 wt% carbon fiber and 2.5 wt% graphite.

Sasmal et al. [55] investigated extensively electrical resistivity of CNT- and CNF-incorporated cement under different variations, such as degree of hydration, content of moisture in the matrix, external voltage field, etc., of nanocomposites [55]. The piezoresistive strain sensitivity of nanocomposites containing CNFs (0.1%) shows better sensing tendency than CNT nanocomposites. The authors' studies brought out the procedure for synthesis and development of carbon-based smart cement composites for both stress and acceleration sensing that can be used as embedded sensors for real-time health monitoring of concrete structures.

Rainieri et al. [56] investigated CNT–cement composites and their self-sensing capabilities in detail, along with new radiofrequency plasma technology for the functionalization of CNTs for incorporation in the matrix [56]. Their analysis of results and theoretical developments provide useful design criteria for the creating of CNT–cement nanocomposites optimized for SHM applications in civil engineering.

D'Alessandro et al. [57] investigated the major problem of dispersion of conductive nanofillers, CNTs in aqueous solutions, the physical properties of a fresh mixture of CNT cement-based composites, the electromechanical properties of hardened composites, and sensing properties of optimized transducers [57]. Their study demonstrated that the simple dispersion technique can produce cementitious high-performance sensors using multiwalled CNTs for SHM applications.

Accurate and reliable sensors are an integral part of the modern intelligent transportation system. Monteiro et al. [58] evaluated piezoresistive properties of a carbon black–cement composite for traffic monitoring applications and found it showed good linearity and repeatability upon compressive cyclic loads [58]. The authors demonstrated that a low-cost carbon black filler provides a remarkable reproducible sensitivity to accurately monitor traffic-like loading with a gauge factor from 40 to 60.

Han et al. [59] investigated a self-sensing CNT–cement nanocomposite via its piezoresistive properties for traffic monitoring by mechanical stress induced by traffic flows well [59]. Their experimental results showed that there is a good corresponding relationship between compressive stress and electrical resistance of the self-sensing CNT–cement composite, including vehicular loadings. Thus, these studies have potential for traffic monitoring use, such as in vehicle detection, weigh-in-motion measurement, and vehicle speed detection. Thus, it can be concluded that self-sensing composites are becoming highly attractive for civil engineering applications to improve the safety and performance of structures. These smart composites show a detectable change in their electrical resistivity with applied stress or strain, and these unique characteristics make them useful for health monitoring of structures [60].

14.6 Summary and Future Outlook

Salient features of research works reported in this chapter are recent findings on cement-based modified nanocomposite materials with applications in ambient energy harvesting, sensor and actuator technology applied to traffic monitoring, and SHM. A brief introduction to the basics of composite materials is described for brevity.

Potential applications of CNTs, CNFs, and graphite in the construction industry represent an innovative discovery. The current literature clearly determines the growing interest of the scientific community around the globe about self-sensing, actuating, smart and energy harvesting pavements, and cementitious materials with nanocarbon fillers. No doubt, future research and applications will be multidisciplinary in nature, such as monitoring concrete and asphalt pavements, weigh-in-motion sensing via self-sensing nanostructured cementitious slabs for traffic and crowd management, monitoring historical structures, and monitoring energy systems such as wind turbine bases and dams. Other important and promising applications will be related to the development of smart cities, such as conducting floors with antistatic capability for data center smart grids, thermally conductive cement-based materials for de-icing of roads or for geothermal applications, smart roads to monitor vehicles with autonomous driving technology, and thermal and hygroscopic sensors for local environmental control of structures, through the combination of conductive nanofillers and phase change materials.

Acknowledgments

AB, AK, and BB would especially like to thank Dr. Chance M. Glenn, Sr., the former dean of the College of Engineering, Technology, and Physical Sciences at AAMU, for his support and, in particular, our fruitful discussions on energy harvesting. The authors also express gratitude to Dr. M. D. Aggarwal for his support and encouragements. AB and BB gratefully acknowledge financial support for this work from the US National Science Foundation grant-RISE/HRD #1546965, and AAMU Title III program. AK acknowledges the support from the U.S. Department of Homeland Security through grant number 2014-ST-062-000060-02.

References

1. Batra, A. K., Aggarwal, M. D., Edwards, M., Bhalla, A. S. (2008). Present status of polymer: ceramic composites for pyroelectric infrared detectors, *Ferroelectrics*, **366**, pp. 84–121.
2. Taya, M. (2008). *Electronic Composites: Modeling, Characterization, and MEMS Applications*, 1st Ed., Cambridge University Press, UK.

3. Guggilla, P. (2007). Studies on pyroelectric materials for infrared sensor application, PhD Dissertation, Alabama A&M University, Normal, Alabama, USA.
4. Satapathy, S., Gupta, P. K., Varma, K. B. R. (2009). Enhancement of nonvolatile polarization and pyroelectric sensitivity in lithium tantalate (LT)/poly (vinylidene fluoride) (PVDF) nanocomposite, *J. Phys. D: Appl. Phys.*, **42**, p. 055402.
5. Sessler, G. M. (1994). Ferroelectric polymer and ceramic-polymer composites, *Key Eng. Mater.*, **249**, pp. 92–93.
6. Whatmore, R. W., Watton, R. (2001). *Pyroelectric Materials and Devices Infrared Detectors and Emitters: Materials and Devices*, Kluwer Academic Publishers, Boston, USA.
7. Guggilla, P., Batra, A. K. (2011). Nanocomposites and polymers with analytical methods, in Cuppoletti, J. (ed.) *Novel Electro Ceramic: Polymer Composites - Preparation, Properties and Applications*, InTech, Croatia, pp. 287–308.
8. Zhang, Q. Q., Ploss, B., Chan, H. L. W., Choy, C. L. (2000). Integrated pyroelectric arrays based on PCLT/P (VDF-TrFE) composite, *Sens. Actuators, A*, **86**, pp. 216–219.
9. Mallick, P. K. (1997). *Composites Engineering Handbook (Materials Engineering)*, 1st Ed., CRC Press, USA.
10. Newnham, R. E., Skinner, D. P., Cross, L. E. (1978). Connectivity and piezoelectric-pyroelectric composites, *Mater. Res. Bull.*, **13**, pp. 525–536.
11. Tressler, J. F., Alkoy, S., Dogan, A., Newnham, R. E. (1999). Functional composites for sensors, actuators and transducers, *Composites Part A*, **30**, 477–482.
12. Batra, A. K., Aggarwal, M. D. (2013). *Pyroelectric Materials: Infrared Detectors, Particle Accelerators, and Energy Harvesters*, SPIE, USA.
13. Nalwa, H. S. (1995). *Ferroelectric Polymers: Chemistry, Physics, and Applications*, Marcel Dekker, USA.
14. Yamazaki, H., Tayama, T. K. (1981). Pyroelectric properties of polymer-ferroelectric composites, *Ferroelectrics*, **33**, pp. 147–153.
15. Batra, A. K., Alomari, A. A. (2017). *Power Harvesting via Smart Materials*, SPIE, USA.
16. Cuadras, A., Gasulla, M., Ferrari, V. F. (2010). Thermal energy harvesting through pyroelectricity, *Sens. Actuators*, **158**, pp. 132–139.
17. Han, B. G., Ou, J. P., Han, B. Z. (2009). Experimental study of use of nickel powder-filled Portland cement-based composite for fabrication

of piezoresistive sensors with high sensitivity, *Sens. Actuators*, **149**, pp. 51–55.

18. Chaipanich, A. (2007). Electrical and piezoelectric properties of PZT-Cement composites, *Curr. Appl. Phys.*, **7**, pp. 537–539.
19. Wen, S., Chung, D. D. L. (2000). Damage monitoring of cement paste by electrical resistance measurement, *Cem. Concr. Res.*, **30**, pp. 1979–1982.
20. Gomg, H., Lie, Z., Zang, Y., Fan, R. (2009). Piezoelectric and dielectric behavior of 0-3 cement-based composites mixed with carbon black, *J. Eur. Ceram. Soc.*, **29**, pp. 2013–2019.
21. Wen, S., Chung, D. D. L. (2003). Pyroelectric behavior of cement-based materials, *Cem. Concr. Res.*, **33**, pp. 1675–1679.
22. Web Access: http://pavehensystems.com/
23. Web Access: http://innowattech.co.il/
24. Web Access: http://www.jreast.co.jp/e/
25. Aggarwal, M. D., Batra, A. K., Guggilla, P., Edwards, M. E., Currie, J. R. (2010). Pyroelectric materials for uncooled infrared detectors, NASA/TM-2010, 216373, USA.
26. Batra, A. K., Alim, M. A., Currie, J. R., Aggarwal, M. D. (2009). The electrical response of the modified lead titanate thick films, *Physica B*, **404**, pp. 1905–1911.
27. Batra, A. K., Bhattacharjee, S. (2011). Energy harvesting roads via pyroelectric effect: a possible approach, *Proc. SPIE*, **8503**, pp. 803519-1.
28. Batra, A. K., Bandyopadhyay, A., Chilvery, A. K., Thomas, M. (2013). Modeling and simulation for PVDF-based pyroelectric energy harvester, *Energy Sci. Technol.*, **5**, pp. 1–7.
29. Bhattacharjee, S., Batra, A. K., Cain, J. (2010). Carbon nano-fiber reinforced cement composites for energy harvesting road, *Green Streets and Highways Conference*, **389**, pp. 258–271.
30. Guide for Mechanistic empirical design of new and rehabilitated pavement structure; final report. NCHRP Project No. 1-37A, Transportation Research Board, National Research Council, Washington, D.C., USA.
31. Batra, A. K., Bhattacharjee, S., Chilvery, A., Stephens, J. (2012). Energy harvesting via pyroelectric transducer, *Sens. Trans.*, **138**(3), pp. 114–121.
32. Batra, A. K., Bhattacharjee, S., Chilvery, A. K., Edwards, M. E., Bhalla, A. (2011). Simulation of energy harvesting from roads via pyroelectricity, *J. Photonics Energy*, **1**(1), p. 014001.

33. Hadi, A. M., Tawil, S. N., Mohamad, T. N., Syaripuddin, M. (2015). Energy harvesting from roads by pyroelectric effect, *ARPN J. Eng. Appl. Sci.*, **10**(20), pp. 9884–9890.
34. Tao, J., Hu, J. (2016). Energy harvesting from payment via polyvinylidene fluoride: hybrid piezo-pyroelectric effects, *Appl. Phys. Eng.*, **17**(1), pp. 502–511.
35. Bowen, C. R., Taylor, J., Leboulbar, E. (2014). Pyroelectric materials and devices for energy harvesting applications, *Energy Environ. Sci.*, **7**(12), pp. 2836–3856.
36. Ghahari, S., Ghafari, E., Lu, N. (2017). Effect of ZnO nanoparticles on thermoelectric properties of cement composite for waste heat harvesting, *Constr. Build. Mater.*, **145**, pp. 755–763.
37. Sun, M., Li, Z., Mao, Q., Shen, D. (1998). Study of the hole conduction phenomenon in carbon firber-reinforced concrete, *Cem. Concr. Res.*, **28**, pp. 549–554.
38. Ji, T., Zhang, X., Li, W. (2016). Enhanced thermoelectric effect of cement composite by addition of metallic oxide nanopowders for energy harvesting in buildings, *Constr. Build. Mater.*, **115**, pp. 576–581.
39. Wei, J., Nie, Z., He, G., Hao, L., Zhao, L., Zhang, Q. (2014). Energy harvesting from solar irradiation in cities using the thermoelectric behavior of carbon fiber reinforced cement composites, *RSC Adv.*, **4**(89), pp. 48128–48134.
40. Lee, J. J., Kim, D. H., Lee, S. T., Kim, J. K. (2014). Fundamental study of energy harvesting using thermoelectric effect on concrete structure in road, *Adv. Mater. Res.*, **1044–1045**, pp. 332–337.
41. Li, Z., Zhang, D., Wu, K. (2002). Cement-based 0-3 piezoelectric composites, *J. Am. Ceram. Soc.*, **85**, pp. 305–313.
42. Xin, C., Huang, S., Chang, J., Li, Z. (2007). Piezoelectric, dielectric, and ferroelectric properties of 0-3 ceramic/cement composites, *J. Appl. Phys.*, **101**, pp. 094110–094116.
43. Lam, K. H., Chan, H. L. W. (2005). Piezoelectric cement-based 1-3 composites, *Appl. Phys.*, **A81**, pp. 1451–1454.
44. Xing, F., Dong, B. Q., Li, Z. (2008). Dielectric, piezoelectric, and elastic properties of cement-based piezoelectric ceramic composites, *J. Am. Ceram. Soc.*, **91**, pp. 779–782.
45. Potong, R., Rianyoi, R., Ngamjarurojana, A., Chaipanich, A. (2013). Dielectric and piezoelectric properties of 1-3 non-lead barium zirconate titanate-Portland cement composites, *Ceram. Int.*, **39**(1), pp. S53–S57.

46. Li, Z., Gong, H., Zhang, Y. (2009). Fabrication and piezoelectricity of 0-3 cement based composite with nano-PZT powder, *Curr. Appl. Phys.*, **9**, pp. 588–591.
47. Jaitanong, N., Chaipanich, A., Tunkasiri, T. (2008). Properties of 0-3 PZT-Portland cement composites, *Ceram. Int.*, **34**, pp. 793–795.
48. Gong, H., Zhang, Y., Quan, J., Che, S. (2011). Preparation and properties of cement based piezoelectric modified by CNTS, *Curr. Appl. Phys.*, **11**, pp. 653–656.
49. Rianyoi, R., Potong, R., Ngamjarurojana, A., Chaipanich, A. (2013). Influence of barium titante content and particle size on electromechanical coupling coefficient of lead-free piezoelectric ceramic-Portland cement composites, *Ceram. Int.*, **39**, pp. S47–S51.
50. Huang, S., Chang, J., Xu, R., Liu, F. (2004). Piezoelectric properties of 0-3 PZT/sulfoaluminate cement composites, *Smart Mater. Struc.*, **13**, pp. 270–274.
51. Jaitanong, N., Yimnirun, R., Zeng, H. R., Li, G. R., Yin, Q. R., Chaipanich, A. (2014). Piezoelectric properties of cement based/PVDF/PZT composites, *Mater. Lett.*, **130**, pp. 146–149.
52. Sun, M., Staszewski, W. J., Swamy, R. N. (2010). Smart sensing technologies for structural health monitoring of civil engineering structures, *Adv. Civ. Eng.*, Article ID 724962 (13 pp).
53. Rana, S., Subramani, P., Fangueiro, R., Correia, A. G. (2016). A review on smart self-sensing composite materials for civil engineering applications, *AIMS Mater. Sci.*, **3**(2), pp. 357–379.
54. Chen, M., Hao, P., Geng, F., Zhang, L., Liu, H. (2017). Mechanical and smart properties of carbon fiber and graphite conductive concrete for internal damage monitoring of structure, *Constr. Build. Mater.*, **142**, pp. 320–327.
55. Sasmai, S., Ravivarman, N., Sindu, B. S., Vignesh, K. (2017). Electrical conductivity and piezo-resistive characteristics of CNT and CNF incorporated cementitious nanocomposites under static and dynamic loading, *Composite Part A*, **100**, pp. 227–243.
56. Rainieri, C., Song, Y., Fabbrocino, G., Schulz, M. J., Shanov, V. (2013). CNT-cement based composites: fabrication, self-sensing properties and prospective applications to structural health monitoring, *Proc. SPIE*, **8793**, pp. 87930-1.
57. D'Alessandro, A., Ubertini, F., Lafamme, S., Rallini, M., Materazzi, A. L., Kenny, J. M. (2016). Strain sensitivity of carbon nanotube cement-based composites for structural health monitoring, *Proc. SPIE*, **9803**, pp. 980319-1.

58. Monteiro, A. O., Loredo, A., Costa, P. M. F. J., Oeser, M., Cachim, P. B. (2017). A pressure-sensitive carbon black composite for traffic monitoring, *Constr. Build. Mater.*, **154**, pp. 1079–1086.
59. Han, B., Yu, S., Kwon, E. (2009). A self-sensing carbon nanotube/cement composite for traffic monitoring, *Nanotechnology*, **20**, p. 445501 (5 pp).
60. Han, B., Ding, S., Yu, X. (2015). Intrinsic self-sensing concrete and structures: a review, *Measurement*, **59**, pp. 110–128.

Chapter 15

Addition of Carbon Nanofibers to Cement Pastes for Electromagnetic Interference Shielding in Construction Applications

E. Zornoza, O. Galao, F. J. Baeza, and P. Garcés
Department of Civil Engineering, University of Alicante, Alicante, Spain
pedro.garces@ua.es

15.1 Introduction

The telecommunication market has been growing exponentially in the past decades. That is the reason why the number of studies on electromagnetic (EM) shielding materials also keeps on growing, with the aim of reducing damage due to electromagnetic interference (EMI), which is produced by radiofrequency systems. EMI shielding refers to the ability of materials to protect against EM radiation, thus avoiding that EM radiation getting through the material or reducing its power. Nowadays, this type of application is gaining importance in our society since in the past years the use of electronic devices sensitive to environmental radiations has

Nanotechnology in Cement-Based Construction
Edited by Antonella D'Alessandro, Annibale Luigi Materazzi, and Filippo Ubertini

ISBN 978-981-4800-76-1 (Hardcover), 978-0-429-32849-7 (eBook)
www.jennystanford.com

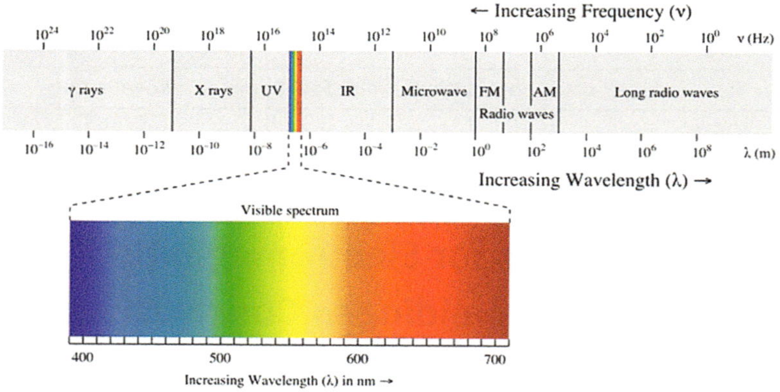

Figure 15.1 Complete electromagnetic spectrum with the visible part extended. This work refers to the highlighted zone (from Wikimedia Commons. Original author: Philip Ronan).

increased exponentially. This is particularly important for high frequencies, such as radiofrequencies produced by mobile phones, which have experienced a higher presence in the environment due to the proliferation of wireless devices. Shielding is especially necessary in electrical facilities underground that may contain power transformer or other electronic components that may be important for power plants and telecommunication facilities [1]. Additionally, this is a controversial aspect due to public health concerns.

Figure 15.1 shows the complete EM spectrum with the visible part extended and the zone under study in this chapter highlighted. On the other hand, Table 15.1 offers the radiofrequency spectrum and the band that is under investigation in this chapter is also highlighted. At present, mobile phones use the frequency range from 5 MHz to 3.5 GHz approximately, which lies in the ultrahigh-frequency (UHF) range.

High-frequency radiation shielding is a barrier against the transmission of EM fields in both ways. A cement matrix behaves like a dielectric material, so it is a bad conductor. That implies the necessity of including another material as an addition, which can be able to conduct electricity in order to achieve an EMI-shielding behavior. That means that we need to produce electrically

Table 15.1 Radiofrequency spectrum. This work refers to the highlighted one

	Abbreviation	Band	Frequency	Wavelength
Extremely low frequency	ELF	1	3–30 Hz	100,000–10,000 km
Superlow frequency	SLF	2	30–300 Hz	10,000–1000 km
Ultralow frequency	ULF	3	300–3000 Hz	1000–100 km
Very low frequency	VLF	4	3–30 kHz	100–10 km
Low frequency	LF	5	30–300 kHz	10–1 km
Medium frequency	MF	6	300–3000 kHz	1 km–100 m
High frequency	HF	7	3–30 MHz	100–10 m
Very high frequency	VHF	8	30–300 MHz	10–1 m
Ultrahigh frequency	UHF	9	300–3000 MHz	1 m–100 mm
Superhigh frequency	SHF	10	30–300 GHz	100–10 mm
Extremely high frequency	EHF	11	300–3000 GHz	10–1 mm

conductive cementitious materials if EMI shielding is required for a particular application.

EM waves consist of two essential components: a magnetic field (**H**) and an electrical field (**E**). Both are perpendicular between them (also called plane waves) in the case of transversal EM (T-EM) waves, as it happens for guided waves in coaxial wires). The impedance of the media (Z), or intrinsic impedance, is defined according to Eq. 15.1:

$$Z = \frac{\mathbf{E}}{\mathbf{H}} \text{ (ohm)} \tag{15.1}$$

A uniform plane wave is a particular solution of Maxwell's equations when **E** and **H** have the same propagation direction, and are perpendicular between them and perpendicular to the propagation direction. From a strict point of view, in practice, a plane wave does not exist, since it would be necessary an emitter of endless extension. However, it is possible to get close to this idea when the receiver is far enough from the emitter or when wave-guiding devices are used. The concept of a plane wave is essential for the study of EMI shielding applications [2].

When plane waves are used, the intrinsic impedance of the media used for the plane wave to travel is expressed according to Eq. 15.2:

$$Z = \sqrt{\frac{j\omega\mu}{\sigma + j\omega\varepsilon}}, \tag{15.2}$$

where ω is the angular frequency; μ, ε, and σ are the permeability, permittivity, and electrical conductivity of the media used for the EM wave to travel, respectively; and $j^2 = -1$.

Magnetic permeability is the ability of a material or media to attract magnetic fields and force them to pass through themselves and therefore to concentrate magnetic flux lines and increase the magnetic induction value. Its calculation comes from the relationship between the magnetic induction created in the material and the intensity of the magnetic field, according to $\mu = B/\mathbf{H}$. Magnetic permeability can be expressed as $\mu = \mu_o \times \mu_r$, where $\mu_o = 4\pi \times 10^{-7}$ (H/m) is the magnetic permeability of vacuum and μ_r the relative magnetic permeability. Relative magnetic permeability is an intrinsic property of each material. In this sense, materials may be classified into three different categories: diamagnetic, such as copper ($\mu_r < 1$); paramagnetic, such as silver ($\mu_r > 1$); and ferromagnetic, such as iron ($\mu_r \gg 1$). In the case of a cementitious matrix with conductive but not magnetic additions, such as the case of carbon nanofibers (CNFs), μ_r can be assumed equal to 1.

Electrical permittivity or dielectric constant (ε) is the ability of the material to acquire a charge dipolar density upon the application of an electrical field. It reveals the presence of atomic and molecular dipoles. Again it is usually expressed as its relative value ε_r with respect to the vacuum (ε_0), according to $\varepsilon_r = \varepsilon/\varepsilon_0$.

Vacuum magnetic permeability and vacuum permittivity are related according to $\varepsilon_0 \times \mu_0 = 1/c^2$, where c is the speed of light in vacuum. Therefore, from Eq. 15.2, and taking into account that conductivity is almost zero for insulating materials, the impedance of air, which is the media used in this work, can be calculated as 377 Ω.

EMI shielding can be achieved by different mechanisms, typically reflection and absorption of high-frequency radiation by a material that acts as a shield or barrier against the penetration of radiation. The shielding ability provided by a particular material increases when the radiation frequency is increased, and it is enhanced by the presence of electrical and/or magnetic dipoles [3–7] if conductivity is ensured at that frequency. Another shielding mechanism is multiple reflection in inner surfaces and the interfaces, but it is a minor contribution when conductive cementitious materials are considered [8].

15.1.1 Shielding by Reflection

The first mechanism that can be explored for achieving EMI shielding is reflection of EM waves. The loss by reflection in the interface between two media (air and shielding material) is related to the characteristic impedance of each. The reflection of a wave in an EM shield is produced as a consequence of the difference in the impedance that exists in the interface between the shield and the original media (typically, the air). This phenomenon, as far as it depends on the impedances (see Eq. 15.2), is a function of the shielding material conductivity, magnetic permeability, electrical permittivity, and frequency.

Reflection is the main shielding mechanism in most materials, but it is especially significant in materials with high electrical conductivity, such as metals. In this type of material, loss by reflection is basically lead by the effective electrical conductivity of the shielding material (higher loss by reflection implies higher reflectivity of the material). On the other hand, effective electrical conductivity of a composite depends on the microscopic characteristics of the addition (CNFs in this case), such as its fractional volume, size, and shape of its particles, and the so-called loss tangent or dielectric loss tangent of the matrix (cementitious paste in this case). Even the preferential orientation of the dispersed particles of the addition in the matrix may affect significantly the effectiveness of shielding. The percolation of the conductive addition (CNFs) is a circumstance that improves the observed shielding, but it is necessary to achieve it to detect the positive influence the addition.

15.1.2 Shielding by Absorption

The second mechanism to obtain EMI shielding is radiation absorption. When an EM wave goes through a shield, its amplitude decreases exponentially inside the shield due to the induced currents that are produced. Inside an EMI shield, the absorption mechanism implies that EM energy is transformed in thermic energy. To enhance the absorption mechanism in a particular material, the presence of magnetic and/or electrical dipoles, which can interact with EM fields, is necessary. Electrical dipoles produce a higher value of the dielectric constant (ε) of the material, whereas magnetic dipoles (such as Fe_3O_4) let the material achieve a high magnetic permeability (μ) value [7, 9, 10].

Absorption is the main shielding mechanism in magnetic materials. The loss by absorption increases if the thickness of the barrier or the EM wave frequency increases.

15.1.3 Shielding by Multiple Reflections

In addition to reflection and absorption, another mechanism is multiple reflections. This refers to the reflection that is produced in several surfaces of inner faces inside the material. This mechanism requires the presence of a high specific surface. An example of a material where multiple reflections can be found is a highly porous material with an internal structure similar to sponges.

Classically, this type of shielding mechanism has been considered almost negligible in most materials, including composites with carbon fibers. However, although there exists a wide knowledge of the behavior of monophasic materials, advances in the field of multiphasic materials or composites have not been developed in the same extent. In a composite that includes a conductive addition, reflections generally occur in the interface between the matrix and the conductive particles. For that reason, multiple reflections may be significant when the size of the particles is much lower than the penetration depth (also known as skin depth). In addition, a rough surface of these particles would imply higher attenuation by multiple reflections. On the other hand, the directions of the reflections are essentially random, so the calculation is rather complicated.

The penetration depth of the wave, or skin depth [1, 11], is defined as the distance needed to observe a decrease of the EM wave amplitude by a factor e (2.718), that is, the depth that the wave has to penetrate in the material to observe an amplitude equal to $1/e$ times its initial value. The intensity of the electrical field diminishes exponentially inside the conductor. On the other hand, skin depth decreases when frequency, magnetic permeability, or conductivity increase, as can be appreciated in Eq. 15.3:

$$\delta = \frac{1}{\sqrt{\pi \cdot f \cdot \mu \cdot \sigma}}, \tag{15.3}$$

where δ is the penetration depth (m), f is radiation frequency (Hz), μ is the magnetic permeability of the conductor (H/m), and σ is the electrical conductivity ($\Omega^{-1}\text{m}^{-1}$). A lower skin depth implies better shielding behavior. Anyway, the wavelength used in the work is much higher than the conductive particle size, that is, the CNFs, so a significant contribution of the multiple reflection mechanism to the overall phenomenon is not expected.

15.1.4 Shielding Effectiveness

Independent of the type of mechanism, loss by EMI shielding is typically expressed in decibels (dB). In general, the power of the transmitted wave will be the difference between the incident power and the sum of the reflected power and absorbed power. The addition of all the losses is the shielding effectiveness (SE), or total shielding, and in the same way it is expressed in dB. SE can be expressed as a function of the ratio between the incident and the transmitted EM power, according to Eq. 15.4:

$$\text{SE} = -10\log_{10}\left(\frac{P_\text{t}}{P_\text{i}}\right) \tag{15.4}$$

where P_t is the transmitted EM power and P_i is the incident EM power.

In addition, the SE can be also defined as a function of the electrical and magnetic field attenuations, according to Eq. 15.5:

$$\text{SE} = 20\log_{10}\left|\frac{\mathbf{E}_\text{i}}{\mathbf{E}_\text{t}}\right|; \quad \text{SE} = 20\log_{10}\left|\frac{\mathbf{H}_\text{i}}{\mathbf{H}_\text{t}}\right| \tag{15.5}$$

where $\mathbf{E}_i$ and $\mathbf{H}_i$ are the incident electrical and magnetic fields, respectively, and $\mathbf{E}_t$ and $\mathbf{H}_t$ are the transmitted electrical and magnetic fields, respectively.

According to these equations, a material with an SE of 1 dB implies an attenuation of the incident wave amplitude of 10.9%, whereas a 20 dB SE would offer an attenuation of 90.0%.

As previously stated, the amplitude of the EM wave decreases inside the material by a factor of $e^{-t/\delta}$, where δ is the penetration depth and t is the distance travelled inside the material.

If the reflectance (R) is defined as the fraction of the incident energy reflected by the material, the absorption (A) is defined as the fraction of the incident radiation that is absorbed by the material, and the transmittance (T) is defined as the proportion of the incident radiation transmitted through the material, the following energy balance can be proposed:

$$1 = R + A + T \tag{15.6}$$

From Eq. 15.6, the shielding by reflection (SE_R) can be directly calculated, according to Eq. 15.7, and the total shielding (SE_t), according to Eq. 15.8:

$$SE_R\ (dB) = -10 \cdot \log_{10}(1 - R) \tag{15.7}$$

$$SE_t\ (dB) = -10 \cdot \log_{10}(T) \tag{15.8}$$

Typically, the target EMI-SE, which is demanded for commercial applications, is about 20 dB [12]. However, for most civil and military applications, the shielding required is around 80 dB, that is, an attenuation of 99.99% of the incident power [13].

It is difficult to find studies on the EMI shielding behavior of cementitious conductive composites, and it is even more difficult if those investigations must include CNFs. On the contrary, there are many studies on polymeric conductive composites with discontinuous conductive carbon fibers, CNFs, and carbon nanotubes. A cement matrix is dielectric; it offers low conductivity. For that reason, conductive particles are included when enhanced electrical conductivity is required in a composite material. However, cement matrix composites offer better shielding than polymeric matrix composites since the conductivity of polymeric composites is typically much lower. Additionally, cement is much cheaper than polymers, so the conductive cement-based composites can be used for applications that require huge volumes of shielding material, such as buildings or other facilities.

As previously mentioned, it is not necessary to reach the percolation threshold of the additions to improve the conductivity of the composite, but it is desirable to introduce a quantity high enough to observe a positive effect in EMI shielding. Also, it has been observed that smaller particles produce higher SE due to an increase in the magnitude of the skin effect [3].

15.2 Experimental

15.2.1 Materials and Specimens

Specimens consisted of cylindrical samples with 100 mm diameter and 10 mm thickness. Ordinary Portland cement type I (according to UNE EN 196) and GANF CNFs supplied by Grupo Antolín-Irausa, S.A., have been used. Due to the high water demand of CNFs increasing amounts of water and plasticizer have been used. Table 15.2 shows the w/c ratio and the percentage of plasticizer for each fabricated samples. The first approach was to avoid the use of plasticizer, but after adding a water quantity equal to the cement content, this criterion was reconsidered since using higher water contents would produce setting problems. Then, for cement pastes containing 10% and 20% of CNFs, plasticizer was included in the formulation. Three replicates of each formulation were prepared and tested.

Table 15.2 Mix proportion of cement pastes

CNFs (%)	w/c	Plasticizer (%)
0	0.35	0
0.5	0.38	0
1	0.42	0
2	0.50	0
5	1.00	0
10	1.00	1.2
20	1.00	2.8

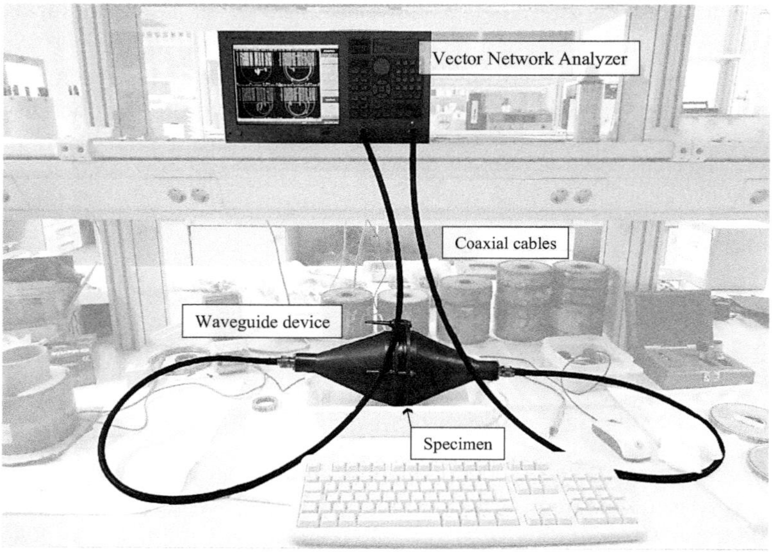

Figure 15.2 Experimental setup for EMI shielding.

15.2.2 Testing Procedures

The EMI shielding offered by cement pastes with CNFs has been evaluated in the frequency range between 0.5 and 1.0 GHz. The SE has been measured with a network analyzer from Agilent Technologies, model E5062A, as shown in Fig. 15.2. A calibration of the device has been run prior to the measurements. The parameters selected to evaluate the EMI shielding of the cement pastes were the reflection and the total shielding, which included reflection and absorption.

15.3 Results and Discussion

Figure 15.3 shows EMI shielding provided by cement pastes containing CNFs due to the reflection mechanism. It can be observed that small additions of CNFs do not significantly modify the value of reflection measured. A minimum of 5% of CNFs is needed to increase the level of reflection. In addition, this increase is only noted

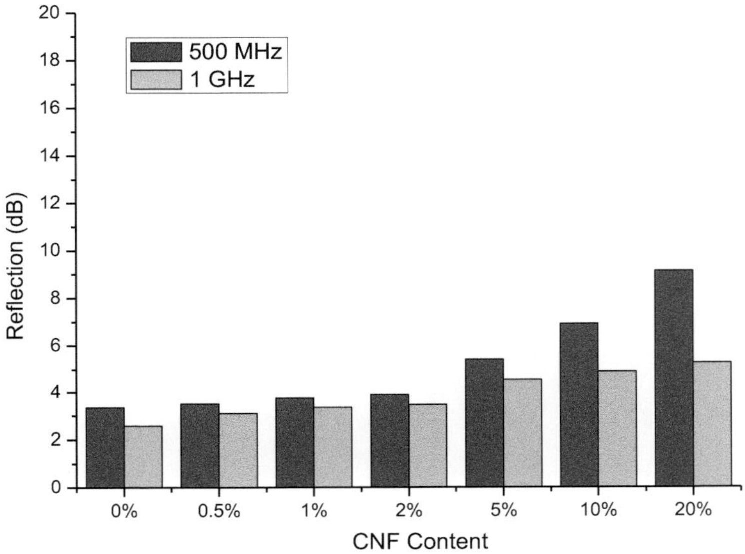

Figure 15.3 EMI shielding provided by cement pastes containing carbon nanofibers due to the reflection mechanism.

for a 0.5 GHz frequency. The reflection obtained for 1.0 GHz does not improve in the same extent since the reflection mechanism is enhanced when the frequency is reduced.

Figure 15.4 shows the overall EMI shielding provided by cement pastes containing CNFs. Once again, it can be observed that small additions of CNFs do not significantly modify the value of EMI shielding obtained. Although an increasing trend is noted as the quantity of CNFs grows, a minimum of 5% of CNFs is needed to significantly increase the level of EMI shielding. In this case, the increase is more evident for 1.0 GHz. The EMI shielding obtained for 0.5 GHz is not improved since both reflection and absorption mechanisms were considered, and the second one is more influential.

Figure 15.5 shows the resistivity of cement pastes containing CNFs. As expected, higher CNF contents offer lower resistivity due to the presence of an electronic conductor in the mix. The consistent reduction of the resistivity is reflected in the increase of the EMI shielding.

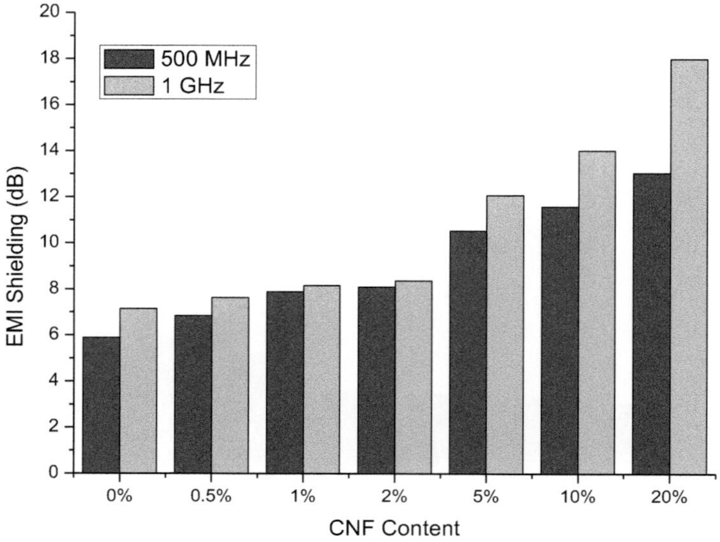

Figure 15.4 Overall EMI shielding provided by cement pastes containing carbon nanofibers.

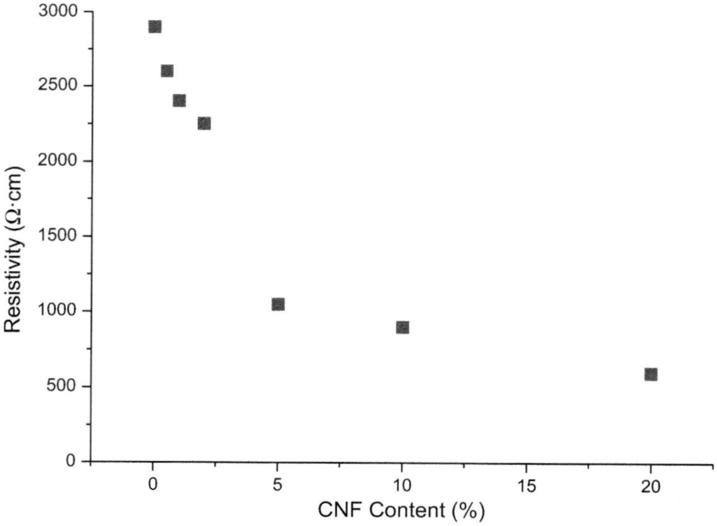

Figure 15.5 Resistivity of cement pastes containing different quantities of carbon nanofibers.

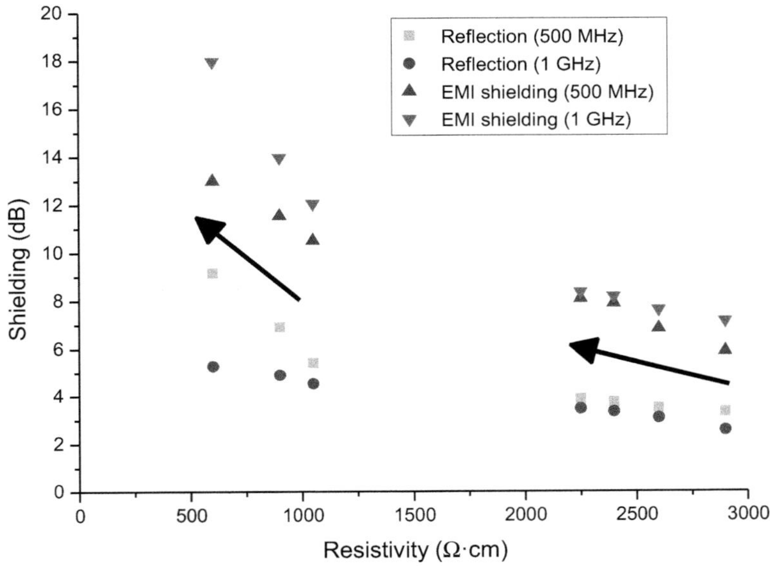

Figure 15.6 Relation between EMI shielding and the resistivity of cement pastes containing different levels of carbon nanofibers.

Figure 15.6 shows the relation between EMI shielding and the resistivity of cement pastes containing different levels of CNFs. Two linear tendencies can be observed: the first one from 0% to 2% of CNFs (right part of Fig. 15.6), and another trend (more marked) from 5% to 20% (left part of Fig. 15.6). According to these results, it seems that an important reduction of resistivity has to be achieved to obtain significant EMI-SE. The values of EMI shielding that have been measured for these pastes lie near 20 dB, which is an acceptable value for most commercial applications, but an addition of 20% of CNFs has to be used. Although other materials, such carbon fibers, behave better than CNFs, according to these results [1], this property of CNF cement pastes could complement other properties shown by these composites, such as heating function, for example, taking advantage of having some EMI-SE [10].

15.4 Conclusions

According to the findings, the following conclusions can be drawn:

- The increase of the CNF content added to cement pastes implies an increase in the EMI shielding.
- The addition of CNFs in quantities up to 2% does not significantly increase the EMI shielding of the cement paste with respect to that shown by the control paste without CNFs.
- The more interesting response regarding the EMI shielding of cement pastes containing CNFs has been obtained for quantities equal to or higher than 5%, but the workability and ease of mixing are dramatically reduced, especially for additions of 10% and 20%.
- The addition of 20% of CNFs to a cement paste with a thickness of 10 mm offered an EMI shielding near 20 dB, which is the minimum shielding usually required for commercial applications.
- As expected, the resistivity of the cement paste is reduced by the addition of CNFs and is related to the level of shielding achieved.

References

1. Chung, D. D. L. (2003). *Multifunctional Cement-Based Materials*, Buffalo, New York, USA.
2. Lu, S.-N., Xie, N., Feng, L.-C., Zhong, J. (2015). Applications of nanostructured carbon materials in constructions: the state of the art, *J. Nanomater.*, **2015**, pp. 1–10.
3. Chung, D. D. L. (2000). Materials for electromagnetic interference shielding, *J. Mater. Eng. Perform.*, **9**(3), pp. 350–354.
4. Wu, J., Chung, D. D. L. (2005). Pastes for electromagnetic interference shielding, *J. Electron. Mater.*, **34**(9), pp. 1255–1258.
5. Cao, J., Chung, D. D. L. (2004). Use of fly ash as an admixture for electromagnetic interference shielding, *Cem. Concr. Res.*, **34**(10), pp. 1889–1892.

6. Fu, X., Chung, D. D. L. (1996). Submicron carbon filament cement-matrix composites for electromagnetic interference shielding, *Cem. Concr. Res.*, **26**(10), pp. 1467–1472.
7. Chung, D. D. L. (2001). Electromagnetic interference shielding effectiveness of carbon materials, *Carbon N.Y.*, **39**(2), pp. 279–285.
8. Galao Malo, Ó. (2012). Matrices cementicias multifuncionales mediante adición de nanofibras de carbono, Universidad de Alicante.
9. Garcés, P., Zornoza, E., Garcia Andion, L., Baeza, F. J., Galao, O. (2010). Hormigones conductores multifuncionales, Editorial Club Universitario, Alicante.
10. Zornoza, E., Catalá, G., Jiménez, F., Andión, L. G., Garcés, P. (2010). Electromagnetic interference shielding with Portland cement paste containing carbon materials and processed fly ash, *Mater. Constr.*, **60**(300), pp. 21–32.
11. Guan, H., Liu, S., Duan, Y., Cheng, J. (2006). Cement based electromagnetic shielding and absorbing building materials, *Cem. Concr. Compos.*, **28**(5), pp. 468–474.
12. Yang, Y., Gupta, M. C., Dudley, K. L. (2007). Towards cost-efficient EMI shielding materials using carbon nanostructure-based nanocomposites, *Nanotechnology*, **18**(34), p. 345701.
13. Wu, J. (2010). Structural and functional polymer-matrix composite materials for electromagnetic applications, PhD Dissertation, University at Buffalo.

Chapter 16

Perspectives and Challenges of Nanocomposites

Antonella D'Alessandro, Filippo Ubertini, and Annibale Luigi Materazzi
Department of Civil and Environmental Engineering, University of Perugia, Via G. Duranti 93, 06125 Perugia, Italy
antonella.dalessandro@unipg.it; filippo.ubertini@unipg.it; annibale.materazzi@unipg.it

Concrete is the most widely adopted material in the field of constructions. The enhancement of its properties or the development of new ones would allow to have a multifunctional and more resistant material for safer structures and infrastructures. Cement-based composites are materials suitable to nanomodifications with addictions of various nature, due to their internal structure with nanosized components. The interaction of the cement matrix with nano-additions can modify the cement reactions, developing novel products for the construction industry with a more controlled and ecologically friendly behavior with respect to traditional concretes and composites.

From an experimental point of view, the choice of the proper method of dispersing the nano-additions in the cement matrix is essential to obtain a homogeneous material, and constitutes the main issue: it is closely related to the peculiarity of the specific

Nanotechnology in Cement-Based Construction
Edited by Antonella D'Alessandro, Annibale Luigi Materazzi, and Filippo Ubertini

ISBN 978-981-4800-76-1 (Hardcover), 978-0-429-32849-7 (eBook)
www.jennystanford.com

filler. Also, the design of the optimal mix design results crucial because literature research demonstrates that a higher addition of nanoparticles isn't always related to an enhancement of the capabilities of the material they are dispersed in. Moreover, the formulation of new functional nano-engineered materials made using tailored hybrid particles can contribute to the development of the use of nanocomposites. The development of micromechanical modeling and simulations surely could support the design of nanocomposites and the prediction of their capabilities.

From the point of view of applications and logistic approach, the main issues of the diffusion of nanomaterials are related to the difficulty of implementation of such innovative technologies by the construction industry and the designers due, first, to their caution in the acquisition of new technologies and, second, to the uncertainties still present about the environmental and safety management of such composites. Indeed, possible hazards resulting from their use are not completely investigated yet, and the literature panorama doesn't demonstrate homogeneous shared opinions related to the risks on the human health and the environmental safeguard attributable to the various types of available nanoparticles. Another issue is related to the economic considerations. Nanoparticles generally exhibit higher costs with respect to traditional raw materials for cementitious materials. However, the price of such novel particles is going to diminish with their diffusion. Moreover, the enhanced or new properties of new nanomodified cement-based materials could compensate for such an issue and contribute to the decrease of the costs.

In conclusion, the potentialities of new nanodevices and nanoparticles derived from nanotechnology are undoubtedly very high. Nanotechnology provides powerful tools and materials with extraordinary properties that can be effectively adopted for specific applications, increasing performance and decreasing costs, if correctly handled. Multifunctional concretes with novel capabilities are also possible to develop positively. So, the advance of cement-based composites modified by use of nanomaterials could result in an important, improving, and sustained impact on the future of engineering and of the construction field.

Index